ANATOMY OF VIOLENCE

For Dorothea, Miriam and Sophia: for their love and patience

Anatomy of Violence
Understanding the Systems of Conflict and Violence in Africa

BELACHEW GEBREWOLD
Helmut Schmidt University/University of the Federal Armed Forces, Germany

ASHGATE

Published by
Ashgate Publishing Limited
Wey Court East
Union Road
Farnham
Surrey, GU9 7PT
England

Ashgate Publishing Company
Suite 420
101 Cherry Street
Burlington
VT 05401-4405
USA

www.ashgate.com

British Library Cataloguing in Publication Data
Gebrewold-Tochalo, Belachew, 1968-
Anatomy of violence : understanding the systems of conflict and violence in Africa.
1. Violence--Africa. 2. Social conflict--Africa.
I. Title
303.6'096-dc22

Library of Congress Cataloging-in-Publication Data
Gebrewold-Tochalo, Belachew, 1968-
Anatomy of violence : understanding the systems of conflict and violence in Africa / by Belachew Gebrewold.
p. cm.
Includes bibliographical references and index.
ISBN 978-0-7546-7528-0 (hardback) -- ISBN 978-0-7546-9746-6 (ebook)
1. Political violence--Africa. 2. Low-intensity conflicts (Military science)--Africa. 3. Africa--Politics and government--1960- I. Title.

DT30.5.G42 2009
303.6096--dc22

2009015710

ISBN 9780754675280 (hbk)
ISBN 9780754697466 (ebk)

Mixed Sources
Product group from well-managed forests and other controlled sources
www.fsc.org Cert no. SGS-COC-2482
© 1996 Forest Stewardship Council

Printed and bound in Great Britain by
TJ International Ltd, Padstow, Cornwall

Contents

Abbreviations

AAA	Addis Ababa Agreement
ACP	African, Caribbean and Pacific States
ACPP	Africa Conflict Prevention Pool (UK)
ADF	Allied Democratic Forces
AFDL	Allied Forces for Democracy and Liberation (DR Congo)
AFRICOM	Africa Command (US)
AGOA	Africa Growth and Opportunity Act
AMIS	African Mission in Sudan
APSA	African Peace and Security Architecture
ARB	*Africa Research Bulletin*
ARPCT	Alliance for the Restoration of Peace and CounterTerrorism
AU	African Union
CAR	Central African Republic
CDNP	*Congrès National pour la Défense du Peuple*
CJTF-HOA	Combined Joint Task Force-Horn of Africa (US)
CPA	Comprehensive Peace Agreement
DFID	Department for International Development (UK)
DPA	Darfur Peace Agreement
DRC	Democratic Republic of Congo
DUP	Democratic Unionist Party
ECOWAS	Economic Community of West African States
EPRDF	Ethiopian People's Revolutionary Democratic Front
EU	European Union
EUFOR RD Congo	European Force in the Democratic Republic of Congo
EUPOL RD Congo	European Police Mission in the Democratic Republic of Congo
EUPOL-Kinshasa	European Police Mission in Kinshasa
FDLR	*Force Democratic de Liberation du Rwanda*
FLNA	National Front for the Liberation of Angola
FLNC	*Front Libération Nationale du Congo*
GB	Great Britain
GNI	Gross National Income
GoS	Government of Sudan

HRW	Human Rights Watch
ICC	International Criminal Court
ICG	International Crisis Group
ICU	Islamic Courts Union
IGAD	Intergovernmental Authority on Development
ILO	International Labour Organization
JEM	Justice and Equality Movement
LRA	Lord's Resistance Army
MCC	Millennium Challenge Corporation
MDC	Movement for Democratic Change
MDGs	Millennium Development Goals
MISAB	Inter-African Mission to Monitor the Bangui Accords
MLC	Liberation Movement of Congo
MONUC	United Nations Organization Mission in DR Congo
MPLA	*Movimento Popular de Libertação de Angola*
NCP	National Congress Party
NDA	National Democratic Alliance
NIF	National Islamic Front
ODA	Overseas Development Assistance
PPRD	People's Party for Reconstruction and Democracy
RCD-ML	Congolese Rally for Democracy – Liberation Movement
RECAMP	(*Renforcement des capacities africaines au maintien de la paix* – Reinforcement of African Peacekeeping Capabilities)
RPF (FPR)	Rwandan Patriotic Front
RRA	Rahanweyn Resistance Army
SADC	Southern African Economic Development
SLA	Sudan Liberation Army
SLM	Sudan Liberation Movement
SNF	Somali National Front
SPLM/A	Sudan People's Liberation Movement/Army
SPM	Somali Patriotic Movement
SSDF	Somali Salvation Democratic Front
SSLM	South Sudan Liberation Movement
SSNM	Southern Somali National Movement

TFG	Transitional Federal Government of Somalia
TNG	Transitional National Government of Somalia
UK	United Kingdom
UNAMID	United Nations African Mission in Darfur
UNITA	Union for the Total Independence of Angola
UNITAF	Unified Taskforce Somalia
UNOSOM	United Nations Mission in Somalia
UNSC	United Nations Security Council
USC	United Somali Congress
WFP	World Food Programme
WMD	Weapons of Mass Destruction
WNBF	West Nile Bank Front
ZANU-PF	Zimbabwean African National Union – Patriotic Front
ZAPU	Zimbabwe African People's Union

Introduction

Soon after Somalia collapsed in 1991 various warlords mushroomed. Since then reconstruction of the country has not been possible and the population has been suffering indescribably. Somalia, as worst-performing state, leads the Failed States Index table of 2008. It is followed by Sudan, Zimbabwe and Chad. Only interrupted by Iraq, another African country continues the list of the failed states: DR Congo. Since its creation in 2004 the transitional federal government of Somalia could not control even the capital, Mogadishu. The government of Somalia has not enjoyed any popularity. In 2006 the Islamic Courts Union (ICU) controlled most of the country but later lost the controlled areas to Ethiopian troops, who were supporting the TFG (transitional federal government). About 700,000 people fled Mogadishu in 2007. In October 2007 Prime Minister Ghedi of the TFG resigned; President Yusuf resigned in December 2008. Islamic radicalism increased dramatically in 2008 and regained control of most of Somalia. Ethiopian troops left the country towards the end of 2008 and early 2009. A new 'moderate' Islamist president, Sheik Sherif Ahmed, was elected in early 2009 and introduced Sharia law to appease the radical Islamists. But the radicals are still fighting the government.

In January 2005 the Comprehensive Peace Agreement was signed between the Sudanese government and the Southern rebels. However, the conflict that started in 2003 in the western part of the country, Darfur, has killed at least 300,000 and displaced about 2.7 million. Despite the AU and UNAMID peacekeeping missions, the population in the region has been suffering dramatically. The ICC arrest warrant in March 2009 resulted in the escalation of violence and expulsion of aid workers from the region.

Zaire, famous for corruption and mismanagement, failed in 1996 when the AFDL coalition advanced, with the support of Rwanda and Uganda from the eastern part of the country, to topple the notorious regime of Mobutu. Between 1998 and 1999 a brutal war took place. Millions have been killed since then. In spite of the various peace agreements in 1999, 2002 and 2003, and in spite of the transitional government established in 2003 and a new government in 2006, violence continued in the eastern part of the country. The arrest of General Nkunda (leader of the rebel group, CNDP) early in 2009 gave a small sign of peace, but the rebels have not yet refrained from violence. Despite the presence of 17,000 MONUC peacekeepers, rape, killing, displacement and exposure to diseases have continued. Even UN peacekeepers have been involved in sexual abuse and gold smuggling. Why did these three states fail?

Many researchers have been trying to explain factors that perpetuate these brutal conflicts. For some researchers, the *intrastate* factors and actors have played the key role. According to Ottaway et al. (2004), African big states such as Sudan, DRC and Ethiopia fail simply because of internal political, ethnic, and religious differences and because of their inability to establish institutions to moderate these differences. For example, Prunier (1995, 2007b) sees the Rwandan and Darfur wars more from intrastate points of view and less from external influences and repercussions. Collier et al. (2005) and Collier (2007) argue that economic factors of conflicts such as resource abundance, deprivation and low economic growth are key conflict factors. Sachs (2005) has suggested that 'big push' (economic aid) would help to solve African economic problems.

Other scholars add other factors to the causes of the state collapse and weakness: unfavourable patterns of ethnic relations, corruption, weak civil society, poor economic policies, lack of state's responsiveness to legitimate public demands, lack of civil rights, incapable leaders, and stifled democracy (Rothchild and Harbeson 2000: 8). For Rotberg, the Horn of Africa, for example, is instable because of civil and interstate wars, displacement, refugees, famine and hunger caused by state weakness and collapse, which further lead to intractable conflicts (Rotberg 2004, 2003). Hence, the specific contexts within which the civil militias thrive are the weakness, failure and collapse of states; porous state borders; and interstate conflicts (Rotberg 2003). Various suggestions to solve these problems have been offered. State-building is considered by the vast majority of the international relations studies to be the solution for political and economic problems (Crocker 2003; Eizenstat et al. 2005; Carothers 2003; Walzer 2004; Fukuyama 2004, etc.).

In the same way, global institutions have been trying to fix these collapsed and weak states. The UN is present in Sudan and Congo to help end the violence. The EU has been actively present, including in the form of military interventions, since the mid-1990s. Global players such as the US and France have been present in those countries in different forms in order to remake those states. But the success of all these actors has been modest.

The aim of my analysis is to illustrate not only the causes and consequences of conflicts in individual states of the region, but also to discuss the correlation between the conflicts and violence of one state with that of the other states. Hence, my hypothesis in this book is: DR Congo, Somalia and Sudan are in a protracted war because of three reasons: first, they are too weak to *regulate* the internal systemic interactions of various actors and factors; second, they are too weak to *resist* pressure from regional and global systemic impacts; third, their internal weaknesses *facilitate* regional and global involvements and interferences. Therefore, many African conflicts are so systemic that peace initiatives, military interventions, development aid and peacekeeping that address only the intrastate level will fail. Hence, a systemic approach is indispensable. The systemic approach attempts to show that there are multi-level actors and factors for African conflicts. For example, poverty does not lead necessarily to conflicts. Amartya Sen's essay on poverty and violence (2008) shows this phenomenon and negates Collier's economistic approach. Similarly, ethnic heterogeneity does not necessarily lead to interethnic conflicts.

But when other conflict factors such as economic marginalization are given, ethnic groups see themselves as homogenously closed systems; elites present their nation or ethnic groups as a closed system. Similarly, states see themselves as closed systems. They construct and conceive others as closed systems of evil seekers *a priori*. But so far there is no evidence that would suggest there is a necessary correlation between violence and ethnic diversity and identity. Even poor ethnic groups can live together in peace and share natural resources like water and grazing land. Violence depends on the conglomeration of various actors and factors.

Though there is an increasing awareness of the role of the *regional* dimensions of conflicts in Africa, the regional dimension is not yet fully integrated into conflict studies. Tull suggests that the neglect of colluding local and regional actors in the Kivus and Rwanda has considerably damaged various peace efforts in Congo (Tull 2007: 114). Indeed the economic interests of states such as Rwanda and Uganda in Congo are widely discussed by various authors (Clark 2002; Nest et al. 2006; Todd 2006; Montague 2002; etc). However, Boas and Dunn, in their analysis on 'why conflicts start in the first place', reject the predominant role given to economic agendas of actors in a conflict. They stress that political, cultural and historical factors should be included (Boas and Dunn 2007: 11). This means, in their attempt to disprove that regional involvement, such as in DR Congo, triggered conflict, they stress that conflicts are primarily internal (Boas and Dunn 2007: 35). 'The conflict in the DR Congo is a Congolese conflict, and to define the movements involved there simply as proxies of either Rwanda or Uganda would be a grave mistake, because it completely overlooks the historicity of the conflict … The important issue here is the relationship between the conflicts, or how localized conflicts become intertwined and regionalized' (Boas and Dunn 2007: 35). But, especially if one takes into account the 1972 and 1981 citizenship laws in DR Congo and the genocide of Rwanda in 1994, one cannot separate local and regional conflicts.

Not only the conflicts in the Great Lakes region, but also those in the Horn of Africa, underline the regional dimension:

> The war between Ethiopia and Eritrea has led both governments to increase their military support to rival proxies in Somalia, thus igniting new rounds of deadly conflict, spreading instability to northern Kenya, re-legitimizing warlords and destroying hopes for internal peace efforts. At the same time, Ethiopia and Eritrea have reduced their support for the Sudanese opposition, thus strengthening the hand of the Khartoum regime and reducing the likelihood of progressive change in Sudan. In addition, Sudan supports insurgent groups in Ethiopia, Eritrea, and Uganda; Somali militias have launched cross-border attacks into Ethiopia and supported Ethiopian oppositionists, while Ethiopian troops have launched assaults into Somalia to create a protective buffer zone; and Uganda has supported the main rebel groups in Sudan. With each new act of

> violence, with each cross-border arms transfer, the regional dimensions of these conflicts deepen. (Prendergast 1999)

Civil militias in one country are actively contributing to the emergence and/ or maintenance of the civil militias in another country. Borders are porous, governments are weak, national security systems are ineffectual. Small arms and light weapons are difficult to control; among cross-border pastoralist communities, arms are acquired overtly and facilitate livestock raiding. Because of the inadequate policing of arms, armed criminality in urban, rural and border areas has internal as well as regional security implications (Griffiths-Fulton 2002).

On the *global* level, there is an increasing expectation that the international community has responsibility to protect the civilians from violence and humanitarian catastrophe. Why doesn't the 'international community' intervene 'in time' to address the causes that make intervention necessary: exploitation of natural resources, arms transfer, economic injustice, corruption, environmental destruction, exploitation, discrimination, marginalization, etc.? Is the international community (if it exists at all!) part of the problem in this case? Various global actors have been pretending to address conflicts and violence in DR Congo and in the Horn of Africa, but it rather seems that they are interested in their strategic interests. Hence, conflict actors and factors are multifaceted.

Tetzlaff and Jakobeit discuss five typologies of wars: the anti-colonial independence wars; between-state conflicts about territory, natural resources and ethnic groups; secessionist wars in ethnically heterogeneous states; anti-regime and internal wars; and regionalized wars (Tetzlaff and Jakobeit 2005: 106). My approach does not negate this typologization; instead it tries to show their systemic interdependence.

The key question of this book is: why have conflicts in DR Congo, Somalia and Sudan been so intractable? To deal with this question, we have to approach these states and their conflicts from internal, regional and global systemic dimensions. It is not adequate to address conflicts only from ethnic or economic perspectives, or to claim that conflicts continue because regional states destabilize each other. In the same way it is not adequate to suggest that conflicts continue because the international community failed to intervene. The success of any global intervention in Africa depends on: Do the intervening actors have the *will* to intervene in African states? Do they have the *power* to intervene? Do they have the *right* intentions when they intervene in Africa? How is this global dimension influenced by the internal and regional actors and factors? These are questions that lead us to the systemic problems of conflict and violence.

Figure I.1 on the next page illustrates this problem. Root causes imply the historical composition of peoples and states during the colonial period as well as during the process of 'state building' afterwards. Trigger causes are the current societal, economic and political actors and factors that trigger conflict; whereas tertiary causes are the external actors that systemically exacerbate conflicts or do not have the *power*, *will* or *right* intention to prevent them.

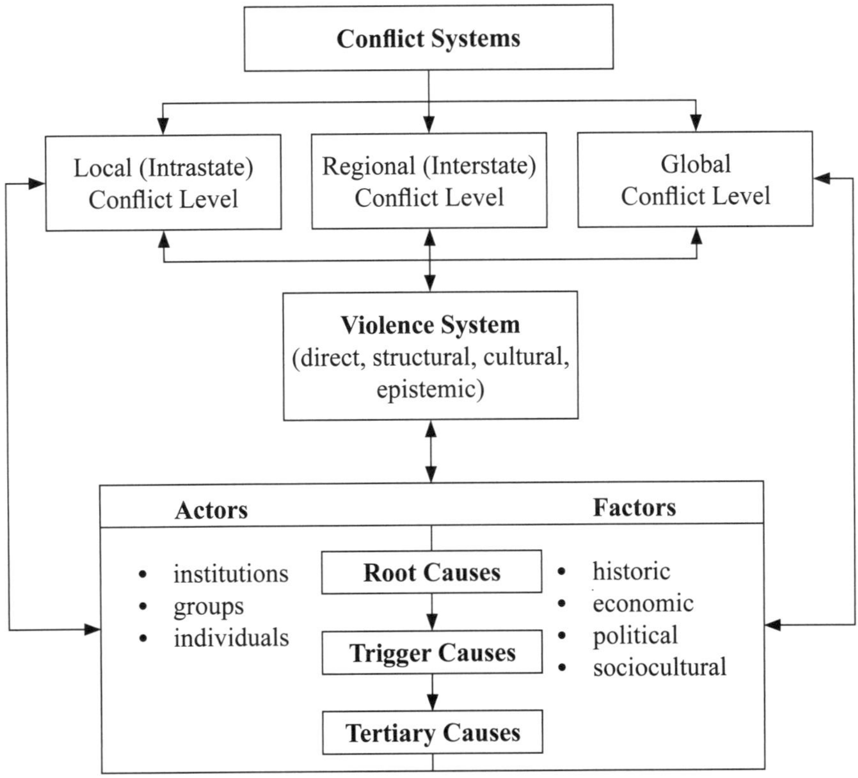

Figure I.1
Source: Belachew Gebrewold/Thomas Spielbuechler

System is to be understood as a number of objects and their relations among each other. This relationship characterizes the system as totality, which, however, can only be differentiated (but not separated) from systemic environment. The three states are state systems that have become conflict systems. Each conflict state consists of intrastate, regional and global conflict systems. Regional and global institutions that are thought to bring about solutions consist of members (elements) which are unpredictable and can change their behaviour depending on the behaviour of other systems or on that of the member states within their system. The external interveners may have the power to influence the system of the intervened-in (weak states), but they neglect the regional and global systems either because they consider it insignificant or because they do not have the will or the power to do it. For example, global players can intervene in Sudan, but they do not intervene in Chad because (though they consider the role of Chad significant) they do not consider it decisive for peace in Sudan; they might intervene in Chad but not in France, which is the backbone of the Chadian regime survival; they can intervene in Sudan, but not in China, a staunch supporter of the Khartoum regime. Therefore, they disregard the different levels of systemic dimension of a conflict.

The crisis and weakness of African states renders them prone to capture by violent non-state actors, such as international terrorist groups in search of wealth, support, recruits and sanctuary. For example, Somalia's collapse was caused by internal, regional and global actors and factors; as a result, Islamic radicalism came up. To address this problem, regional and global actors are acting against it. What we see here is a vicious circle. Therefore, it is important to underline that the behaviour of global systems and the behaviour of units are reciprocally constituted and constrained. This means that the behaviour of the global players contributes to state weakness, failure and collapse. State weakness, failure and collapse causes the global players to act against it. The dynamics and intensity of conflict and violence depend on this vicious circle. Hence, how can this vicious circle be terminated or reduced? What I suggest is system transformation, as will be shown in the concluding chapter: transformation of systemic isolationism or separation into systemic differentiation and interdependence.

The first chapter discusses what system is and how African states function or dysfunction as systems. However, the objective is not to analyse system theory as such, instead just to clarify the concept of system in order to show the phenomenon of conflict as system and how different state and non-state actors and factors interact and constitute a conflict system. Chapter 2 discusses Africa in the international security system and shows how the global actors are present in Africa. The objective of the chapter is to demonstrate the impact of the intersystemic interpenetration on African states' systems. Chapter 3 demonstrates how the systems of states and security become conflict systems. In this chapter various theories of conflict and violence are discussed. Chapter 4 discusses DR Congo as a case study and shows the intrastate, regional and global conflict systems there. Chapter 5 discusses Somalia from an intrastate systemic point of view, whereas Chapter 6 deals with Sudan again, from an intrastate systemic point of view. Chapter 7 deals with the regional and global conflict systems in the Horn of Africa, with a main focus on Somalia and Sudan. The reason why the regional and global conflict systems of Somalia and Sudan are dealt with in a single chapter is that because of the interconnectedness of regional actors in those states, their conflicts on the regional level are interconnected. The intrastate dimension of each case study is divided into three main sections: society, economy and politics. The societal, economic and political dimensions of each intrastate conflict system are penetrated by regional and global actors and factors making the regional and global conflict systems. The conclusion, therefore, suggests that conflict and violence can only be ended or reduced if such system is transformed.

Chapter 1
The System of the State

> State is a human community that (successfully) claims the monopoly of the *legitimate* use of *physical force* within a given territory. (Max Weber 1972)

> War-making and state-making are organized crime. (Charles Tilly 1985)

African States Systems

Various scholars of state theories or international relations have been adopting some aspects of the system theory into their respective fields. Deutsch, in his analysis of political and state systems, suggests that systems may have different characteristics than the components or subsystems they comprise (Deutsch 1974: 156). And he defines system as 'a collection of recognizable units or components which hang together and vary together, in a manner regular enough to be described' (Deutsch 1974: 229). In his 'world-systems' theory, Wallerstein underlines that it is not about various systems (such as economies etc.) *of the* whole world, instead about these systems *as a* world. It is about 'spatial/temporal zone which cuts across many political and cultural units, one that represents an integrated zone of activity and institutions which obey certain systemic rules' (Wallerstein 2006: 17). World-systems theory shows the limits of reducing complex situations to simpler variables, and complicates and contextualizes all so-called simpler variables in order to understand real social situations (Wallerstein 2006: 19).

> States … exist within the framework of an interstate system, and their relative strength is not merely the degree to which they can effectively exercise authority internally but the degree to which they can hold their heads high in the competitive environment of the world-system. All states are theoretically sovereign, but strong states find it far easier to 'intervene' in the internal affairs of weaker states than vice versa, and everyone is aware of that. Strong states relate to weak states by pressuring them to keep their frontiers open to those flows of factors of production that are useful and profitable to firms located in the strong states, while resisting any demand for reciprocity in this regard. (Wallerstein 2006: 55)

This systemic understanding has effected considerable impacts on the theory of states and international relations. In the global war on terror and state collapse, failure and weaknesses, state-building is a goal-oriented process and political project that attempts to create reliable organizations and interdependence of relationships (Fukuyama 2004; Carothers 2003; Crocker 2003; Eizenstat et al. 2005; Hippler 2004).

On the international level, studies on international relations discuss the 'international structure' (Waltz 1979; Mearsheimer 2006) or 'international system' (Clapham 2003; Buzan and Little 2000; Emmott et al. 1997; Hanson 2006; Buzan et al. 1998; Buzan and Waever 2006; Jackson and Sorensen 2003); and states are systems (Bull 1995). For Buzan and Gonzalez-Pelaez (2005), system is a 'sustained and structured pattern of interaction among agents'. For Neuman 'system' is 'characterized by its interconnectedness and the sensitivity of each unit to change within and among other member units' (Neuman 2005: 4). For any analysis of international relations as system it is indispensable to discuss state as system or subsystem of the international relations system.

This system implies order. Hobbesian strong state as well as Lockean division of powers in the state aim to create order in the state. Kant's Enlightenment suggests a 'way out' of 'immaturity', which is a certain state of our will that makes us accept someone else's authority to lead us in areas where the use of reason is called for (Kant 1900). These concepts imply a state is an organized or 'civilized' system or order. On the national level the state fulfils its task by organizing the interests and relationships of the citizens. According to Kant's Enlightenment, the liberating reason of humankind organizes international relations and enables perpetual peace through creating organizations like the United Nations to end the state of immaturity prior to the organized system. If a state is not able to organize itself in an ordered structure, it can be designated a 'failed state' or failed system. Hence, a successful state system is indispensable to establishing international relations as an organized system.

For understanding conflict or state systems, it is important to discuss whata 'system' is. According to von Bertalanffy:

> …there exist models, principles, and laws that apply to generalized systems or their subclasses, irrespective of their particular kind, the nature of their component elements, and the relationships or 'forces' between them. It seems legitimate to ask for a theory, not of systems of a more or less special kind, but of universal principles applying to systems in general. (Bertalanffy 1968: 32)

According to system theory, 'system' means a number of objects and their relations among each other. This relationship characterizes the system as totality, which consequently differentiates itself from its environment. Environment in this context means all other systems beyond the borders of the focal system. Though other systems have an impact on the focal system, each system owns its characteristic dynamic in which all its elements contribute to self-reproduction. Luhmann suggests that the modern society differentiated into political, economic, scientific, educational, etc., subsystems and their respective environments are building up an unprecedented degree of complexity, and their boundaries do not correspond to the common territorial frontiers. But at the same time the political subsystem (the sovereign state) continues to maintain its territorial frontiers, while other subsystems, and with this the society itself, are spreading over the globe (Luhmann 1990a: 177–8).

According to Luhmann, system *is* the difference to the environment. Systems characterize themselves by constructing differentiation between themselves and the environment through operations (differentiation) or events (*Ereignisse* in German) (Stark 1994: 38). Therefore, system is the connection between operations that differentiate the system and environment. System as 'form' is characterization as well as differentiation: *characterization* is on the side of the system and *differentiation* is on the side of the environment. Based on the systemic principle of *recursivity*, the difference between system and environment necessitate specific operations that continuously constitute this difference. The difference emerges through operations: system–environment differentiation is reproduced by which short-term operations systemically interconnect subsequent operations. The wholeness of the system exists because of the 'interconnectibilities' (*Anschlussfähigkeiten* in German) of the system-specific operations that are produced in the system itself.

System is characterized by:

- varieties of objects, which can be perceived as units from the perspective of the observer;
- institutions, interconnections between individual units that, as system structure, keep the system together as totality;
- system-specific environment, while the environment is understood as a multitude of objects whose interconnection with the focal system can be differentiated empirically and theoretically from the system structure (of the focal system) (Stark 1994: 10).

According to Parsons, social system is about coordination or mutual relationship of social positions (not individuals) (Parsons 1969). This means, social system as general system of action is an integrative function through which structures and processes are related to system stability and system transformation in a given environment (Parsons 1964: 3–6). This function implies certain contribution of a system to another system (Parsons 1968). Functions are related to the problematic existence of systems in an environment that does not belong to the system. The environment is on the one hand the source of the means for the functioning of the system, on the other hand a threat to the existence of the system; and since the system has to maintain an identity towards the environment, the structural border between the system and environment should not be destroyed. Hence, system/environment observation is the basis for the functioning of the system (Parsons 1964: 27–31).

For Willke, each system is a self-referential structure in which it as observer observes its own observing through reflection of consciousness (Willke 2005: 67).

Luhmann understands this act of differentiation (through observation) as system formation (Luhmann 1999: 7). Autopoiesis, or self-reproduction, of the system is possible only through this self-referentiality and differentiation. ‘Systems constitute and maintain themselves by creating and maintaining a difference from their environment, and they use their boundaries to regulate this difference … Boundary maintenance is system maintenance’ (Luhmann 1999: 17). According to Maturana, autopoiesis not only determines by itself the relations of its elements but also it produces its elements by the elements of the system (Maturana 1987). Autopoietic system suggests that system-specific operations are reproduced by system-specific operations; the integrity of the system is reproduced by the integrity of the system, so that the difference between system and environment is continuously reproduced. One of Luhmann’s theses suggests that all operational systems are operationally closed because any operation for maintenance of the system–environment differentiation has to be on the system side of the differentiation (self inward observation of the outward observing system) (Luhmann 1993: 53). According to Luhmann, autopoietic systems reproduce all elementary units through the network of the elements themselves which the systems consist of, and through this reproduction they delimit themselves from the environment. However, internal interdependencies of a system are not greater than system/environment interdependencies.

Therefore, it is about how to organize the intrasystemic as well as intersystemic relations. System theory deals with organization and interdependence of relationships, or configuration of parts. The system concept attempts to disprove the reductionist theory, which focuses on single parts or elements. This is important for analysis of conflict as system. System theory maintains that system is composed of regularly interacting elements that form a new whole which has properties that cannot be found in the elements (Luhmann 1984, 1999, 2004). The relationship of elements creates ‘emergent properties’ which may not be found in any analysis of the parts, i.e. the ‘wholeness’ can’t be seen in them. Relationships of elements in the system are goal-oriented processes. This suggests that in the states system, state as system and states’ relations as system are essentially interrelated, that neither part can be understood separated from the other part.

System is the selected unit (focal system), whereas environment is the totality of the other (excluded) unit. But as constructivism suggests (based on relativity and contingency), this reality could be differently cognized than it is, and each cognition is construction (Guggenberger 1998: 35). Even if the reality as such is not dependent on cognition, the *acceptance* of the reality is dependent on constructivism, i.e. the accepted reality is reconstructed through the act of acceptance. Construction of a new reality as an objective and pre-existent system is especially relevant for discussing ‘state’ or ‘state system’. While discussing the act or process of construction of the state it is important to discuss *what* the state is built for.

This ‘what for’ implies to *meaning* of the system according to system theory. The meaning of the system through its referential structure forces the next step to selection in its relation to the environment (Luhmann 1999: 61). The meaning of the state system is crucially in formation of national identity and interests. The intention of meaning is self-referential because it takes care of the self-realization of the system. Meaning in general and boundaries constituted in meaning in particular guarantee the irrevocable nexus of system and environment in a form distinctive to a meaning (Willke 2005: 6). State identity is essentially dependent on this. No meaningful system can conclusively lose itself in its environment or in itself because there are always implications that refer back over the boundary (Luhmann 1999: 62). This takes place while defending national identity and interests. Such a conception of state system is crucial for a conflict system.

According to system theory, the understanding of the meaning of the system depends on the understanding of communication (relation) and action (elements) in the system (Guggenberger 1998: 49). Communication and action cannot be separated and they form a relationship of complexity (Luhmann 1999: 137–8). Complexity is 'selection drive' (*Selektionszwang* in German) from plenty of various and mutually interdependent factors that create environmental interconnectedness. According to Willke, complexity implies that system has to react to the demands of its environments. This means, it is about system–environment relation in which the abundance of possibilities as well as dangers in relationship with the environment becomes a problem for the system. Such abundance of possibilities results in contingency (varieties of alternatives the system has got) (Willke 2006: 30). For example, traditional societies have lower contingency than the modern capitalist and democratic societies. Retention (of the existing) is the negation of contingency (Willke 2006: 33–4).

For communication and action of the system the relationship to the environment is constitutive in system formation because everything that happens belongs to a system and always and at the same time to the environment of other systems (Luhmann 1999: 176–7). This type of relation results in interpenetration: an intersystem relation between systems that are environments for each other means that systems interpenetrate each other (Luhmann 1999: 211–15). In the complex interpenetration, each effect of a system has infinite causes, and each cause has infinite effects. At the same time each cause can be combined with or replaced by other causes because of the various differences of effects that can happen. Therefore, each causal process can be split or traced back into the infinite (Luhmann 1974: 18; Luhmann 1990b: 70). This effect–cause mutual interaction is a kind of systemic self-creation (Willke 2005: 31).

In interpenetration and complexity an important phenomenon for conflict is the so-called 'negative double contingency': A will not do what system B wants if system B does not do what system A wants, leading to a vicious circle (Luhmann 1999: 389). With increasing contingency, conflict potential increases too (Willke 2005: 30). Double contingency creates the problem of mutual intransparency in the interaction between two autopoietic systems. However, if both parties would like to establish contact, interest in negation of the negativity (mutual intransparency) increases (Luhmann 1984: 172). Hence, double contingency suggests that, first, system elements are bound to the system structure through positive and negative sanctions; second, that inputs into the system influence the focal system (Parsons 1964: 36–41). Therefore, the task of the ordered and organized social systems is to solve the problem of double contingency

System is characterized by asymmetrization: to make its operations possible, a system chooses points of reference that are no longer put in question and which must be accepted as given. This is especially true for national identity that is dogmatized and intolerant of any revision. Very often, national identity is not something that can be reflected critically; it becomes a kind of a supernatural entity beyond the sphere of criticism, because critical reflection implies insecurity and transience. The state system and conflict system are closely related.

The nature of the state system is based on this systemic change of the state from constructed system into pre-existent system. Studies of Benedict Anderson's 'imagined community' (2000), Michael Billig's 'banal nationalism' (1995), the 'pathological homogenisation' of Rae (2002), etc., show how the state system changes through nationalism. Through a stream of collective memories, social mobilization of group awareness based on existing symbols and institutions, a nation regenerates itself, or even becomes a supranational church (Deutsch 1979: 27). Nationalism is the last instance that tries to secure the order of the system. Order is the balance and stability of the systems (Luhmann 2004: 44). In this order, the external threat or the environment as possible danger indirectly creates internal cohesion. This is significant for the conflict system.

The System of the State

State is the system of tension between legitimate power monopoly (ideal state) and organized crime (real state). According to Max Weber's famous definition, 'state is a human community that (successfully) claims the monopoly of the legitimate use of physical force within a given territory' (Weber 1972: 822). State is simply 'institutionalization of power' (Nettl 1994: 9). The important question in this section will be 'what constitutes state as system?' Before we deal with this question, let us see what the system of the state from the point of view of system theory is. Willke suggests that in the form of politics the state creates its own principles. Since secularization dynamics and the awareness of contingency have undermined the external (such as supernatural) justifications of politics, the politics has to create its own legitimacy by replacing dangerous contingency by formal necessity as a form of operation. Social systems as communicative and symbolically constituted systems – in the face of lacking religious plan of salvation or objective truth – are exposed to the power of contingency determined by changeable and selectable alternatives and operations. Consequentially, social systems change their modes of operation from unity to difference, from the continuation of the always identical to the contingency of particular identity. It is about how the system (state) sees and understands itself in comparison with the other systems or other sides of its world or environment (Willke 1993: 216). State is neither a supra-societal and autonomous reality nor a mere means of power elites and their interests; instead it is a specific function of a political system for a differentiated society as well as it is preconditions of operations in a self-legitimating politics by excluding arbitrariness or contingency (Willke 1993: 219). The notion of territory is a good example here.

Territory becomes a 'homeland', 'fatherland', or 'motherland' that has to be commonly defended by those who share similar obligations for its protection and because it defines 'them'. They sacrifice their blood and taxes for its defence because it 'clearly' distinguishes 'them' from 'others' (Goemans 2006: 27). Specification of homeland requires focal principles: natural frontiers (by reference to nature and topographical features: mountains, rivers, seas), common culture (pre-existing principles: language, religion, ethnicity, 'nationality'), prior historical formation, and cartography (conventions and maps) (Goemans 2006: 32–40).

> The political organization of space is not limited to notions of the state but it equally, perhaps even more importantly, impacted at the local and micro levels of daily behaviour and practices ... Territory constitutes an important component of our individual, group, and national identities, not simply because our state territories are delimited by fixed boundaries but because territory has a symbolic dimension which determines our attachment and affiliation to particular spaces and places, attachments which are thought consciously – through political and territorial socialization ... Since we reject the absolute notion of deterritorialization (in favour of a continuing process of reterritorialization and changing territorial configurations of power) we must, by definition, reject the notion of a borderless world. (Newman 2006: 87)

For the system of the state, national 'memory' is crucial. Deutsch asserts that 'memory is the storing and recall of past information that constitutes the *self*', and on the group level it is the common memory of the group-self as well as a self-steering system or an autonomous *political self*. Therefore, 'identity is the recognition and awareness of one's own memory and of one's self' (Deutsch 1974: 178). As Billig neatly put it, 'every nation must have its history, its own collective memory. This remembering is simultaneously a collective forgetting: the nation, which celebrates its antiquity, forgets its historical recency. Moreover, nations forget the violence which brought them into existence...' (Billig 1995: 38).

Identification with territory has substantial impact on national, group and individual behaviour, identities and practices and it determines their attachment and affiliation to particular spaces and places, attachments which are thought-consciously – through political and territorial socialization – based on historical and mythical events in the nation's history (Newman 2006: 96). 'Territory is important, not so much because land contains important natural resources that can be translated into tangible assets, but because it plays the more important role of defining one's social, spiritual, and communal world' (Walter 2006: 289).

'Territorial attachments and people's willingness to fight for territory appear to have much less to do with the material value of land and much more to do with the symbolic role it plays in constituting people's identities and providing a sense of security and belonging' (Walter 2006: 288). Further, Walter correctly argues that:

> measuring the value of land strictly in terms of its tangible assets and making these assets the basis of a settlement is unlikely to bring long-term peace. To date, policymakers have tended to focus disproportionately on issues related to border length and placement, and the distribution of natural resources, rather than dealing with the deep psychological bonds that may exist among individuals at the local level. (Walter 2006: 294)

Such psychological bonds have huge impacts on nationalism and violence.

In the state system state-building can become a pathological homogenization (Rae 2002). This homogenization is a process of unification through denial and destruction of contingency. It even becomes a simplifying hubris that violently controls the powers (Willke 2006: 29). Willke suggests that state theory has to keep in mind that its assumptions and self-evidences are contingent, and because of rationalization, secularization and positivism there is no basis for belief in absolute forms and order. This formalized and ordered operation (organization) produces asymmetries, i.e. it radically excludes the 'external'. Confronted by insecurity and complexity, state system is tempted to look for and establish lasting security and closed permanent operation procedures. But the state has to observe itself ironically: not as an absolute operator, rather a contingent observer (Willke 1993: 226–9).

State is an autonomous collectivity that can institutionalize its sovereignty. But the established states are characterized not only by sovereignty towards outside, but also by state's sectoral autonomy, quality of law enforcement, and social goal-attainment of the state as a social system (Nettl 1994: 25–9). Usually, the state is defined through its function or its capability to function. The pluralist paradigm of the state maintains that the function of the state is to achieve consensus and social order through 'continuous exchange of demands and responses by social groups and government and a continuous sequence of bargaining processes' (Alford 1994: 65). Goldsmith and Posner maintain that the 'existence of the state depends on the psychology of its citizens' (Goldsmith and Posner 2005: 4). Besides the fact that state is an abstraction, the distinction between the state and the influences on it are difficult on the grounds that the negotiations and treaties in the name of the state are conducted by individual leaders who ultimately comply with or breach them. Goldsmith and Posner conclude that states do not have agency; as a consequence the theory of the state must be founded on the theory of the individual choice because state actions depend on individuals' beliefs and actions (Goldsmith and Posner 2005: 4). This is why state interests are not always easy to determine because they depend on the preferences of the state's political leadership. Therefore, states' interests can vary by context, and contrary to realists' assertions, states' interests are not limited to security and wealth (Goldsmith and Posner 2005: 6).

Based on empirical arguments, state is often defined as a process of exploitation and an instrument of the ruling class (Alford 1994: 66; Offe 1994: 104–5). This is mainly because state is a collection of individuals, each with his or her own particular interests (Levi 1994: 149). Renan argues that common interests are not enough to build a nation because common interests may lead to, for example, trade agreements, whereas nationality consists of feelings, and it is body and soul at the same time (Renan 1995: 55). According to Hintze, state is an organization of (for) *war* driven by the necessities of defence and offence. Therefore, state is an organization that cannot be determined solely by economic and social relations and clashes of interests (Hintze 1994: 202). Gourevitch similarly suggests that political development of a state is shaped by war and trade. Whereas the former shows the share of states' power in the international states system, the latter implies the share of wealth of a state in the international economy (Gourevitch 1994: 204).

International anarchy is a threat as well as opportunity for the state: a threat to be conquered, occupied, annihilated or made subservient; an opportunity because the threat of anarchy can lead to power accumulation, dominion, empire, glory, total security (Gourevitch 1994: 215). Gourevitch concludes that both international relations and domestic politics are interdependent.

According to Schmitt, the task of the state is to materialize the law. Through this task the state becomes the centre of the law–state–individual relations. Whereas the law is the pure norm, the state is the link between this ideational world (law) and the empirical world of the citizens. The individual as an empirical being will be conceived and consummated by the law and the state (Schmitt 2005: 10). Since the law is the ideational norm of the state, the law is not in the state; instead the state is in law. This means the state is to be derived from law, and its essential characteristic is to be viewed in its relation to law. Therefore, the state is the construct of law and mediator of the law: state's meaning consists of the task to materialize the law; and it is the instrument through which the law impacts on the reality or the empirical world. The state makes use of the empirical world in order to reform the very empirical world (Schmitt 2005: 52–76). This means that the empirical world also has an impact on the state. But this depends on the legitimacy of the state.

Weber defines legitimacy as the willingness of followers to accept orders given to them as properly to be obeyed. Legitimacy of the state implies its ability to provide a territorially centralized form of organization (Weber 1972: 122–30). This ability of the state is further the source of its autonomy (Mann 1985). States need bureaucracies nurtured by interest and the capability to pursue collective goals in a predictable and coherent way (Evans 1994: 411). Carniero concludes that state is not a product of 'genius' or the result of chance, but the outcome of a regular and determinate cultural process that takes place if the appropriate conditions existed as a response to certain specific cultural, demographic and ecological conditions (Carneiro 1970). Two of the key causes for the rise of the state are economic factors (industrialization) and military factors (such as expansionism and war) (Mann 1977). Renan maintains that the creation of the state is based on the act of committing violence but forgetting it (Renan 1995: 45). Violence is forgotten as violence, but remembered as patriotism to preserve the nation. Through externalization of the evil and threat a nation is a large community of solidarity sustained by violence towards external threats or domestication of internal disturbances through collective consent and desire to carry on the common life of all (Renan 1995: 57). For Tilly, war-making and state-making are organized crime. As a consequence, state is a double-edged protection: one is comforting, and the other is ominous. The state is a shelter against danger as well as it tries to avoid the damage that itself threatens to deliver (Tilly 1985). Constructed by and for war, state on the one hand concentrates all functions of the society in its own hands (sovereignty) and it exercises powers over all the people within its borders (territorial integrity), and on the other hand it is an abstract organization not identical with either rulers or ruled, neither man nor a community but an invisible being with an independent *persona*, as if it were a person in making contracts, owning property, defending itself (Creveld 1999: 416).

A functioning state presupposes a centralized authority that is able to defend territory from external aggression and to regulate conflicts within (between elites, between economic groups, and between socio-cultural or ethnic groups). Therefore, 'the "state" comprises a set of differentiated, autonomous, and public institutions which are territorially centralized and claim jurisdiction over a given territory, including monopoly over coercion and extraction' for the sake of social order (Smith 1994: 63–4). Order is an organizational performance in relation to a complexity that can be managed (Willke 1997: 306). This complexity is crucial for state in international system.

State Systems and the International System

In many colonized African states the colonial powers incorporated different ethnic groups unequally into a divisive system of colonial power (Smith 1994: 73), such as in Rwanda where the Tutsis were dominant, or Sudan where the Arabs have dominated the politics and economy. Smith suggests some important thoughts for states affected by conflicts due to ethnic factors: states that have a divided ethnic core (lack of a dominant ethnic core that gathers all other ethnic groups together and integrates them into the state-making project) and rival ethnic pasts that are weak are less well developed than their ethnically secure counterparts (Smith 1994: 80). This aspect is one of the characteristics of many of the African states because, in many cases, the political authority is more personal than institutional. But Jackson and Roseberg 'conclude that international society is at least partly responsible for perpetuating the underdevelopment of the state in Africa by providing resources to incompetent or corrupt governments' which are not institutionally bound and controlled to use those resources effectively and properly. Therefore, the role of the external factors is considerably significant for understanding the persistence of Africa's weak states (Jackson and Roseberg 1982). Consequently, African states systems cannot be separated from the international states system. This leads us again to the system theoretical concept of system–environment (state-international system) interpenetration.

According to Bull, the essential attributes of the international system are: plurality of sovereign states; a degree of interaction among them, in respect of which they form a system; and a degree of acceptance of common rules and institutions, in respect of which they form a society (Bull 1995: 225). States are and become what they are through their relations towards inside (self-reproduction) as well as outside (system–environment communication). Because of these systemic reasons, states are in continuous, unfinished and unpredictable process. In this context Campbell suggests that the nature of identity of states is performative: it is in permanent need of reproduction. State is an imagined political community which consists of territoriality and the many axes of identity. States are never finished as entities. They are an unfinished world because they depend on the continuous creation of excluding order towards outside and including identity towards inside. The performative nature of identity, i.e. the permanent need for reproduction, is indeed acknowledged to some extent but the fact that the political community is imagined based on territoriality and the many axes of identity has not been given due consideration (Campbell 1998: 12–20). In this misperception the sources of danger will be misrepresented as given or pre-existent realties independent from the context of intersystemic causes. Realism in international relations misrepresents this intersystemic phenomenon by essentializing the victimhood of the self (own nation) and the evilness of the other states.

As von Glasersfeld argues, realists believe that a continuous and well-structured reality exists in itself. This reality determines our experiences and becomes the object of our knowledge. What we consider as objective reality emerges basically from the fact that our own experience is confirmed by others. Things that are perceived not only by us but also by others are considered in daily experience as well as in our knowledge system as real. Intersubjective repetition of experiences is the best guarantee for objective reality (Glasersfeld 2003: 33). This intersubjective (which we can call also intersystemic) repetition implicitly or explicitly recognizes the adversaries or agents as system.

States are systems founded primarily on such intersystemic misrepresentation than international law. Commonly accepted criteria for statehood based on international law are a permanent population, a defined territory, government, capacity to enter into relations with other states, independence, sovereignty (Murphy 2006: 32; Crawford 2006: 45–89; Dixon 2006: 106–8). But here we have to point out two important aspects: first, though the international law pretends to set common and defining characteristics of state, states are in continuous creation. States have to permanently define themselves towards the inside as well as the outside. This means, ultimately, it is not only the international law which defines the state; instead states themselves have also to define themselves. Second, even if a state is not legally recognized, it can politically exist, defend its territorial integrity and independence, provide for the preservation and prosperity of its citizens, administer its services and determine the jurisdiction and competence of its courts of justice (Crawford 2006: 38). Based on its financial system, security instruments, political system etc., state defines itself as a system operating in complexity (communication towards outside) as well as in contingency towards inside.

Based on the Weberian definition of the state, an orderly modern state depends not on abolishing of hierarchy; instead it depends on its capacity to grant the subsystems' autonomy and responsiveness and protect them through appropriate mechanisms. Through this type of system–subsystem relations hierarchy becomes an effective organizational principle even for complex systems (Willke 1997: 314). According to Dunleavy, a state is a set of organized institutions or subsystems with a level of connectedness or cohesion (Dunleavy 2000: 611). A state operates in a given spatial territory, inhabited by a substantial population, which is organized as a distinct society. A socially accepted function of a state's institutions is to define and enforce collectively binding decisions on members of society. Their existence creates a public sphere differentiated from the realm of private activity or decision-making; and the state must be ultimately both absolute and legitimate (Held 1989: 14). The concern of the international security system is how an institutionally failed African state system can be rebuilt for the benefit of international security (Rotberg 2003, 2004, 2005; Walzer 2004; Fukuyama 2004).

The Systemic Problems of the African States

Ayittey suggests that the basic problem of Africa lies in the fact that there are three Africas that are constantly clashing.

> The first is traditional or indigenous Africa that historically has been castigated as backward and primitive. Yet it works – albeit at a low level of efficiency. Otherwise, it would not have been able to sustain its people throughout the centuries. Today it is struggling to survive. The second Africa is the modern one, which is lost. The third is the informal sector, a transitional sector between traditional and modern. Most of Africa's problems emanate from its modern sector. They spill over onto the traditional, causing disruptions and dislocations and claiming innocent victims. Most Westerners generally have difficulty dealing with and reconciling these two Africas. (Ayittey 2005: 19–20)

Eleven of the twenty critical states in the world, according to the Foreign Policy Failed States Index of 2008, are in Africa; this is caused by various factors: corruption and militarized politics and resources-based conflicts in Nigeria; the devastated democratic institutions and international isolation of Zimbabwe; the politically divided and economically poorly performing Ivory Coast; the politically and economically devastated DR Congo; Sudan, marred by conflicts based on socioeconomic and cultural marginalization; post-1998 undemocratic Eritrea and Ethiopia; authoritarianism in Togo, Cameroon, Guinea, etc. These are just a few of the cases where conflicts, bad governance and undemocratic institutions are interacting. Most of the African states are characterized by state *dirigisme*; administrative ineptitude; overspending, wasteful practices, extravagance with public funds; many failed political and economic grand initiatives (Ayittey 2005: 307–16). Similarly, the Foreign Policy Failed States Index gives some reasons for state failure: 'rampant corruption, predatory elites who have long monopolized power, an absence of the rule of law, and severe ethnic or religious divisions'.

However, it has to be noted that the *Foreign Policy* magazine concentrates only on internal factors that cause state failure or weakness. Such analysis is misleading because it does not take into account the role of the external factors or the systemic dimension of state failure in the international system. The role of global competition for resources such as oil, arms transfer, alliances with non-democratic African governments, etc., are not given due consideration in the analysis as causes of state failure.

According to Jean-François Bayart et al. (2001), there are five symptoms that characterize African political and economic situations: first, the relegation of the sub-Saharan Africa in diplomacy, economics and finance (i.e. loss of diplomatic importance with an economic and financial crisis, caused by the devaluation of Africa in the estimation of great powers); second, the failure of democratic transition; third, the continuation and spread of armed conflicts in most parts of Africa; fourth, the recomposition of the subcontinent around new foreign influences and powers such as China and India; and fifth, the growing implication of African and Western and Eastern economic and political entrepreneurs based in Africa in activities that may be considered illegal or criminal according to Western criteria (Bayart et al. 2001: 2–9). Undemocratic institutions emerged at least partly based on the 'idea of the state' of the African elites (Clapham 1996: 45). In the process of state-building in Africa, the 'idea of the state' resulted in undemocratic culture.

Making a Democratic System out of Africa

Democracy is more than mere rituals of voting and elections. It is the plurality of opinions, freedom of expression, multi-party political system, political competition, free and universal multi-party elections, fundamental and human rights, rule of law and accountability of the rulers that constitute democracy. This is even more than mere majority rule because even a majority can be dictator against the ideational, ideological, racial, ethnic, lingual and religious minority (Tocqueville 1956: 112–27). Thompson has summarized the constituent elements that are considered to be the most important for democratization: democratic consolidation needs a credible opposition, a strong civil society, strong economies, separation of the state and ruling party, regime change through democratic elections, addressing the challenges of ethnic mobilization, dealing with threat of the military, and establishing political culture (based on shared political ideas, attitudes and belief that underlie a society) (Thomson 2007: 236–44).

The quality and extent of democracy can be judged on the possibility of freedom and equality; the possibility of associating and communicating in public spheres, 'informed by liberal presuppositions, and governed politically by representative institutions based on wide suffrage and contested elections' (Mackie 2003: 1). Habermas suggests that democracy is a deliberative politics and a communicatively generated power. This is about public deliberation of matters of mutual concern to all (Benhabib 1996: 87). Since democracy is a public deliberation about common concerns, it is a symbol of freedom. But at the same time, it also points to anxiety and agony (Perry and Moran 1994: 283). Pointing to this problem of democracy, Birch (1993: 47–8) argues that if we define democracy as the 'rule of the people', we will have troubles because we have to define 'people' and 'rule'. Similarly Robert Dahl (1989) argues that it has become self-evident that democracy means 'rule by the people' but who ought to comprise 'the people' and what does it mean 'to rule'? (Dahl 1989: 3). According to Benjamin Barber, 'democracy is the regime within which the struggle for democracy finds legitimacy – legitimates itself' (Barber 1996: 357). Birch concludes that 'we cannot arrive at an objective and precise definition of democracy' (Birch 1993: 48).

Any democratization process is dependent on the concept of the state. This means a clear idea of the state and its people are decisive for defining democracy.

> …'idea of the state' encompasses not only its explicit institutions and goals, but also the often unarticulated collections of assumptions, identities and traditions which shape its behaviour, and at the same time confer on its government such legitimacy as it possesses. In cases where the idea of the state is almost universally shared, any threat to the security of the incumbent government, by other than legitimate constitutional means, can come only from outside, and the state's foreign policy can be regarded as representing the domestic political community as a whole. Such threats as it has to contend with will then largely be determined by the extent to which its national mission congruent with those of its neighbours and the major global powers … The foreign policies which African states followed after independence therefore depended most basically on the extent to which their ideas of the state were shared, firstly among their populations, secondly with their neighbours and African states as a whole, and finally with the dominant states of the international system. (Clapham 1996: 45)

Various African leaders tried to substantiate through various Africanist ideologies that there is already democracy in Africa. *Pan-Africanism* of Nkrumah, *African Socialism* of Nyerere, *African Humanism* of Kaunda, *Négritude* of Senghor, *Authenticity* of Mobutu, etc., tried to demonstrate that there is already democracy in Africa, and argued state-building as well as national political culture in the African states have to be based on African conceptions of democracy. Various researchers have shown that democracy was already there in Africa when the colonizers came (Ayoade 1986; Molutsi 2004; Kabongo 1986; Mazula 2004; Magang 1986). But what are African democratic concepts? The 'village democracy' (palaver democracy) in which elderly *male* members of the village come together and conduct a long parley until they agree or solve a problem?

Kabongo argues that democratic systems functioned in Africa in the past and are functioning in the present, therefore democracy is not *intrinsically* alien to African people (Kabongo 1986: 35). Kabongo suggests that 'a more pertinent question at this juncture to ask oneself is why the Western type of democracy has been so difficult to implement in the African context over the years, and what kind of democratic mechanisms are more suitable…' (Kabongo 1986: 35). Kabongo attributes the failure of the Western democratic system in Africa to 'artificiality' and 'sophisticatedness' of the former (Kabongo 1986: 36). But in my view the problem emerges in the idea of the state often leading to poor governance and undemocratic institutions.

African State System Against the Backdrop of Quality of Governance and Democracy

Governance as the art and process of decision-making is the most important parameter to judge the quality of democratization, the performance of the state and the overall system of the state. In July 2007 the World Bank published a study on the worldwide governance indicators. The study measured six dimensions of governance:

1. *Voice and accountability:* the extent to which a country's citizens are able to participate in selecting their government, as well as freedom of expression, freedom of association, and a free media.
2. *Political stability and absence of violence:* perceptions of the likelihood that the government will be destabilized or overthrown by unconstitutional or violent means, including domestic violence and terrorism.
3. *Government effectiveness:* the quality of public services, the quality of the civil service and the degree of its independence from political pressures, the quality of policy formulation and implementation, and the credibility of the government's commitment to such policies.
4. *Regulatory quality:* the ability of the government to formulate and implement sound policies and regulations that permit and promote private sector development.
5. *Rule of law:* the extent to which agents have confidence in and abide by the rules of society, and in particular the quality of contract enforcement, the police, and the courts, as well as the likelihood of crime and violence.
6. *Control of corruption:* the extent to which public power is exercised for private gain, including both petty and grand forms of corruption, as well

as 'capture' of the state by elites and private interests' (Kaufmann et al. 2007).

Positive effects of these six criteria seem to be self-evident. But if we look at them carefully, government effectiveness does not necessarily mean that everyone profits from the economic performance and efficiency of government policy. Many people can live in poverty in spite of the effectiveness of the government and its economic policies. Similarly, there are various mechanisms to make a state look stable. A state can still exist in spite of vivid violation of human rights or interethnic or similar conflicts, and it can be stable; but this does not mean that a state is peaceful. Even where we cannot observe physical or direct violence in which civilians are killed, women are raped, children are abused like in DR Congo or Sudan, there are cases in which structural violence takes place like in the post-civil war Angola, in corrupt Nigeria, or in Gabon, where 15,000 (out of 1.5 million) people hold 80 per cent of the nation's wealth. Here, there is no physical violence but an appalling structural violence. Good governance study should seriously include this phenomenon of structural violence.

At the same time it is obvious that these indicators of good governance have impacts on political as well as economic development. However, it is not easy to draw conclusions whether there is a positive correlation between democracy and economic growth. Of course, in the case of Botswana one could argue that democratically succeeding governments played a positive role in the economic performance of Botswana (Molutsi 2004; Rotberg 2004). Mauritius, with its US$13,700 GDP per capita, shows that there is a correlation between democracy and economic growth. Diamond suggests that Africa lags behind economically because it lags behind in governance (Diamond 2004: 267). But if we see the economic growth in China, there is no necessary correlation between economic growth and democracy.

We can see at least three phenomena when we observe democratization and the idea of the African states within the international system: first, because of the national pride within Africa it is difficult for Western countries to assert pressure, and many African politicians act ignoring international opinion; second, though donor agencies know that corruption is the bane of development and democracy in Africa, they are not always honest in their demand for democracy and good governance; third, the global players have been cooperating actively with various corrupt and undemocratic rulers in Africa, for example Biya of Cameroon, Eyadema of Togo, Mobutu of Zaire/Congo, Arap Moi of Kenya, Meles Zenawi of Ethiopia, Idriss Déby of Chad, etc., have been supported by the West. Diamond maintains that highly authoritarian and corrupt governments, in countries such as Cameroon, Angola, Eritrea, Guinea, and Mauritania, received levels of aid equalling or even well exceeding the African average of US$20 per capita. Almost all the authoritarian regimes, including those under international pressure for bad governance (such as Kenya and Zimbabwe), received aid well above the global average (US$11 per capita) in 2000 for low- and middle-income countries. 'Africa needs a truly new bargain: debt relief for democracy and development for good governance' (Diamond 2004: 272, 278). Because of bad governance in Africa, African states and governments have got varieties of designations: 'vampire state' (Ayittey 1999: 157–8, 2005: 239); 'criminalised states' (Bayart et al. 2001); 'shadow state' (Reno 1998).

The discussion about state, governance and democracy leads to the notion of government. What is government? What are its benchmarks? Karl Deutsch defines government as the direction and self-direction of large communities of people. This self-controlling or self-steering of the community (city, state or nation) requires mastery of knowledge or techniques. According to Deutsch, there are four elements that constitute government which reflects the type of governance: first, the manner of staying in control or in power; second, the basic nature and state of the country or organization, which is being controlled; third, the limits and opportunities to cope with; and the results wished to be attained. Deutsch maintains that the art of government lies in combining these four kinds of knowledge, and acting upon them (Deutsch 1974: 8–9). Most of the governments in Africa do not seem to be able to combine the four kinds of knowledge and implement them.

Gyimah-Boadi concludes that the general tendency in Africa to appoint incompetent persons to key bureaucratic and technocratic positions reflects the immaturity of African democratic systems and processes (Gyimah-Boadi 2004: 7–12). But Chabal suggests that some internal factors have caused the democratization process: 'the erosion of the legitimacy of the one-party state; the decline in all aspects of state capacity; the failure of development; the depth of economic crisis; and the strength of political protest and/or pro-democracy movements' (Chabal 1998: 291). According to Alex Thomson the democratization process in Africa in the early 1990s came to happen because of the state's loss of authority, i.e. its coercion power diminished, its co-option abilities starved by lack of resources; a new international political environment emerged which reduced aid from international financial institutions and foreign governments; rejuvenation of civil society (churches, trade unions, ethnic associations, women's organizations, professional bodies, etc., began to play a role). These developments were furthered by the wave of democratization on the global level, such as in Eastern Europe since 1989 (Thomson 2007: 232–6). Similarly, according to Chabal, there were external factors that have been at least as important as the internal factors for the democratization process: first, emergence of a more conservative outlook on North–South relations in the West (especially by the Thatcher and Reagan administrations) regarding aid to Africa, which was accorded a lower priority, and support for one-party states; second, the widespread imposition of structural adjustment programmes; third, the collapse of communism and the end of the Cold War (Chabal 1998: 293).

These developments towards democratization lead us to the question of what democracy or a democratic government is. Karl Deutsch has suggested that under a democratic government we understand that on the one hand the majority (directly or indirectly) makes or confirms laws and elects or confirms the government, its officials and its policies. But on the other hand the minority that disagrees today with these policies and laws should have the possibility to become a majority tomorrow. A political system can be designated as democratic only if a minority can remain free to express its views, to agitate for them, to organize, and try to win converts to its side. According to this democratic concept, the minority must have this freedom not only in its own interest but also in the interest of every member of the majority. The benefit of the rights of the minority for the majority is that if minority views are silenced, the majority cannot compare its ideas, cannot learn new ones, and cannot change its action. Democracy means a possibility for the majority as well as the minority to switch back and forth between majority and minority roles or situations (Deutsch 1974: 20).

> If so, they may find themselves in a minority on one specific issue but will get their way as a majority on some other, or they be now in a minority whose view may win majority support later. When the members of a minority group find themselves outvoted permanently and on most or all issues about which they care, then they may no longer see the prospect for political give-and-take – of reciprocity and change of roles – as realistic. When this happens, the minority status of the group has become diffuse (oriented to many issues [or almost all issues]) and permanent; their identification with the larger democratic community will be weakened, and their feelings of legitimacy and loyalty toward it may become severely strained. The outcome of such a development may be an effort on the part of the minority at secession, revolution, or rebellion… (Deutsch 1974: 20–21)

From the background of the flexibility of the border between minority and majority, a legitimate and stable democracy in the eyes of the entire population is gauged by the performance of the political system, which consists of effectiveness (making an unlikely outcome more likely to happen) and efficiency (how the benefits outweigh the costs) (Deutsch 1974: 230). The flexibility between the state of minority and majority leads us to the fundamental quality of democracy. Democracy is a process, not an end state. Democracy is not an apocalyptic promise which takes place after the evil is defeated. Democracy is a point of departure, the way as well as the goal. Mansfield and Snyder suggest that democratization is a rocky transitional period where democratization suffers reversals (Mansfield and Snyder 1995: 5). Further they conclude that violence is concomitant to the democratization process or democratic transition (Mansfield and Snyder 1995: 12–15).

Crippled Democratization and Prevalent Violence in the African States' System

Poverty and inequality are seen as the main causes of state failure in Africa (Rotberg 2003, 2004, 2005). Moreover, they challenge the establishment and consolidation of democracy in Africa. On the institutional level, lack of accountability consists of lack of answerability (that the public officials have to inform and explain to the public what they are doing); centralized governments, personalized powers, ethnicized politics and lack of transparency and accountability are widely accepted as causes of state weakness and failure (Kpundeh 2004: 121–3). Tetzlaff and Jakobeit designated democratization in Africa as an 'incomplete journey' because of the blocked democratic transitions, military coups, totalitarianism and military dictatorships (Tetzlaff and Jakobeit 2005: 153–95).

Lack of democracy has led many states to weakness or collapse. Zaire/DR Congo, Somalia, Sudan, Zimbabwe, Kenya and Chad are just a few examples. The one-party state in Africa did not give a chance to the articulation and functioning of majority–minority relations. According to Alex Thomson (2007), the causes of or arguments for one-party state (centralization) in Africa were: 'multi-party system would be socially divisive' (K. Nkrumah); 'no opposition existed in Ivory Coast' (Houphouet-Boigny); 'socialist ideology needed one-party system' (Sekou Touré); one-party system was most appropriate for democracy in Africa (J. Nyerere). Many African and non-African intellectuals argued that there is no reason why Africa should imitate the Western political culture (multi-party system).

In many countries bad governance is augmented by ethnicism. While discussing the African state system this is an aspect which cannot be disregarded, be it Zimbabwe or Ethiopia, or Kenya, Somalia, Sudan or Chad. Most African countries suffer from this sociopolitical disease. Berman et al. suggest that ethnic pluralism is and will remain a fundamental characteristic of African modernity that must be recognized and incorporated within any project of democratic nation-building (Berman et al. 2004: 3). 'There is little doubt that the wave of "democratisation" in Africa since the 1990s has seen an increase than decrease in the visibility of ethnic politics and conflict' (Berman et al. 2004: 9). Further, for African states system it is important to take into account that a rigid focus on ethnicity as the main cause of corruption and 'clientelism' is not necessarily true, because even within a single ethnic group there are subcategories based on kinship or membership of an extended family. Hence, Berman et al. suggest that competition for power between competing elites of the same ethnic communities play a decisive role in the process of governance and democratization. Elites of ethnic communities are never homogenous. There are symbols, grievances and expectations that are marshalled by elites to foster political consciousness of some of the members of the ethnic group while excluding some members and elites in the same community (Berman et al. 2004: 10).

Since ethnically divided societies in all African countries were perceived as a danger to state integrity, strong leadership has been demanded as indispensable for economic development and political stability. But a single-party structure encouraged corruption and an exploitative bureaucratic bourgeoisie and subordinated 'peripheral' state institutions (parliaments) to the core executive (government) (Thomson 2007: 110–13). 'In a political environment where executives were so powerful, laws became arbitrary. Politicians and bureaucrats felt disinclined to obey the constitution if their private interests were threatened.

Laws became less binding on those who ran the state, while those in civil society were still expected to conform' (Thomson 2007: 114). Cheryl Hendricks argues that even if ethnic identities are socially constructed, they become real for their bearers and have consequences. In the case of southern African, they have become an 'institutionalised difference' (Hendricks 2004: 113, 126).

A stable democratic society, rule of law, and economic development can happen in Africa only if there is a strong and independent civil society, a private sector, and a political society. Civil society means more or less autonomous associations, not state-intervened, 'that capture the private concerns of social forces and condense and transmit them to the public sphere of politics' (Mazula 2004: 183). A political society takes place if state and political parties adjust to the social and economic norms, market economy and civil society, multi-party parliament, independent media, and proliferation of non-governmental organizations (NGOs). Hence, the contributions of civil society are: to help pry open authoritarian systems, limiting the power of the states and challenging abuses of authority, monitoring elections and enhancing the credibility of the democratic process, educating citizens and building a culture of tolerance and civic engagement, incorporating marginal groups and enhancing responsiveness, providing alternative means for material development, opening and pluralizing the flow of information, building a constituency for economic and political reforms. African civil societies are severely constrained by an extremely weak material base, and rely on the state and international donors. Because of the state dependence the civil societies compromise their autonomy and are co-opted by undemocratic regimes, which further distorts their accountability (Gyimah-Boadi 2004: 100–108). Instead of fighting corruption, they have become part of the corrupt state system.

Corruption and State Systems

Africa's image is shattered by corruption, poor economic performance and weak competitiveness. Economic data suggest that just 3 per cent of global foreign direct investment flowed to sub-Saharan African in 2005, whereas most of the money has gone to only a handful of resource-rich countries like South Africa, Nigeria and Angola. Only 22 per cent of African households benefit from power supply. However, the number of mobile phone subscribers in Africa, which soared by over 50 per cent a year, is hoped to boost the African economy (Green 2007: 48).

Corruption in Africa increases the costs of goods by up to 20 per cent, and deters investment and impedes growth by 0.5 per cent (Lockwood 2006: 66). Corruption and undemocratic institutions in most African states have resulted in the continent's current indebtedness of US$300 billion. Nigeria's US$400 billion in oil revenue since the early 1970s, crude oil output of 2.1 million barrels per day and about 184 trillion cubic feet of natural gas reserves should have made Nigeria one of the most stable and rich countries in Africa. However, Nigeria carries a burden of US$35billion foreign debts, 60 per cent of its population lives below the poverty line, and life expectancy is 47 years. Nigeria's rulers between 1960 and 1999 stole US$400 billion in oil revenues. This economic exploitation and corruption in Nigeria was accompanied by a civil war that killed 1 million people, 30 years of military rule and six coups (Perry 2007: 26; Smith 2006). This structural injustice and exploitation has left two-thirds of the country's 140 million people in poverty, a third in illiteracy and 40 per cent without a safe water supply because of environmental damage from the spillover of more than 1.5 million tons of oil over 50 years in the Niger Delta, which is one of the most polluted places on earth. It is estimated that corruption costs Africa US$148 billion per year.

Gabon's Libreville ranks among the top 10 most expensive cities in the world; 15,000 people in Gabon hold 80 per cent of the nation's wealth; four-fifths of the country consists of rainforest and it has coastal waters full of fish, but nearly all of the country's food is imported from Europe (Perry 2007: 28). States such as Kenya, Angola and Chad are also affected by the high level of corruption in Africa. Angola's economy was expected to grow by 27 per cent in 2008. Gabon's wells are slowly drying up but it is one of the major oil producers. Angola is the second-largest oil producer in sub-Saharan Africa, and production is growing 25 per cent a year, mainly because of China (Perry 2007: 25). China is always ready to offer money to corrupt countries such as Angola if its oil interests are at stake, as it offered in 2004 US$2 billion credit without particular requirements of democracy and good governance, and announced in 2005 and 2006 an additional US$3billion in loans. 'Well-connected businessmen and unscrupulous government officials grow impossibly rich, and the ruling elite uses its wealth and largesse to consolidate its own power.' In spite of the US$10 billion in oil revenues in 2005 alone, 70 per cent of Angolans still live below the poverty line (Perry 2007: 25–6). There are systemic causes for this.

The system theory has shown us that system and environment are interpenetrating. The internal self-reproduction of a system cannot be totally separated from the environment or other systems, or international relations in this case. Some of Africa's challenges, such as patronage politics, corruption, neopatrimonialism, ethnicization of political alliances, elections marred by violence, rigged elections or pseudo-multi-party systems, manipulations of the constitution and frequent postponement of elections within the state system, cannot be separated from some factors penetrating from the international system. The international system (as will be shown in Chapter 2 in detail) hinders in some cases the emergence of any viable democratic institutions. At the same time it has to be stressed that the emergence of viable democratic institutions in many African states depends not only on the goodwill of the national political actors, but also on a genuine will of the international actors or international system in Africa. Unfortunately, the international system has been contributing to anti-democratic systems in Africa.

African states are in an anti-democratic international system. Therefore, while discussing problems of democracy and the weak state system in Africa, it is important to take into account the global systemic problems of democracy and governance. On the global level there are very often policies that are selective and at times selfish and contradictory. Selectivity refers to the concentration of criticism on some fraudulent elections in Africa (such as in Zimbabwe 2002, 2008) while virtually ignoring others, as in Zambia (November 2002) and Madagascar (December 2001). In the 'fight against global terror' external powers like the US are guided by selfish interests rather than engagement in global democracy or good governance. Since the bombings of the US Embassies in Kenya and Tanzania in 1998, Kenya has become one of the key US allies in Africa. In the context of the US-led 'war on terrorism' and the search for Islamic militants in Eastern Africa, the US as well as the UK ignored then President Daniel Arap Moi's poor human rights record in favour of renewing a military cooperation agreement that allowed British troops to use bases in Kenya. In the Great Lakes region, on the one hand, there have been cases of pushing for peace in the DRC but without openly criticizing friendly states such as Yoweri Museveni of Uganda and Paul Kagame of Rwanda for their involvement in the war (Williams 2004: 48–9). Between 1965 and 1997 Mobutu was supported by various international actors in spite of the fact that he economically depleted and undermined all democratization moves.

Oil producer regions of Africa belong to the 'strategic national interest' of the US. Warships patrol off West Africa and there are demands for a permanent military base in the region, such as a separate US African Command. Whereas Nigeria supplies 10–12 per cent of US oil imports, the Gulf Guinea will supply 20–25 per cent of total US oil by 2010 (Perry 2007: 24). Angola supplies 47 per cent of Africa's oil exports to China, followed by Sudan (25 per cent), the DRC (13 per cent), Equatorial Guinea (9 per cent), and Nigeria (3 per cent) (Broadman 2007: 81). Russia, Japan and India are increasingly interested in the oils of Equatorial Guinea, Cameroon, Chad and DRC. The Gulf of Guinea will earn US$1 trillion from oil by 2020 (Perry 2007: 24). These countries are neither politically stable, democratic nor institutionally strong. Their weakness and failure as a state system cannot be separated from the international system.

Successful and Failed African State Systems in the Democratization Process

Chabal suggests that between 1989 and 1994 most African states began to step towards multi-party democracy and away from a single-party political system (Chabal 1998: 290). Political developments such as democratization since the beginning of the 1990s have been dubbed a 'second liberation' caused by the end of some single-party political systems and military dictatorships, such as in Ethiopia and Eritrea starting in 1991 (Gyimah-Boadi 2004: 6). Even Mobutu claimed to be moving towards the democratization process in 1992. Apartheid came to an end in 1994. Liberal democratic constitutions in countries such as Benin (1990), Mali (1992), South Africa (1994), Ghana (1993), Malawi (1994), Nigeria (3rd Republic in 1993, 4th Republic in 1999), Ethiopia (1995) and Eritrea (1997) began to change the democratic face of Africa. Civil society and parliaments began to play in political arenas. Gyimah-Boadi argues that some basic conditions for the establishment of rule-based governments and states are increasingly introduced; official arbitrariness is being reduced; human rights are increasingly enjoyed; corruption is addressed more and more; constitutional documents are becoming the normative point of reference for African politicians; mass media is expanding, though confined to urban areas (Gyimah-Boadi 2004: 6–7).

There are some democratically *well-performing* countries in Africa. Though the ethnic violence between Inkatha/Zulu and ANC supporters marred the first democratic elections campaign of 1994 in South Africa, the elections in 1999 happened peacefully. According to Marks, some of the positive factors that contribute to the peace between ANC and Inkatha are that the Apartheid's 'divide et impera' strategy came to an end; even the formerly hostile anti-Inkatha youths began to approach the traditional Zulu leaders; a good diplomatic progress between the ANC leaders and the Inkatha; lack of external support for Inkatha; international recognition of the new South Africa; and the high level of sympathy that especially Mandela enjoyed on the international level (Marks 2004: 194–99). The emergence of a new party (Congress of the People – COPE) in 2008 is a chance to further democratization not only in the state of South Africa as such but also in the African National Congress itself.

Because of an active free press, an independent judiciary, human rights respected by the authorities and lack of corruption, Botswana and Mauritius are seen as exemplary democracies in Africa. Botswana is usually considered as one of the few African success stories politically as well as economically. Some factors that contributed to democracy and development in Botswana are: historically, the Protectorate established in 1891 was negotiated between the British government and individual chiefs; ethno-cultural homogeneity; 'getting the politics right', i.e. continuous, overlapping policies while successors following the footsteps of the predecessors with stable party and leadership (Seretse Khama period 1965–1980; Ketumile Masire 1980–1998; Festus Mogae since 1998) (Molutsi 2004: 160–70). In Namibia, even if SWAPO has dominated Namibian politics since independence in 1990, Namibia is a democratic state respecting human rights.

There are a good number of democratically *promising* countries. When Ghana's President Rawlings stepped down in 2000, President Kufuor of the opposition NPP (New Patriotic Party) was elected, first in December 2000 (with 57 per cent of second-round votes) and again in December 2004 (with 53 per cent of second-round votes). The NPP dominates Ghanaian politics, not only on the governmental level but also on the legislative level. John Atta Mills from the National Democratic Congress was elected in December 2008. Similarly, presidential elections that took place on 5 March 2006 and parliamentary elections, held in March 2007, witnessed promising democratization records in Benin. Proliferation of political parties, a vibrant independent press, powerful trade unions, civil society, some 5,000 local NGOs, including human rights groups operating freely without government interference, characterize the politics in Benin. In Zambia, by regional standards, neither the democratization process nor human rights record is particularly bad. In the 2006 elections, a serious but democratic challenge from Michael Sata's

Patriotic Front confronted Levy Mwanawasa's Movement for Multi-party Democracy (MMD), but in the end Mwanawasa secured 43 per cent of the popular vote against Sata's 29 per cent. However, in October 2007, Mwanawasa threatened those who opposed plans for a new constitution with treason charges, which is an undemocratic democratization.

Senegal used to be one of the stable countries in Africa; however, the tumultuous elections in February 2007 have shown that this could change. The tradition of democracy in Senegal has been a rare model, but fights between supporters of President Abdoulaye Wade and those of his former prime minister, Idrissa Seck, before the elections on 25 February 2007 overshadowed the Senegalese democratic records of remaining stable despite its religious and ethnic heterogeneities. In the weeks before polling for the February elections, police used teargas to disperse protesting oppositions, critics of Wade received death threats, and votes were rigged.

Uganda's politics since 1986 was known as the so-called 'no party' political system, or Movement System. However, Uganda's 1995 constitution provided for political participation. The Movement System was endorsed by 91 per cent in a referendum in 2000. But since the turnout was low and the pro-multi-party side had limited opportunity to present its case, it cannot be designated as entirely democratic. The 2001 elections were marred in some places by violence and intimidation, which sent Dr Kizza Besigye into exile for four years in South Africa. Museveni orchestrated a parliamentary move in August 2005 to lift the constitutional two-term limit. Consequently, in the February 2006 elections – which the EU Observation Mission and the Commonwealth Observer Group concluded as free and fair despite identifying significant flaws in the campaign process – Museveni won 59 per cent of the vote, whereas Dr Kizza Besigye of the main opposition group, Forum for Democratic Change (FDC), gained 37 per cent, and now controls 37 seats out of 215 parliamentary seats. However, Uganda still has a reasonably free media and the human rights record improved enormously after Museveni came to power in 1986.

Though large numbers of different ethnic groups comprise the population and despite the Chama Cha Mapinduzi (CCM – 'Movement for the Revolution') remaining the overwhelmingly dominant force, Tanzania is one of the peaceful, stable and relatively democratic countries in Africa. However, sometimes violent conflicts between the CCM and Civic United Front (CUF) (with a strong power base on Zanzibar – most notably the island of Pemba) have been affecting not only the political stability but also the democratization process. Violence, intimidation and serious allegations of rigging overshadowed the 1995 and 2000 elections. The rest of the opposition seems to be weak and divided. In May 2005 CCM's Kikwete comfortably won the December 2005 election, securing 80 per cent of the vote.

Sierra Leone's presidential and parliamentary polls on 11 August 2007 were designated as free, fair and credible – at least partly thanks to the great job done by Christiana Thorpe, Chief Electoral Commissioner – as a continuation of the democratization process that started in 2002 when a 10-year civil war, which began in 1991 and killed 50,000 people, came to an end. Whereas Rwandan politics is still strongly ethnicized, the 'power-sharing politics' of Burundi is trying to overcome the Hutu–Tutsi hatred (Lemarchand 2007: 1–20), and large numbers of Tutsi have been joining previously Hutu parties.

It seemed originally that Madagascar had recovered from the presidential elections of December 2001, which pitched the country into crisis when the supporters of Ratsiraka and Ravalomanana became embroiled in a period of civil unrest that lasted for several months. But Madagascar's High Constitutional Court carried out a recount of the first-round votes and declared Ravalomanana the winner in April 2002. Neither the December 2006 presidential elections (which granted President Ravalomanana a second, five-year term) nor the legislative elections of 15 December 2002 (which enabled President Ravalomanana's party to control 102 out of 160 seats) led to a similar crisis. But the opposition leader, Andry Rajoelina, staged a coup attempt in March 2009. Democratization is still very unstable on the island.

There are *crippled democracies* in Africa, mainly by corruption. Angola is one of the highly corrupt states in Africa. New election laws were passed in April 2005, and the legislative elections took place in September 2008, in which MPLA won 81.6 per cent, UNITA won 10.4 per cent, and other minor parties shared the rest. Presidential elections are expected for September 2009. Angola and Nigeria have some similarities: dominant oil economy and corruption. According to Abgaje, even the 1999 Constitution of the Fourth Republic of Nigeria is fundamentally flawed and fraudulent 'given the fact that the final version of the document was authored by a few military officers in the countdown to the Fourth Republic' (Abgaje 2004: 209). Moreover, the constitution bestowed the president legislative powers; it is centralistic, i.e. it favours federal government over states and local governments (Abgaje 2004: 209–12). Abgaje concludes that 'an enduring legacy of the many years of undemocratic rule (colonial, military, and civil) to which Nigeria and Nigerians have been exposed is a deep-rooted militarization of the socio-cultural landscape' (Abgaje 2004: 220).

Abgaje suggests that key challenges for Nigeria are: managing ethnic conflicts, regional, and religious tension; managing Nigeria's size, resource endowment, and economic development; addressing corruption and rent-seeking practices (Abgaje 2004: 205). Mustapha underlines an important aspect that inter-ethnic as well as intra-ethnic sectarian processes threaten democracy and Nigerian unity (Mustapha 2004: 257). Even the April 2007 presidential elections in Nigeria were widely considered to be undemocratic. Olusegun Obasanjo paved the way for the victory of his favoured candidate Umaru Yar'Adua in the elections on 21 April 2007 (Perry 2007: 26).

Egypt, Libya, Morocco and Ethiopia are *bypassed by the democratization* process. Egypt's undemocratic politics can be deduced to the fear of terrorism. However, not only Islamist parties like the Muslim Brotherhood are banned or all religious parties are prohibited under the constitution, but secular parties are also subject to restrictions. By taking advantage of the fear of terrorism, President Mubarak has been in power since 1981. Libya is an authoritarian state led by Colonel Muammar Abu Minyar al-Gaddafi. Morocco is a constitutional monarchy with undemocratic records, but the political system is evolving from a strongly centralized monarchy to a parliamentary system. The election on 7 September 2007 was considered as transparent and the entire election campaign professionally organized, but the voter turnout of about 37 per cent was an historic low. For years Ethiopia was one of the African darlings of the West. Since the general elections on 15 May 2005 the public support for opposition parties became very clear. This support resulted in deterioration of the political atmosphere and the arrest of a number of opposition leaders, civil society representatives and journalists for an alleged role in stimulating violent protests in November 2005. In Kenya the unrest triggered after the election on 27 December 2007 not only killed hundreds, displaced thousands and ignited interethnic conflict, but it also substantially damaged the democratization process in Kenya and its image on the international stage.

Countries such as Burkina Faso, Cameroon, Côte d'Ivoire, Togo and Guinea have gone *astray from democracy*. Burkina Faso's Compaore, who believes only he guarantees political stability and economic progress, stood for a third term in the presidential elections on 13 November 2005; he won with 80.3 per cent of the vote against 11 other candidates. In the name of this stability, opposition parties are often harassed. After the 22 July 2007 elections in Cameroon, a country in which opposition to Mr Biya is limited, Biya's Democratic Rally of the Cameroonian People Party held 149 of the 180 seats in parliament. The leader of the opposition SDF party, John Fru Ndi, called the polls 'a sham' marred by fraud, and fears that the ruling party might try to use its two-thirds majority in the assembly to amend the constitution to allow Mr Biya, already head of state for 25 years, to seek a new term in 2011. Moreover, in the municipal polls, the governing party won 303 out of 363 communes. Togo has been known as one of the dictatorial states in Africa, especially up until President Eyadema Gnassingbe died on 4 February 2005. His son, Faure Gnassingbe, won the presidential election in April 2005, which the opposition condemned as rigged, but the African Union and ECOWAS accepted this result and urged Gnassingbe to include members of the opposition in the new government. Parliamentary elections were held on 14 October 2007, and the next ones will be held in 2012. Togo had one of the most appalling human rights records in Africa under Eyadema.

Central African Republic (CAR) and Chad are two examples of uncontrolled states and non-extent democracies. CAR is politically unstable, affected by conflicts in Sudan and Chad, and the government authority barely extends outside the capital. It is dependent upon French army support, not on the democratic legitimacy of the population; human rights are abused by the security forces, which are rarely under the full control of state authorities. In Chad Idriss Déby stood for a third term as president in elections on 3 May 2006 – after having successfully orchestrated the removal of presidential term limits through a referendum in June 2005 – and won a new mandate with 64.67 per cent of the vote, which was boycotted by the main opposition parties.

The incursions by rebels in Guinea's border regions with Liberia and Sierra Leone since September 2000 have claimed more than 1,000 lives and caused the displacement of thousands of the population. In late 2001 President Conte proposed to extend his presidential term, which contributed to further escalation of the conflict. Even though the agreement in March 2002 between Guinea, Sierra Leone and Liberia on measures to secure mutual borders and to tackle insurgency eased the relations between these neighbouring countries, the detainment and suppression of oppositions within Guinea have remained an unresolved internal problem. Elections boycotted by the opposition parties in 2003, assassination attempts against the president in 2005, crippling general strikes since 2006, and the declaration of a state of emergency on 13 February 2007 have made Guinea one of the most out-of-control states in Africa. Captain Moussa Dadis Camara staged a military coup, following the death of President Lansana Conte on 22 December 2008, and declared himself President of the National Council for Democracy and Development.

In Ivory Coast, after the conflict divided the country into north and south, a buffer zone between the two sides was created. On 27 March 2007 the two sides agreed to appoint the prominent rebel leader, Guillaume Soro, to the position of prime minister. On 30 July 2007, when President Laurent Gbagbo set foot in the north for the first time since rebels occupied it in 2002, he declared, 'The war is over!' For Soro, Gbagbo's presence in Bouake, the former rebel stronghold, sealed the reunification of the country. In the Failed States Index of 2008, Ivory Coast is situated between Afghanistan and Pakistan, which is quite gloomy.

Somalia is a political *vacuum*. Somalia collapsed in 1991. In 2000, the Transitional Government of Somalia, and towards the end of 2004, the Transitional Federal Government of Somalia, were created, but not successfully – mainly because of the influx of arms from regional states such as Ethiopia, Eritrea, Yemen, Syria, Saudi Arabia, Iran and many other Middle Eastern countries, which are supporting various factions, not only financially but also with weapons (Menkhaus 2006/7: 74–106). There is not a functioning democracy and the human rights situation is appalling. The new government has installed Sharia law.

Zimbabwe and Eritrea have taken a path from democracy to *dictatorship*. With unemployment at 80 per cent, by July 2008 Zimbabwe's inflation had reached 341 million per cent, more than 80 per cent of the population is living below the poverty line, and life expectancy is just 35 years, Zimbabwe is probably the worst-performing country in the world. Hundreds of thousands of Zimbabweans were made homeless in the government's Operation Murambatsvina, or 'urban clean-up', campaign in May 2005, which set out to destroy slums in the cities where opposition to the government and informal economy flourished. Economic mismanagement, official corruption, ethnic favouritism and political intimidation all issued a bad certificate for the government of Mugabe. The 'clean-up operation' brought forth very short-term solutions, but the fundamental problem of lack of democracy and devastating economic mismanagement are still there, and the 'clean-up' operation was opposed by about 70 per cent of Zimbabweans (Bratton and Masunungure

2007: 43–4). Draconian media laws curtail the media, access to information and restrict protection of privacy, because media, like foreign interventions, is seen as a threat to national security (Ronning 2003: 196–221). Mugabe has got enough support from the Eastern world (in particular China) and many African states. African leaders are not willing to condemn Mugabe (Dowden 2007: 15). Neither the 'big player', South Africa, nor the SADC is willing to put pressure on Mugabe, the once anti-apartheid hero. In the meantime Zimbabwe has become the symbol of authoritarian state ideology and the democratic agenda of a multifaceted civil society (Ronning 2003: 221). In March 2002 Mugabe stole the elections; he and his media denigrated Tsvangirai as 'tea boy'; opposition supporters were hunted down; 'party cards' determined the safety of most Zimbabweans, to criticize the president was forbidden; and the commander of the defence force, General Zvinavashe, declared that the army would not support anyone other than Mugabe (Meredith 2003: 225–36). It is still unclear whether the formation of a coalition between the MDC and ZANU-PF in February 2009 will change the situation.

Eritreans hoped that after independence in 1993 their country would become one of the most democratic and prosperous countries in Africa. However, Eritrean politics are completely militarized; the constitution ratified in 1997 was not implemented on the grounds that Eritrea's priority is the war with Ethiopia, not democratization (Mengisteab and Yohannes 2005: 131–60); the economy has been destroyed and the country is increasingly isolated, both regionally and internationally. It is estimated that every month between 400 and 600 Eritreans flee poverty and conscription through the dangerous Sudanese deserts, mainly towards Europe (BBC News 25 March 2007).

DR Congo and Sudan are failed states as well as *failed democracies*. In the DR Congo, the transitional government was sworn in on 30 June 2003 and was supported by European Union troops and UN peacekeepers. One of the key objectives of the transitional government was to prepare elections for 2005, but belligerent leaders, corruption and mismanagement threatened the population throughout the transition period. Ongoing conflicts and institutional weaknesses caused the postponement of national elections from June 2005 until March 2006. Moreover, the elections preparation faced huge challenges because there were 33 presidential candidates, 8,650 parliamentary candidates for 500 parliamentary seats, and 267 registered political parties. In October 2006 Joseph Kabila was elected president with 58.05 per cent of the vote in the second round and was inaugurated on 6 December 2006 for a five-year term. Now Kabila's PPRD controls 114 of 500 seats, the largest number of any political party, and controls more than 200 through its political allies. Similarly, the Senate (108 seats), elected on 19 January 2007, is controlled by the PPRD taking the largest share of seats (22), with the MLC (party of Jean-Pierre Bemba) obtaining 14 parliamentarians for a five-year mandate. President Kabila and PM Gizenga announced a government of 60 ministers on 5 February 2007. This high number of ministries shows that Congolese politics is about power-sharing, not about political efficiency. The democratic future of the Congo will depend not only on internal developments, but also on the impact of the politics in the region as well as global competition for the natural resources of DR Congo.

Since early 2003 the Arabization campaign of the Khartoum government in Darfur has killed more than 300,000 black Africans, displaced more than 2.7 million and sent more than 200,000 as refugees into Chad. The mutual destabilization between Sudan and Chad has displaced at least 140,000 Chadians and 235,000 Sudanese refugees and exposed them to the deserts in eastern Chad. Because of the very slow process in and deliberate hampering of the implementation of the Comprehensive Peace Agreement of January 2005, in the first half of October 2007 the SPLM/A threatened to leave the Transitional Government.

Conclusion

Above we have discussed the state of the African states in their attempt to organize themselves as political systems. Some have failed, while others have been performing promisingly. But their success as well as failure has to be seen in the systemic context. Tilly suggests that 'states form a system to the extent that they interact with each other regularly, and to the degree that their interaction affects the behaviour of each state' (Tilly 2001: 162). But this does not mean that the international system is taken for granted (Buzan and Little 2000: 90) or exogenously given (Wendt 1992: 396); instead the system is also constituted by the units, or states. These units can of course be *differentiated* from the system as well as from other units but *cannot be separated* from them because they constitute each other intersubjectively. Wendt has shown this intersubjective interaction convincingly.

As the international system influences the behaviour of the states in intersystemic interpenetration, the political culture within the states influences their behaviour in the international arena. This means that state leaders' decisions to rely upon either peaceful diplomacy or military force as the means to resolve international disputes are influenced by the political institutions and norms of political competition and conflict resolution within states. It is more or less widely accepted that democracies are less likely to resort to the aggressive threat or use of military force against all other states: not only are two democratic states very unlikely to become engulfed in military conflicts with each other, but democratic states are also less likely to initiate crises and wars regardless of whether an international opponent is democratic or not (Huth and Allee 2002: 2). The last part of this sentence is simply not true. The US invasion of Iraq has disproved it. At any rate, it is true that all states in Africa that are in conflict with their neighbours are undemocratic: Ethiopia, Eritrea, Somalia, Chad, Sudan, Rwanda, Uganda, etc. Democratic leaders seek negotiated settlements in disputes, and they should have a democratic consent from the national parliament and similar institutions before they wage a war against another state. This means that powerful democratic domestic political institutions exert constraints on democratic leaders because of the political accountability (Huth 2002: 7–8). Democratic leaders will weigh the relative advantages of negotiated compromise, military conflict, and continuing diplomatic deadlock quite differently. The internally as well as regionally unstable African states lack democracy; they destablize their neighbours and contribute to their own destabilization (as a backlash). This results in the interpenetration of system–environment instabilities of the state system internally, regionally and globally. Dörner suggests that 'we have to know that in complex systems we do not do just a single thing, instead, whether we like it or not, we do plenty of things. We have to know how to deal with side effects. We have to know that side effects of our decisions and determinations appear in areas where we never thought about' (Dörner 1989: 307). In the system–environment interaction, a system cannot be analysed in isolation, but instead only in discourse and within the context of its environment. The state system is an institutionalization of relationship with environment as well as autopoietically within the system.

The international system has been contributing to the current weak African states' system for decades, or even centuries. The African intrastate system and the international system are inseparably interrelated. This systemic interpenetration integrates Africa into the international security system. The international system, first, has integrated African states into the global system; second, has contributed to their dysfunction; third, they reflexively contribute to its dysfunction. However, it has to be emphasized that, as will be shown in the following chapters, African states are not mere victims of the international system; instead they have become an essential part of it.

Chapter 2
Africa in the International Security System

Since 11 September 2001, international security policy cooperation has come to the top of the agenda. State collapse, state failure and weak political and socioeconomic institutions in various African countries are considered to be breeding ground for rising threats to international security. Somalia, Sudan and DR Congo are among the top six failed states of the world (Failed States Index 2008). Moreover, global actors like the EU, some of the EU member states, the US, China, India, Russia and Brazil are consolidating their presence in Africa to secure their strategic interests. This chapter shows the impact of intersystemic penetration on Africa.

As Latham et al. argue, there is a silent intersection between the global and local through structures and relations that involve global and local actors. At the same time, the local and the global actors have fundamental impacts on the regional actors. They shape one another through specific junctures, joining diverse structures, actors, ideas, practices and institutions with varying ranges in a common social and political frame (Latham et al. 2001: 6). Barnett suggests a reconceptualization of the global and the local, since closely watched developments in local circumstances can lead to global conclusions (Barnett 2001: 64). This means avoiding seeing one particular location as the natural starting point for analysis and for tracing the causal chains because of the multiple interactions of domestic and international politics (Barnett 2001: 65).

The international security system discussed here is the system of multifaceted mutual interactions of international agents and structures. In the global system of competition, the security agenda of one global player impacts first on its globally powerful rival or rivals, and then significantly impacts the weak nations. To show these global systemic impacts, the main global actors in Africa are discussed in this chapter.

The International Security System

International security system implies two aspects: first, an organized international regulation or reaction to address international security threats. The international system is a kind of international integration, while integration is conceived as an attainment of sense of community or the creation of a security community (Haas 1977: 224). Intensified international security policy cooperation with Africa implies that Africa has to be integrated into the international security system. For example, various G8 summits have discussed how to solve Africa's economic hardships, how to stabilize Africa politically, how to prevent Africa from becoming a possible breeding ground for terrorism, how to prevent and resolve violent conflict on the continent, and how to undertake peace support operations in accordance with the United Nations Charter (Ramsbotham et al. 2005: 329–30).

There are various debates in international relations theory about whether the system influences the units' behaviour, or the units influence the system. David Singer argues that the behaviour of the state is not dictated by the international system (Singer 1960), whereas for Waltz, the system constrains the behaviour of constituent units and operates independently of the units (Waltz 1979). However Wendt, from a constructivist approach, suggests that the system and units are mutually constituted, and it is not possible to talk about one without the other (Wendt 1992). They are intersubjectively constituted and constrained (Buzan and Little 2000: 39–45). This chapter tries to show the intersubjective interaction among the big global players and the weak states.

Various current studies attempt to show that it is the conflict-ridden, weak, failed and collapsed states that will have negative impacts on global security. Ayoob suggests that there has been a 'serious underrating of the potential impact of conflict and disorder in the global periphery … on the international security agenda…' (Ayoob 2005: 32). Further, Ayoob points out that in international relations theory, some have argued that the issues of the conflict-ridden Third World are so marginal for the economically, politically and technologically powerful states that it is not imperative for them to be overly concerned with conflict and disorder. Conflicts, migrations, drug trafficking, terrorism, WMD, economic interdependence, and natural resources in the south have rendered the south to dominate the international system and its security agenda in the twenty-first century. In this system the domestic activity of the states, regional balances of power, global structures, international norms, and great power policies have impacts on the evolution and course of both the state-making enterprise and regional conflicts (Ayoob 2005: 33–4).

Since explicit interstate wars have been continuously diminishing and intrastate conflicts have not, international relations theory increasingly suggests that 'the threat posed by internal war is almost always greater than that posed by outside state' (David 2005: 77). Starting from a realist point of view, David argues that since people are driven by fear, honour and greed in pursuit of security, glory and gain, it 'is more accurate to assume a world of international order and domestic anarchy' (David 2005: 78). For Holsti, though the legacies of outside powers have been playing some role in conflicts, 'the problem of war in the Third World … states is not one of the relations between states, but of relations within states' (Holsti 2005: 108).

On the one hand, Acharya suggests that conflict analysis after the Cold War 'will need to pay more attention to systemic, as opposed to domestic or local, causes of international conflict' (Acharya 2005: 160). According to Acharya the superpower competition and alliance system during the Cold War had been contributing to the prolonging of regional wars, to the ineffectiveness of global and regional institutions created after the Second World War (Acharya 2005:

165–7). On the other hand, however, Acharya concludes that prospects for conflict and disorder in the post-Cold War period may have less to do with changes to the system structure than with the developments at local and regional levels (Acharya 2005: 189). This chapter argues that the international level has substantial impact on the local as well as regional levels, and vice versa.

Buzan et al. suggest that international systems are the largest conglomerate of interacting or interdependent units and there are no system levels above them. Currently, this level encompasses first the entire planet; second, the international subsystems, meaning the group of units within the international system that can be distinguished from the entire system (such as AU); third, units, i.e. actors composed of various subgroups, organizations, communities (such as states, nations); fourth, subunits, i.e. organized groups of individuals within units (such as lobbies); and fifth, individuals (Buzan et al. 1998: 5–6).

The definition of 'region' by Buzan and Waever is relevant for my analysis of conflicts in the Horn of Africa and in DR Congo. According to them, the region:

> refers to the level where states or other units link together sufficiently closely that their securities cannot be considered separate from each other. The regional level is where the extremes of national and global security interplay, and where most of the action occurs … Each regional security complex is made up of the fears and aspirations of the separate units (which in turn partly derive from domestic features and fractures). Both the security of the separate units and the process of global power intervention can be grasped only through understanding the regional security dynamics. (Buzan and Waever 2006: 43)

Buzan et al. define security complex as a set of units whose major processes of securitization, de-securitization, or both are interlinked. However, their security problems cannot reasonably be analysed or resolved apart from one another (Buzan et al. 1998; Buzan and Waever 2006). Buzan et al. suggest that securitization may serve short-term interests of internal and regional power consolidation, but in the long run it produces the international equivalent of autism and paranoia. On the intrastate level, securitization stifles civil society, creates an intrusive and coercive state, cripples (eventually) the economy, increases military expenditure, and maximizes the intensity of the security dilemma with neighbours that do not share the ideological project or that are not part of the security complex. To avoid excessive and irrational securitization is thus a legitimate social, political and economic objective of considerable importance for individual states as well as regional and intercomplex cooperation (Buzan et al. 1998: 208).

For Buzan and Waever, 'interstate security dynamics in Africa are often simply spillovers of domestic dynamics, particularly refugee flows, expulsion of foreigners, and civil wars. In a sense, security interaction in Africa is generated more by weakness than by strength, as when imploding states inflict spillover on their neighbours. Intervention by neighbours in domestic turbulence is fairly common' (Buzan and Waever 2006: 229). Buzan and Waever made a very significant observation about how conflicts in various parts of Africa have spillover effects:

> Spillover interactions between neighbours can create what might look like regional patterns, but these patterns have no obvious boundaries, and they are more often chains of discrete events rather than coordinated patterns of alliance and rivalry. Thus, for example, Somalia and Ethiopia, Ethiopia and Sudan, Sudan and Uganda, Uganda and Rwanda, Rwanda and Zaire/DR Congo, and Zaire/DR Congo and Angola have all played into each other's civil wars, creating a potential chain … There is no such interplay between Somalia and Sudan, or Sudan and Rwanda, or Rwanda and Angola, let alone between Somalia and Angola. The general pattern is that each country sits at the centre of a set of security interactions connecting it to its immediate neighbours, but with limits of power meaning that these individual patterns have not as a rule linked significantly into wider patterns of security interdependence. (Buzan and Waever 2006: 232)

However, it is important to note that the interconnectedness of the systems of conflicts is more than mere spillover effects. Various conflict levels enhance each other, and as a consequence we can only differentiate but not separate them. Moreover, as will be shown in the following chapters, it is not true that Somalia and Sudan, Angola and Rwanda, etc., do not influence each other directly. In my systemic conflict analysis I will demonstrate that there is a substantial mutual influence beyond immediate neighbours. Even if they do not immediately intervene militarily in a non-neighbour country in a structured way, they influence each other. This is clearly shown in the case study chapters.

The system of international relations is a system of national interests. Buzan and Gonzalez-Pelaez, while discussing the international security system, define 'system' as a sustained and structured pattern of interaction among agents (Buzan and Gonzalez-Pelaez 2005). According to Thompson, nations are more inclined than individuals to follow their own interests because of which only pretension of morality and hypocrisy exist instead of the morality itself (Thompson 1977: 21). For Dunn, international politics is about 'power relationships that exist in a community lacking an overriding authority' (Dunn 1977: 14). According to Kaplan, the international system determines the behaviour of the states. 'The materials of politics are treated in terms of systems of action. A system of action is a set of variables so related, in contradistinction to its environment, that describable behavioural regularities characterize the internal relationships of the variables to each other and the external relationships of the set of individual variables to combination of external variables' (Kaplan 1977: 111), suggesting a kind of systemic role. Role-based conflicts are a characteristic element of the international system. 'Roles come into conflict when actors existentially participate in different systems and when they must choose between the sometimes conflicting demands that their different roles place upon them' (Kaplan 1968: 384).

As constructivists argue, our knowledge of security is based on our knowledge of the world. If we mean to know the 'real' threats of security, we are talking about 'epistemic realism' in the sense that the world of insecurity comprises threatening objects whose existence is independent of our ideas or beliefs about them (Campbell 1998). In foreign and security policies, defining the foreign as an ontological danger is the *sine qua non* of the state. National identity is a discourse of danger: what we fear makes what we are. Our knowledge of insecurity consists of 'secularized eschatology' based on the conviction of the innocence of the Self and the threat of the Other and of the Otherness. A key argument of constructivists is that 'we do not ask whether a certain issue is in and of itself a "threat", but focus on the questions of when and under what conditions who securitises what issue' (Buzan and Waever 2006: 71).

Security policy is containment of domestic and international contingency and delineation of borders: a) by isolating the threats outside of the social body (state), and b) by domesticating the internal threat (Campbell 1998: 156). Campbell suggests that security policy depends on the ability to create new dangers if the old ones are not sustainable anymore, and on identifying the ethical boundaries of identity endangered by threats to order and morality that constitute national security. According to Campbell, it is an evangelism of fear. The dogmatization of identity is underpinned by the belief in a stable and pre-discursive and non-performative identity of state and imagined community (Campbell 1998: 179–97).

Various security studies suggest that the enemy is not always reality, instead a process and a notion. If we see the enemy as a process that can emerge based on our attitudes towards the Other, the Other is securitized and socially constructed: the Other as enemy is intersubjectively established as an existential threat with a saliency sufficient to have substantial security effects (Buzan et al. 1998: 25, 31). Societal security refers to collective identities and action taken to defend their 'we identities'. Threats to identity and construction or reproduction of 'us' are functionally and existentially interdependent (Buzan et al. 1998: 120). This suggests that security is the process of organization of violence towards the inside as well as the outside. But for realists, enemies are not constructed. Security threats are real. George W. Bush said in his second-term inauguration speech, 'life is fragile, and evil is real, and courage triumphs'.

Structural realists argue that there is a perpetual struggle for power among states in a pervasive condition of anarchy (Waltz 1979). Some suggest that in the international system 'the weaker members of the system, such as the Third World countries, simply do not possess the capabilities needed to affect the system structure' (Acharya 2005: 162). However, I argue that in a system all elements of the system influence the system structure. Failure or collapse of the weak states, illegal migration, poverty, AIDS, all have an impact on the global system. For example, the military build-up of powerful states such as Russia as well as the state collapse of Somalia have an impact on the global system.

The competitive and conflictive interaction among the global players has been structurally including the weak nations in the conflict system and even transforming them from weakness to global significance without resulting in improvement of the human rights record, economic performance, rule of law, good governance, etc. The competition between the West and Russia, the West and China, and within the West itself while pursuing solely their national interests, has given leverage to undemocratic and corrupt regimes in Africa. Chad, Sudan and Angola are just a few examples. Because of such competition in global politics we can conclude with certainty that it is not only the global system that determines the behaviour of the states (units), but also the units that certainly influence the global system.

Besides the competing powers of the international system, the international legal system is also a component of the global conflict system. The states of the international system that act with legal effect as a competent, international organ determine with certainty the subjects of the system (Crawford 2006: 20). Sudan has been arguing that the Darfur war is an internal matter (because Sudan is a sovereign state) and any intervention without the consent of Sudan is a violation of Sudanese sovereignty. Sovereignty is, in addition to independence and security, the symbol and substance of the state (Deutsch 1979: 202). Sovereignty implies 'full and unchallengeable power over territory and all the persons from time to time therein', and a state's 'sovereignty over its territory is absolute and complete' (Dixon 2006: 144). Therefore, the victims of the Darfur war are not only victims of the Sudanese government, but also of the international legal organ from which the Sudanese state derives its sovereignty or inviolability.

It is important to note in the sovereignty or inviolability debate that global players deal with similar cases in completely different ways. Kosovo became independent on 17 February 2008 because most of the major global players were in favour of it, whereas Somaliland declared its independence in 1991 but no global actor or institution has recognized it, in spite of the fact that Somaliland is the only economically and politically well-performing and relatively peaceful region of Somalia. Indeed, recognition does not solve all problems, but it could be one important factor. Crawford suggests that recognition as an institution of state practice can resolve uncertainties as to the status of the state, but this does not mean that the status of the state is, in principle, dependent of recognition (Crawford 2006: 27–8). This means recognition itself does not determine personality under international law because the state is entitled to the rights and subject to the general duties of the international system anyway.

However, recognition of the state by the international system has internal as well as external consequences for the state, for its citizens as well as for its neighbouring states. At least from the point of the declaratory theory of international relations, recognition is nothing more than acknowledgment of pre-existing legal capacity (Dixon 2006: 119), since the political existence of the state is independent of recognition as long as it has a defined territory, a permanent population, effective government, and the capacity to enter into relations with other states (Murphy 2006: 32; Crawford 2006: 45–6; Dixon 2006: 106–8). The 'constitutive theory of recognition' contends that 'only when other states decide that such conditions have been met, and acknowledge the legal capacity of the new government, is a new state actually constituted' (Murphy 2006: 32–3), and the act of recognition is 'a necessary precondition to the existence of capacities of statehood or government' (Dixon 2006: 120). In the case of Somaliland, the non-recognition by international law is certainly not going to prevent possible conflict between Somalia and Somaliland. Somaliland will maintain its sovereignty and Somalia will reclaim Somaliland as its territory. The non-recognition will for sure fuel conflict more than recognition because Somalia will be boosted by the stance of the international law. Moreover, the non-recognition of Somaliland by global powers has more global systemic aspects. Global players such as Russia and China have got their own separatists. Whereas the West supported the independence of Kosovo, Russia was against it, not only because Kosovo became independent from its closest ally, Serbia, but the issue of Chechnya is still there. Therefore, the non-recognition of Somaliland has to be seen in the global context as well.

Superpower competition during the Cold War made Africa part of the international security system. Various African countries served as proxies for the West as well as the East. Many African political elites profited from that competition, whereas that competition destroyed the long-term political and economic structure in those African countries. Many Africans feared the end of the Cold War would lead Africa from the world's attention into oblivion deprived of the leverage provided by superpower competition. Clapham argues that that has not been the case. 'Africa attracts a far greater share of global notice than its apparent importance warrants, not only from the United Nations (…) but even from a body such as the G8' (Clapham 2005: 275). Clapham underlines that the landmark year of 1989 shows Africa in a particularly unfavourable light in the global agenda, and the continent is increasingly seen as a collection of interlinked problems and threats (Clapham 2005: 275). Dictatorship, civil war, AIDS, refugees, and famine enhanced by environmental degradation in Africa are a 'scar on the conscience' of the rest of humanity, as Tony Blair maintained. The 'something-has-to-be-done' to save this 'dark continent', not only to save it but also to protect the rest of the world from the spillover effects. On the one hand, there is the emergence of a new security agenda, but on the other hand there are significant limitations on the capacity and indeed the willingness even of a Western government that gives Africa an exceptionally high profile to carry through on its promises (Clapham 2005: 275–7). Who are the global actors and what are their impacts on Africa?

Global Actors in Africa

Rational theory maintains that states act rationally to maximize their interests. Accordingly, the preferences of state interests are consistent and complete. Goldsmith and Posner acknowledge that states act sometimes irrationally because their leaders make mistakes, or their institutions fail (Goldsmith and Posner 2005:

7). Goldsmith and Posner point to three weak points of rational choice theory approaches: that rationality is primarily an attribute of individuals; that there may occur cognitive errors while taking decisions and acting on the international stage; and that, as constructivists argue, states' interests can be influenced by international law, institutions and the behaviour of other states (Goldsmith and Posner 2005: 8–9). Are the global actors behaving rationally in Africa?

The EU in Africa

On 12 December 2003 the European Union issued a strategy paper to address security threats such as hunger, malnutrition, AIDS and many other diseases, poverty, conflicts in a number of countries, terrorism, proliferation of WMD, state failure and organized crime. Responses to these threats include cooperation in fighting terrorism, policies against proliferation, dealing with regional conflicts, 'putting failed states back on their feet', acting before a crisis occurs, conflict prevention, cooperating with partners such as Africa through effective multilateral systems leading to a fairer, safer and more united world (European Security Strategy 2003). Multilateralism consists of an intricate web of states, regimes, treaties and organizations, i.e. a multi-level governance to improve people's access to peace, security and development. In the case of failed, collapsed or weak states, this includes nation-building, preventing and settling conflicts in the EUs neighbourhoods, controlling migration flow (Biscop 2005: 24–40).

The EU's Migration Policy towards Africa

In 2005, 22,939 illegal migrants were registered on the Italian island of Lampedusa, and between January and August 2006, 14,567 newcomers arrived there. Similarly, in 2005, 11,781 African illegal migrants arrived in the Spanish Canary Islands, and between January and July 2006, 17,058 new arrivals were registered. According to Spanish authorities, up until the end of 2006, more than 31,000 migrants arrived at the Canary Islands, and about 6,000 died or went missing in the Mediterranean and Atlantic in the same year. On the Maltese side, there were 1,822 illegal migrants in 2005, and 1,502 between January and August 2006. Refugee trends show that in 2005, 14,855 by 154 boats; in 2006, 18,096 by 341 boats; in 2007, 11,749 by 270 boats; in 2008, 31,700 by 397 boats, arrived at Lampedusa (BBC News 23 January 2009).

European boats, planes and helicopters are patrolling the shores of Senegal, Mauritania and Cape Verde and the routes to the Canary Islands to stop the increasing flow of African illegal migration to Europe within the framework of Frontex (European Agency for the Management of Operational Cooperation at the External Borders of the Member States of the European Union), established by Council Regulation (EC) 2007/2004 of 26 October 2004. The main tasks of Frontex are coordinating operational cooperation between member states in the field of management of external borders; developing a common integrated risk assessment model; helping member states train their national border guards; developing research relevant for the control and surveillance of external borders; assisting them technically and operationally at external borders; and supporting them in organizing joint return operations. The EU considers these measures as a preventative operation to locate and identify any illegal boat and to immediately return it, if found, to African waters (Laitinen 2007). In addition, some European states have already been arming North African states such as Tunisia, Egypt, Algeria, Morocco and Libya, bilaterally, to reinforce their coast defence capacities and prevent the influx of African migrants. Spain has been equipping itself with a new radar system, day/night vision cameras, a tear gas diffusion system and detection wires.

The Declarations of the Euro-African conferences on Migration and Development, of Tripoli in November 2006 and of Rabat in July 2006, designed measures on how to manage African migration to Europe. According to the declarations, migration management depends on partnership with countries of origin, transit and destination. Good and effective governance, sustainable environmental policy, integrating Africa into global trade, peace, and coherent international policies of development cooperation, to address underdevelopment especially in rural areas, political conflicts, HIV/AIDS and other similar diseases, and unchecked population growth are some of the key components of the declaration. In Paris in November 2008 the AU and EU developed a three-year programme on migration, with its main focus on the economic development of Africa.

The EU's Development Policy towards Africa

On 20 July 1963 and 29 July 1969, Yaoundé Convention I and Yaoundé Convention II, respectively, the EEC and Africa signed agreements to promote the increase of exchange between the associated states and member states; to reinforce their economic relations and the economic independence of the associated states; and to contribute to the increase of international trade through preferential trade arrangements, such as the duty-free access of specified African goods into the European market, financial and technical assistance (through the EDF), and investment. The objective of the *first* ACP-EEC Convention signed at Lomé on 28 February 1975 and of the *second* Lomé Convention signed on 31 October 1979 was 'to promote trade between the Contracting Parties, taking account of their respective levels of development, and, in particular, of the need to secure additional benefits for the trade of ACP States, in order to accelerate the rate of growth of their trade and improve the conditions of access of their products to the market of the European Economic Community'. According to the *third* ACP-EEC Convention signed at Lomé on 8 December 1984 and the Fourth ACP-EEC Convention signed at Lomé on 15 December 1989, both parties agreed to promote and expedite the economic, cultural and social development of the ACP states and to consolidate and diversify their relations in a spirit of solidarity and mutual interest based on the fundamental principles of: equality between partners; respect for their sovereignty, mutual interest and interdependence; the right of each state to determine its own political, social, cultural and economic policy options; and security of their relations based on the 'acquis' of their system of cooperation. In the Cotonou Agreement of 23 June 2000, the parties agreed to promote and expedite the economic, cultural and social development of the ACP states; to enable a stable and democratic political environment; reduce and eradicate poverty, sustainable development, regional and subregional cooperation and integration, gradual integration of the ACP countries into the world economy and international trade; enhance human rights and fundamental freedoms, respect for fundamental social rights, democracy, rule of law and transparent and accountable governance, and peace and security through peace-building policies, conflict prevention and resolution.

The Africa-Europe Summit under the aegis of the OAU and the EU in Cairo between 3 and 4 April 2000 declared a plan of action as a new strategic dimension to the global partnership between Africa and Europe, which consists of: supporting regional economic cooperation and integration within Africa; integrating Africa into the world economy through trade, private sector development, investment; providing resources for development, such as ODA; addressing infrastructural problems and strengthening the industrial base; development and transfer of research and technology; debt relief and providing additional resources for poverty reduction strategies; promote human rights, democratic principles and institutions, good governance and the rule of law; addressing the root causes of migration; creating and maintaining a vibrant civil society; eradication of the root causes of refugees and displaced persons; provide assistance to refugees and displaced persons and to participate in their voluntary return and in their reintegration; peace-building, conflict prevention, management and resolution, disarmament, demobilization and reintegration; fight against terrorism; combat the problem of illicit trafficking in small arms and light weapons; assisting and rehabilitating landmine victims, promoting mine awareness and assisting and developing the national capacities and skills in de-mining; implementation of the Treaty on the Non-Proliferation of Nuclear Weapons; post-conflict assistance; addressing challenges to sustainable development and poverty eradication through education, fight against endemic, parasitic and infectious diseases, cooperation in the field of prevention of natural disasters and setting up disaster prevention and preparedness mechanisms, implementing food security policies; combating drugs trafficking and money laundering, and cultural cooperation between Africa and Europe.

The EU Strategy for Africa of 2005 suggests that creating an economic environment is indispensable for achieving the Millennium development goals (MDGs): the most important factor to create this environment is that Africa needs to achieve average growth of at least 8 per cent per year. The success of this goal further depends on the development of South–South, North–South and multilateral trade, and increasing the competitiveness and productivity of African agriculture. This needs further regional, continental and intercontinental interconnectivity. At the *regional level*, the EU would support (sub)regional integration and development strategies and programmes of different regional economic communities (RECs). Since, according to the EU, trade is a key factor for Africa's development, the EU expressed its commitment to increase aid for trade to pay €1 billion per year by 2010. Moreover, asymmetric and flexible implementation of economic partnership agreements (EPAs) with Africa's regional groupings since 2008, and improving African access to European and regional markets, have been considered decisive to achieve the MDGs and a gradually increasing involvement of Africa in global trade (Consilium 2005a). The EU is already Africa's biggest trading partner, with 45 per cent of Africa's foreign trade, accounting for €144 billion in 2000, which is five times' more than intra-African trade (€29 billion a year) (Consilium 2005b).

The EU envisages helping make basic social services available for the poorest people by allocating 0.7 per cent of its gross national income (GNI) to development aid by 2015 to materialize the MDGs. Compared with 2006, this commitment should result in an estimated additional €20 billion in official development assistance (ODA) per year by 2010 and an additional annual €46 billion per year by 2015. The EU also agreed to allocate at least 50 per cent of this amount to Africa. This increased development aid is meant to create a positive economic environment through support to macroeconomic stability by promoting pro-poor growth; creating integrated regional markets (South–South trade) to use trade as a policy tool to reduce poverty, and on the intercontinental level an increased market access through North–South trade. Whereas the EU's development aid to Africa in 1985 was €5 billion, in 2003 it totalled €15 billion.

Support of the New Partnership for Africa's Development (NEPAD)'s African Peer Review Mechanism (APRM) is considered as one of the key measures to implement good governance in Africa. This process demands participatory democracy and accountability, reliable and strong parliaments and civil society. These objectives are believed to be materialized only through development assistance by the EU and other external actors. But the stance of many African heads of state and government on Zimbabwe has shown that APRM is a paper tiger. In the SADC region, with the exception of Zambia and Botswana, member states were supportive of Mugabe (mainly Angola, South Africa, DR Congo and Namibia) or reserved, such as Malawi, Mozambique and Tanzania. Generally, most of the other African countries were on the side of Mugabe. But even within the EU itself, there has not been a coordinated position against Mugabe. France was very often on the side of Mugabe, whereas the UK and EU in general have been against Mugabe. Similarly, there were contradictions between the EU and French positions on Chad when the EU was preparing its EUFOR Chad/Central African Republic Mission in late 2007 and early 2008.

The EU's Political and Security Strategy towards Africa

The EU pledged to allocate €300 million for the period 2008–10 to strengthen African capabilities for the prevention, management and resolution of conflicts, security sector reform and disarmament, demobilization and reintegration, assistance to the counterterrorism centre and the African Centre for Study and Research on Terrorism (CAERT) based in Algiers, and stemming the illicit flow of small arms and light weapons. The African states are expected to join counterterrorism, in return for technical assistance, information and support to the AU Anti-Terrorism Centre. Capability-building includes strengthening the AU's African Standby Force by building on existing activities by member states through training, technical, planning and logistical support, military and civilian, and strengthening fragile states. The EU member states declare their will to adopt the EU Code of Conduct on arms exports, discouraging transfers that contribute to instability, tackling illegal trafficking, border management controls and an international arms trade treaty. Further, the EU conducts in Africa its own crisis management missions, including potential deployment of EU battle groups.

Western Africa is a key region for Euro-African relations. There are various reasons that make the region significant for the European Union: first, two influential EU states (France and the UK) colonized the region; second, these former colonizers still have vital economic interests in the region; third, the region is one of the leading migrants senders and migrants' transit route to the EU; fourth, the UK intervened militarily in Sierra Leone, and France has military bases in various states of the region (its former colonies); fifth, the region is volatile because of arms transfer (Vines 2005) mainly caused by bad interstate relations and devastating civil wars in Liberia, Sierra Leone and Ivory Coast. A Joint EU-ECOWAS Declaration on Proliferation of Small Arms and Light Weapons (SALW) on 24 April 2007 suggested cooperation to prevent the transfer of SALW, their ammunition and other related material to non-state actors. They agreed to develop and actively implement efficient mechanisms of border management and control, and to cooperate further at the continental, regional and subregional levels, including the implementation of the ECOWAS Small Arms Control Programme (ECOSAP). The ECOWAS security policy in the region, such as the plan to transform the moratorium on SALW into a binding convention, conflict prevention and stemming arms transfer (Vines 2005), was supported by the allocation of €10 million as of February 2005 from the 9th EDF (Consilium 2005b).

Somalia, as the collapsed state of Africa, has regional as well as global implications as a possible terrorist haven. Even if no EU member state is interested in intervening directly in Somalia, the EU has been attempting to take part in rebuilding it politically and economically. The conclusions on 23 April 2007 of the Council of the European Union on Somalia reaffirm the EU's support for the Africa Union (AU) Mission in Somalia (AMISOM), which began with the deployment of the Ugandan contingent in March 2007. The Council pledged to contribute €15 million to finance AMISOM, and to provide advisory and planning support to the AU.

On 18 July 2005 the EU expressed its willingness to support the African Union mission in the Darfur region. This support consisted of military as well as civilian components. The civilian component supported the AMIS II Police Chain of

Command by providing the AU with experienced senior police advisors; support for training of CIVPOL personnel through providing a capacity for in-mission training by a group of EU trainers; support for the development of a police unit within the Secretariat of the AU. This support project was allocated €2,120,000 supplied by the EU. The military component consists of the provision of planning and technical assistance to all AMIS II levels of command, including the logistic support coordination structure; provision of military observers, in the context of the AMIS II enhancement plan; training of African troops and observers forming part of AMIS II enhancement, strategic and tactical transportation; aerial observation. This component was allocated €1,970,000 supplied by the EU (EUR LEX 2005).

As of February 2005 the 'consolidated EU' package in support of AMIS II in Darfur committed €92 million to AMIS I and II from the African Peace Facility. For humanitarian assistance, the EU allocated a total of €72 million to meet the needs of people affected by the conflict; €26 million was allocated as of April 2005 for relief to the Sudanese refugees that have fled to Chad as well as local host communities. The AU requested another €70 million for the AMIS budget until mid-2006 (Consilium 2005c). All of these measures are meant to support the AU's ongoing establishment of an African Peace and Security Architecture (APSA), including the creation of the African Stand-by Force (ASF) (Consilium 2006).

When the crisis was exacerbated in the Great Lakes region in general, and in the Democratic Republic of Congo (DRC) in particular, after the genocide in Rwanda in 1994, the EU sent its Special Representative Aldo Ajello. The military operation 'Artemis' conducted in summer 2003 in Ituri, DRC, was the peak of the EU's involvement in the region. In Ituri, Eastern DRC, in 2003 the EU intervened militarily with 1,500 troops, as well as provided support to the transitional national government by providing €2.5 million. The European Union conducted a military operation in the Democratic Republic of Congo in 2003, named *Artemis* in accordance with the mandate set out in UNSCR 1484 (2003), and deployed its forces to operate in accordance with the objectives set out in the 'Framework for EU action in response to the crisis in Bunia' (eastern Democratic Republic of Congo) approved by the Council (Council Joint Action 2003/423/CFSP). The EU contributed €25 million to the African Union to sustain its military operation in Burundi. For the Demobilisation and Reintegration Programme for the Greater Lake Region, the EU contributed €20 million, just as the EU had already financed a disarmament programme in Congo Brazzaville with €2 million. In the DRC, by allocating €8 million, the EU supported the establishment of an integrated police unit (IPU), and this project was carried on with a further budget of €4.4 million for 2005. To prepare for the elections that took place in 2006, the EU allocated €9 million to establish and implement a strategy of elections security to be implemented by the Police National Congolaise with the assistance of the 'Mission de l'Organisation de Nations Unies en République Démocratique Congo' (MONUC) (Consilium 2005b).

Nevertheless, states such as the UK and France pursue their own foreign policies towards Africa, which do not always reflect the principles of the EU. Such conflictive competition and policy divergences have considerable impact on the economic development and political security of the African states. In the following sections we shall see the impact of such divergences on peace in Africa and on the EU's credibility there. Some undemocratic regimes in Africa are unconditional allies of France and the UK.

France and Africa

France, since the start of the colonial period, has determined the political path of many African states. Bernard Debré, French Minster of Cooperation from November 1994 to May 1995, suggested that France should play an active role in the democratization process in Africa because, if Africans are left by themselves in their democratization, if they are not encouraged, assisted, helped and stabilized in their democratization process, they will be more unstable and institutionally weak rather than democratic (Martin 1995: 17–18). However, France realized that it could take advantage of the EU to play an active role in the international arena (Guyomarch et al. 2002: 105), to advance its interests in Africa, and to secure power and influence through Africa on the wider international stage, to seek an active Third World policy, and to establish national grandeur (Guyomarch et al. 2002: 106; Kroslak 2004: 64; Utley 2002: 134).

Guyomarch et al. maintain that 'France supported the development of a European foreign policy only in so far as it was consistent with these overriding priorities' (Guyomarch et al. 2002: 107). France was the prime mover behind trade and aid agreements (such as Yaoundé I of 1963, Yaoundé 1969, and Lomé I–IV between 1975 and 2000 and Cotonou Agreements of 2000) reached between the EC and developing countries since the Treaties of Rome (Guyomarch et al. 2002: 110–11). Through the Lomé Convention between the EEC and the 46 African, Carribbean and Pacific (ACP) countries in February 1975, France indirectly secured its influence in Africa while sharing the cost with its European allies and rivals on the continent, and it keeps economic advantages without paying too high a price (Kroslak 2004: 73). It was about how to make France a world actor independent from the two superpowers. In the international power and security system, France tried to advance its influence, its international rank and status during the colonial as well as the Cold War period (Utley 2002: 129).

For strategic, political and economic reasons, African officials and their families continued to receive special treatment from President Mitterrand. Those special treatments included higher education in France's best schools for the children and extravagant shopping trips for African 'first ladies' (Kroslak 2004: 61–8). Non-democratic regimes were supported financially and were rescued whenever their internal security seemed to be endangered. The idea of the 'grandeur de la France' has been more important than human rights or good governance in French Africa. Mitterrand's son Jean-Christophe was involved in illegal arms dealing and money laundering (Kroslak 2004: 64–8). Factors crucial for Mitterrand in particular and France in general to maintain special relations with Africa include:

the sentimental bonds linking Paris to Francophone Africa through 'civilization' and dissemination of French culture; the projection of French identity and values overseas; the desire to maintain France's status and influence in international politics; the fear of Anglophone encroachment in the French domain in Africa; and protection of its citizens in Africa, currently about 240,000 French nationals living in Africa (Hansen 2007).

However, some imperatives began to push France to change its Africa politics by the mid-1990s: France came to acknowledge that so far the French African politics failed; crisis in the Great Lakes region, not only despite but also because of the French alliance with Rwanda (with whom France concluded a military cooperation agreement in 1975), came to damage the image of France as well as its interests. France has been known for supporting and equipping corrupt and undemocratic governments and defending them militarily from rebellion. After the Rwandan genocide, France established safe havens for Hutus, including perpetrators of genocide, to protect friends of France from public scrutiny and accountability (Utley 2002: 130–2).

DR Congo is the symbol of France's failure in Africa. France failed: tactically, because Kabila was backed by the United States and Anglophone African countries; morally, because France had been supporting the notoriously corrupt Mobutu until his demise; and geo-politically, because Zaire was an essential element in the French presence on the continent (Kroslak 2004: 67). Moreover, the UK has been attempting to gain influence where France failed, such as in DR Congo, whereas France has been attempting to gain ground where the UK has difficulties in Africa such as in Zimbabwe.

Towards the end of Mobutu's regime, France was increasingly isolated in the Great Lakes region. The US as well as Europeans embraced the advancement of Laurent Kabila in 1996; and those African states who were against Mobutu began to condemn the French policy in the region. The US scaled up its ties with African states in the region through various high-ranking visits: Secretary of State Warren Christopher in October 1996, followed by Hillary Clinton in March 1997, Madeleine Albright in December 1997 and Bill Clinton in 1998. These developments alarmed France.

In October 2007 France and Morocco agreed that France's Alstom would build a high-speed rail link between the cities of Tangiers and Marrakech in Morocco whose first section, 200km from Tangiers to Kenitra, expected to be operational in 2013. For France, which is Morocco's main export destination, this is a considerable business deal, worth €2 billion. According to further business deals, Morocco will buy 18 French Rafale fighters and 140 armoured vehicles. In July 2007 Sarkozy proposed the idea of creating a Mediterranean union between southern European and North African countries to improve cooperation security, economic development, energy policy and immigration, and the union was established in July 2008. During his visit to South Africa in February 2008, Nicolas Sarkozy suggested renegotiating France–Africa relations. He added that 'France hopes for the Renaissance of Africa with all her heart and soul, she hopes for it for the sake of everything that unites her with Africa.' According to Sarkozy, it is unthinkable for France to be drawn into domestic conflicts (Ambafrance 2008).

France is still an unconditional supporter of the Chadian president, whose regime ranked as one of the most corrupt in the world in 2007 (ranked 172 out of 179 surveyed countries) and in 2005 it was the most corrupt besides Bangladesh. The Zaghawa minority, less than 3 per cent of the total population, dominates Chadian politics. It is because of French military support that the Chadian government survived the rebel attack in 2006, as well as in late 2007 and early 2008. France organized the delivery of munitions to the Chad army through the intermediary of Libya to defend the presidential palace (ICG 2008c: 29). France has a dilemma in Chad: to renegotiate the traditional relations with Africa or to defend its allies. At least in the Chadian case, France has opted for the second. President Déby violated the Chad Constitution in June 2005 when he changed the law to enable him to stand for a third term as president. In April 2006, his move resulted in the killing of hundreds as rebels fought government troops on the outskirts of N'Djamena; and in May 2006 President Déby won an election boycotted by the opposition. France argued that it is its obligation to protect an ally and 'legitimate' government from 'external' aggression. In his speech in South Africa, Sarkozy defended the undemocratic and corrupt president of Chad as 'legitimate'.

In the mid-1990s France developed and introduced the 'RECAMP' (Renforcement des capacities africaines au maintien de la paix – Reinforcement of African Peacekeeping Capabilities) programme to help Africans address their security problems on the continent. In 1997 the United States, Great Britain and France recognized the need to coordinate their military cooperation programmes in Africa. As a consequence they issued a P3-Declaration on 22 May 1997, an initiative to support African peacekeeping capacities. In 1998 France presented the RECAMP concept in Paris at the Franco-African summit. In 1999 France created *Ecole du maintien de la paix* (EMP) in Zambakro (Ivory Coast) as a first and special training centre for the preparation of peacekeeping operations. From *Direction de la coopération militaire et de défense* (DCMD), the successor of *Mission militaire de coopération* (MMC), the centre was given about €1 million per year for its maintenance, in addition to the originally allocated amount of €2.4 million, until the centre was relocated towards the end of 2002 to Koulikoro (Mali) as a consequence of the tense relations between France and Ivory Coast. There are also other similar institutions, such as *Collège interarmées de défense* (CID) in Paris and *Écoles d'état-major* in Libreville (Gabun) (Koepf 2005: 75–6).

RECAMP, according to France, is meant to provide a solution to the peacekeeping needs expressed by African security mechanisms, the African Union and African states. Its principles include full compliance with the targets of the UN Charter; joint decisions with the Africans; opening of the RECAMP programme to the whole African continent and to any state or organization willing to participate; bilateral cooperation and measures taken by subregions, partners and especially Europeans; freedom of each state to determine the nature and rate of its contribution; consistency between civilian and military players; focus on the current features of peacekeeping operations and the international context (RECAMP 2005). The tasks consist of crisis planning, support, formation of the peacekeeping force, financial support, command of forces, advice and expertise to the subregion and troops. These tasks can be categorized into *cooperation, training* and *commitment.*

1. *Cooperation* consists of political, politico-military (command) and military levels while setting up crisis prevention and management teams in Africa within subregional organizations with the European Union and in association with the UN. On the *political* level it backs up the reinforcement of continental and regional political institutions in partnership with the European Union; supports the implementation of defence instruments, and coordinates between the member states of the EU; and participates in the training of the military and civil servants of the subregional organizations under European or national cooperation programmes. On the *command* level it develops a detailed plan for operation and reinforces the capacity to set up strategic- and operative-level staff; participates in the training of staff officers. On the *military* level it trains officers for peacekeeping operations in military academies in Africa or in Europe; runs repeated military training at the subregional level (Koepf 2005: 76). The peacekeeping training centre in Mali concentrates mainly on peacekeeping where France supports the training of 150–200 peacekeepers every year from all over Africa. Moreover, there are region-specific national training centres in Africa as well as peacekeeping-specific training centres in France for Africa.
2. The RECAMP training programme is a two-year programme, including three main components: a major exercise is staged within one of the subregional organizations in view of managing a crisis at the subregional level; intermediary trainings are ensured in the other subregions by pre-positioned armed forces in close cooperation with subregional organizations and neighbouring states; and off-program exercises. RECAMP supports African or non-African exercises that strengthen African peacekeeping capabilities.
3. The RECAMP's main commitment is to support the implementation of the standby African force placed under the authority of the African Union as well as the rapid development of peacekeeping and security entities instituted by African subregional organizations.

RECAMP provided equipment, logistic support and set-up for MISAB in 1997 in the Central African Republic, a mission made up of six contingents (Chad, Gabon, Mali, Burkina Faso, Senegal and Togo). In 1999 RECAMP provided assistance and logistic support for the setting up of a multinational battalion (Benin, Gambia, Niger and Togo in Guinea Bissau). Until 2005, within the framework of ECOWAS, two trainings took place – RECAMP I and IV – one each within the frameworks of ECCAS and SADC. RECAMP Cycle I (1996–98) took place in February 1998 in the region between Senegal, Mauritania and Mali. In addition to the troops from eight African states (Senegal, Mauritania, Mali, Ghana, Guinea, Guinea-Bissau, Gambia, Cape Verde) and 600 French soldiers, some units from the US and the UK participated. It was sponsored by the US, the UK and Belgium. RECAMP Cycle II (1998–2000) took place in Gabon in January 2000 involving Chad, Cameroon, Gabon, the Central African Republic, Congo-Brazzaville, Equatorial Guinea, Burundi und Sao Tome e Principe, and was supported financially and logistically by the US, the UK, Belgium, Spain, Italy, Portugal and the Netherlands. The training RECAMP Cycle III (2000–02) *Tanzanite* of February 2002 involved 16 African states: DR Congo, Angola, Tanzania, South Africa, Kenya, Namibia, Botswana, Zimbabwe, Zambia, Malawi, Mozambique, Madagascar, Lesotho, Swaziland, Mauritius and the Seychelles. It was sponsored by France, the US, Canada, Japan, the UK, Germany, Belgium, Spain, Portugal, Sweden, Denmark and the Netherlands. RECAMP Cycle IV (2002–04) *Bénin 2004* involved Senegal, Mali, Ghana, Guinea, Cape Verde, Benin, Burkina Faso, Ivory Coast, Niger, Nigeria and Togo. It was financially supported by the same states as 2002, but Denmark was replaced by Austria (Koepf 2005: 77). RECAMP V (2005–06) training *SAWA* took place in November 2006 in Cameroon. The participating countries were Cameroon, Guinea, Sao Tome and Principe, Gabon, DR Congo, Angola, Rwanda, Burundi, Congo Republic, Chad and the Central African Republic. France pledged $1million to support Cameroon's plans to build a police peacekeeper training school, a facility that will be able to train 1,000 officers a year from across the region as a step to building African capacity in the non-military sphere of APSA (Vines and Middleton 2008: 28).

When the Anglo-French summit took place at St Malo in December 1998, it initially seemed that the post-imperial competition between the UK and France would be substituted by a coordinated EU policy towards Africa. In November 1994, the UK's John Major endorsed French President Francois Mitterrand's proposal made at the Franco-African summit that 1,000–1,500 African troops should be trained, equipped and financed for peacekeeping duties by France and other European powers and by the EU itself. Nevertheless, contradictory policies and competition between France and the UK have continued.

The UK in Africa

Competition between France and the UK has characterized European relations towards Africa. Though officially both France and the UK declare that they will coordinate their Africa policies, the reality looks different. Both states consolidate their traditional ties with their respective former colonial countries. That is, the UK is not less inclined to pursue its interests in Africa, in some cases leading to undermining the efficiency of the European Africa policy and contributing to insecurity and the undermining of democratization.

In 1998 Tony Blair and President Chirac of France agreed to improve Europe's own defence capabilities in order to advance mutual self-interest and moral purpose in defending the values Europe cherishes in its own national interest (safety): values of liberty, the rule of law, human rights and an open society. For this the principle of non-interference must be qualified in important respects to prevent and punish genocide. The New Labour Government came up with a 'new' idea of a 'third way' between liberal internationalism and national interests in its commitments in Africa to act out of 'ethical' commitment and moral outrage (Abrahamsen and Williams 2001: 252). Potential security 'threats' and 'risks' are dominating Britain's Africa policy. From this background, Blair called Africa 'a scar on the conscience of the world' in urgent in need of international support (Williams 2004: 42).

However, 'Britain's post-Cold War attempts to promote peace in Africa have been selective, inconsistent, under-resourced' (Williams 2004: 46). Even New Labour's Africa policy is selective because it criticizes fraudulent elections (as in Zimbabwe) while other similar cases are ignored (as in Zambia). In its global war on terror the British government has been coalescing with non-democratic states with poor human records, such as Ethiopia and Kenya (during President Arap Moi). Yoweri Museveni of Uganda and Paul Kagame of Rwanda contributed in Eastern DR Congo to the brutal war, but the UK was still their unconditional friend. They committed serious human rights abuses in DR Congo and have been involved in massive illegal exploitation of DR Congo's economic resources. Unfortunately, undemocratic and corrupt African governments have been directly or indirectly encouraged or tolerated by the West, including Britain (Porteous 2005: 284). In Nigeria, since Britain's oil and other commercial interests were consistently considered paramount, the democratic principles and human rights do not play a central role.

Blair's New Labour created the Department for International Development (DFID), whose priority was to reduce poverty in Africa and elsewhere in collaboration with the Foreign and Commonwealth Office (FCO) and the Ministry of Defence (MOD). Further, the so-called Conflict Prevention Pool in the UK was created to facilitate conditions for the African Peacekeeping Training Support Programme through an integrated institutional structure and resources. The British government proclaimed to prevent conflicts through addressing the root causes: fighting poverty and promoting sustainable development; good governance; curbing the flows of small arms and light weapons; preventing the trade of conflict goods. At the St Malo (France) summit in December 1998, France and the UK declared a more coordinated Anglo-French policy in Africa. Even after the era of John Major, despite various initiatives from the Labour Party for Africa, the UK's Africa policy was not without contradictions.

New Labour has been one of the arms sellers to Africa, which the Campaign Against the Arms Trade estimated to have exceeded US$200 million in 2003. In 1999, for instance, the government granted 970 single individual export licences in the small arms category, including to Eritrea, Kenya and Zimbabwe, despite democratic deficits (Williams 2004: 49). The FCO colluded with the military consultancy firm Sandline International, which brought 30 tonnes of arms and ammunition into Sierra Leone in contravention of the UN arms embargo (Porteous 2005: 287). Similarly, Avient, a company run by a British businessman, supplied military assistance to the DRC's air force while the civil war was going on. Western business people, including UK nationals and residents, have been directly or indirectly involved in resources exploitation in the west African civil war countries (Porteous 2005: 295). On the one hand, British mercenaries were involved in the Ivory Coast's ongoing conflict, and on the other hand British troops supported the UN force to stabilize Freetown and its environs since 2000. Such contradictions between developmental and strategic goals in Africa have been growing sharper and more numerous since 9/11.

Tony Blair had made African development, peace and security capacity a lead issue during the UK's presidency of the G8 in 2005. If it were about money, then with the approximately $1 trillion that Africa has had in aid in the last 50 years, most of its problems would have been solved (Easterly 2006). Instead, the misallocation of funds, lack of prioritization, deep-rooted corruption and wastage swallow the money because the system fundamentally fails to promote national economic growth and development (Taylor 2005: 301–6). Paradoxically, 9/11 has contributed to global policy contradictions.

Nevertheless, though the UK has not established ambitious programmes like France, the British Military Advisor and Training Teams (BMATTs) in Ghana, Kenya, Sierra Leone, South Africa and (until 2001) Zimbabwe are not insignificant. The British Peace Support Teams (BPSTs) based in Kenya and South Africa have been preparing African troops for peacekeeping operations (Koepf 2005: 79). Moreover, the UK's conflict prevention initiatives have become increasingly significant.

The Africa Conflict Prevention Pool (ACPP), which was established in April 2001 as a joint initiative of the UK Department for International Development (DFID), the Foreign and Commonwealth Office (FCO), and the Ministry of Defence (MOD), attempts to find long-term solutions to Africa's problems related to conflicts within the context of the UK Sub-Saharan Strategy for Conflict Prevention. Africa Conflict Prevention Pool (ACPP) is about capacity-building while working together with the African Union (AU) and the subregional organizations contributing to the African Peace and Security Architecture (APSA), including disarmament, demobilization and reintegration (DDR). The UK has been supporting the capacity-building of African countries to contribute to peace support operations (PSOs).

ACPP began with an annual budget of £50 million; it rose to £60 million in 2004, and has probably risen further to £64.5 million by the end of 2008. The money has been spent in various areas, such as security sector reform (SSR); disarmament, demobilization and reintegration (DDR); enhancing African peace support operations capacity; political dialogue and peace processes; NGO/civil society grassroots peace-building programmes; support to the African peace and security architecture; media projects in conflict-affected areas; research and conflict analysis (DFID 2006: 6).

Between 2001 and 2004, the ACPP contributed about £25 million on DDR programmes in Sierra Leone, Liberia, the Great Lakes (DRC, Rwanda, Burundi, Angola) and Sudan. In western Africa between 2001 and 2005, ACPP funded various activities: 2,000 African staff officers have received foundation and advanced training in peace support operations between 2004 and 2005; over 500 West African civilian police officers have received training in peace support operations; training for the ECOWAS Standby Force; military, diplomatic and civil society personnel from across West Africa have received training in a broad range of peace-building skills – including grass roots peace-building, negotiation skills, DDR planning, human rights and international humanitarian law, election monitoring, gender, HIV and peace support operations; it hosted conferences on AU peace and security strategy, civil society and peace support operations (PSOs) in West Africa. The UK has provided more than £14 million in support and additional training for the African Union Mission in Darfur (AMIS) (DFID 2006). African PSOs received around £15 million in 2006–07, whereas security sector reform (SSR) was allocated £24 million (DFID 2007). In 2007–08 the ACPP spent just over £60 million, the main areas being: 37 per cent security sector reform, 12 per cent DDR; 22 per cent PSOs; 22 per cent peace-building; 3 per cent small arms and light weapons (DFID 2008: 10).

The future role of Germany, under whose leadership the EU's military intervention (EUFOR RD Congo) took place in 2006, will further determine the future and features of the EU's Africa policy. Germany is pursuing a position of being an important global player. In its Africa policy, the UK is cooperating with the US rather than with France or Germany. But the Common Foreign and Security Policy of the EU reiterates that the EU member states must coordinate their foreign policies. This does not always seem to be the case. In the 1990s the British–American cooperation in Africa seemed to undermine the French interests. Clinton as well as Blair supported the new regimes in Ethiopia, Eritrea and Rwanda; they supported the toppling of Mobutu; whereas France had been supporting Mobutu, and the Rwandan regime responsible for the genocide of Rwanda. For more than a decade, the US presence in Africa has been re-expanding.

US and Africa

The US has been proclaiming that its objective is to unseat non-democratic governments, advance human rights and consolidate transitions in the new democracies (Roberts 1992: 303), with no compromise on the proud dictators who pervert all principle and debase men whom they have first oppressed (Cook and Moos 1954: 210). George W. Bush, in his speech at West Point on 1 June 2002, said the US is 'in a conflict between good and evil, and America will call evil by its name' and 'moral truth is the same in every culture, in every time, and in every place'. According to Bush, this moral truth consists of respecting human rights, freedom, and dignity; fighting poverty, hunger and disease; making peace, fighting terrorism, etc. Accordingly, the US mission has been to persuade the rest to follow its paths instead of living in or progressing towards 'tyranny', not only by sparing the humble but also by delivering the oppressed. Indeed, such noble objectives are theoretically pursued, but practically, the US policies in different parts of the world have been undermining the same 'moral truths', as will be shown below.

Conflicts and instability in Africa have become an increasing concern for global security in general and for that of the US in particular. Spreading democracy in Africa seemed important out of a globalist approach as well as out of democratic realism. But the 9/11 attackes resulted in democratic contradiction: spreading democracy as a solution to global terrorism, but at the same time cooperation with undemocratic allies increased as well. If such allies are producers of oil, democratic realism and globalism are increasingly replaced by security realism and economic realism.

Overall, in Africa US aid quadrupled from $1.3 billion in 2001 to $5 billion in 2008. There are different causes for this. The events of 11 September 2001 and new oil discoveries in Nigeria and Gulf of Guinea have intensified US economic interests in West Africa. Currently Africa supplies the United States with 15 per cent of oil imports; Africa could be supplying the United States with 20 per cent by 2010 (Lake et al. 2006: 30). High levels of natural gas or oil exploration and production investment are taking place in various parts of Africa, such as in Nigeria and Angola. There are currently about 60 billion barrels of proven reserves and $50 billion of investment by foreign companies in this decade in West Africa's Gulf of Guinea (such as Equatorial Guinea, Gabon, Sao Tome and Principe, Cameroon) and in Chad and Sudan. Exploration is commencing in Ethiopia and off the shores of Kenya and Namibia. In the countries of the Gulf of Guinea, nearly $5 billion had been invested until 2006 by major US investors like ExxonMobil, Amerada Hess Corporation, and Marathon Oil.

The threat of Islamic fundamentalism and terrorism in eastern, western and northern Africa compels the US to be present in Africa more and more. In the same way the military basis of the US in the Horn of Africa, especially in Djibouti and Ethiopia, seems to shape the security policies of the region. The Combined Joint Task Force-Horn of Africa (CJTF-HOA) aims to disrupt and defeat transnational terrorist groups posing an imminent threat to the United States and its allies in the Horn of Africa region. CJTF-HOA is focused on denying safe havens and external support and material assistance for terrorist activity within the region. Additionally, the force is prepared to counter the re-emergence of transnational terrorism in the region through civil–military operations and through support of international organizations working to enhance the long-term stability of the region.

Lyman and Morrison (2004) maintain that the US administration must deal with threats from Africa by adopting a more holistic approach to fight terrorism. Economic distress, ethnic and religious fissures, fragile governance, weak democracy and rampant human rights abuses create an environment in which terrorists thrive (Lyman and Morrison 2004: 76). Since the explosion of two massive bombs outside of the US embassies in Dar es Salaam and Nairobi on 7 August 1998, the consciousness of the necessity of focusing on the threats that could emerge in Africa has risen. Radical Islamism and anti-Americanism in Nigeria is growing steadily, especially after the US invasion of Iraq in 2003. Radical Islamist groups have been focusing on Nigeria, a close ally of the West, a key source of oil for the West and an area where there is already armed violence between armed militias and the government, following Osama Bin Laden's exhortation early in 2003 to do so.

Walt maintains that various regimes consider their partnership with the US as an essential protection against a variety of internal and external challenges and a means to gain greater leverage over key international outcomes; these regimes hope to gain direct influence over US policy deliberations and foreign-policy initiatives because US leaders would consult them before launching major initiatives (Walt 2006: 190–91). Many African states expect that US competition (conflicts) with China, Russia, North Korea, Iran, Iraq, Libya, Syria, Venezuela, Cuba, or 'rogue states', will have beneficial results for them.

Clinton's approach to Africa was to increase the number of capable states in Africa to define the challenges they face, manage their resources to effectively address those challenges, and build stability and peace within the borders and their subregions (Steven Metz in Rotberg et al. 2000: 71f). After Clinton took office, various US policies for Africa were initiated: US-Africa Economic Cooperation Forum; Overseas Private Investment Cooperation; ACRI (1997); African Centre for Security Studies (ACSS) (1998); Military Medical Exercises in Africa (MED-FLAG), FLINTLOCK (promoting regional cooperation); Joint Combined Exchange Training (JCET); foreign military financing (FMF); foreign military sales training (FMS-T); direct commercial sales (DCS); international military education training (IMET); Excess Defence Articles (EDA); Humanitarian Assistance Programmes-Excess Property (HAP-EP). In 2000, the AGOA (African Growth and Opportunity Act) was established to offer tangible incentives for African countries to continue their efforts to open their economies and build free markets.

US attempts to continue professional military education for Africa; to building expertise in areas of defence, law enforcement, economics, political organization; to support regional peacekeeping cooperation programmes; to establish rapid US and European response to humanitarian disasters – these initiatives are meant to lead the world community towards effective peace-building in Africa (Metz in Rotberg et al. 2000: 75–8). However, these initiatives cannot be and should not be viewed outside of the global context. That is, when we see them in the global context of competition they will be easily compromised: conflicts in Africa continue and undemocratic regimes establish their power further.

On 13 September 1996 the African Crisis Response Force (ACRF) was established; it was changed in 1997 to the African Crisis Response Initiative (ACRI). Between 1997 and 2002 9,000 African soldiers from eight countries (Ghana, Kenya, Malawi, Uganda, Benin, Ivory Coast, Mali and Senegal) were trained. In September 2002 the US presented during the G8 Conference in Berlin the successor of ACRI, the African Contingency Operations for Training and Assistance (ACOTA), to 'train the trainers' and to support the African troops financially and logistically (Koepf 2005: 78; Ramsbotham et al. 2005: 328–9). Furthermore, the TSCTI (Trans-Sahara Counterterrorism Initiative), replacing the Pan-Sahel Initiative of 2002, was formally approved by the US government in early 2005. It is envisioned as a five-year programme based on counterterrorism, democratic governance assistance, a public diplomacy component, and military assistance. The military assistance involves training, advising and assisting regional forces, and establishing institutions to promote better regional cooperation, communication and intelligence-sharing. EACT (East Africa CounterTerrorism Initiative) was created in June 2003 to address corruption, terrorism, money laundering, terror financing, etc., in eastern Africa. The Department of State and the National Defense University's Africa Center for Strategic Studies (ACSS) collaborates with the AU and the regional security mechanisms and African states through running counterterrorism workshops.

The Millennium Challenge Corporation (MCC) was created in October 2004; its objectives include *reducing poverty* through economic growth (economic growth to reduce poverty through investments in areas such as transportation, water and industrial infrastructure, agriculture, education, private sector development, and capacity-building); making and implementing *good policies* (assistance based on their performance in governing justly, investing in their citizens, and encouraging economic freedom); *country ownership* (high-level engagement and leadership by the partner government, as well as civil society and other domestic stakeholders); and *focus on results* (enable sustainable progress even after the funding under the compact has ended). Further, the US Command, AFRICOM, was created in October 2007 to promote a stable and secure African environment in support of US foreign policy, to enable Africa to have greater security capacity, and to provide African states with military training and assistance.

State failure is not always a homegrown issue, but is also a by-product of the international security system. The US has at times structurally contributed to the maintenance of 'order' through state terrorism, anti-democratic politics and bribery (Johnson 2004: 284), which at the end would lead to state failure and even more terrorism. On the one hand there are initiatives like MCC or AGOA, but on the other hand there had been obvious cooperation with undemocratic states like the former Zaire, Kenya and Ethiopia in its global war on terror. Further, its competition with states such as Russia and China will have further negative consequences for Africa, for the US and for the whole world.

China and Africa

The increasing influence of China in Africa has made the US as well as other global actors nervous in their Africa policy. African oil and other natural resources have caused an economic scramble for Africa. Chinese presence in Africa will probably have the most decisive impact on Africa's politics and economy by creating a chain reaction in the international security system.

Both China and Africa saw themselves as the Third World, or the South. China supported Africa by supporting various liberation movements with arms and money, food and medicines, and by training the liberation movements' commanders in the theory and practice of guerrilla war (Snow 1994: 186; Taylor 2006: 33–4). China argued that the struggle of the African people is the struggle of the Chinese people, because of which these two distant regions of the world are similar, or even identical (Snow 1994: 285). But China was not necessarily supporting liberation movements, but instead its allies without taking into account how representative they were as long as they were not allies of Russia. That is, the impact of the intensification of the Sino–Soviet dispute cannot be disregarded in the Chinese engagement in Africa.

China took advantage of the liberation movements, such as in southern Africa, to promote an anti-hegemonic agenda (Taylor 2006: 34–5). Through economic aid and building infrastructures, such as a railway between Tanzania and Zambia between 1970 and 1975, China tried to strike at the Europeans (Alden 2007: 15). Chinese involvement in Africa was not only out of common ground between China and Africa, but it was mostly a camouflage act against Russia, the US and the Europeans (Snow 1994: 293). Even during the Cold War Russia and China were supporting opposing liberation movements: while China was supporting FLNA, Russia was supporting the MPLA in Angola; while China was supporting ZANU in Zimbabwe, Russia was supporting ZAPU. Hence, to a great extent, Chinese relations with Africa have been conditioned by its relations with the United States as well as Russia, since bilateral interaction in world affairs is conditioned by a combination of indigenous and exogenous variables (Shambaugh 1994: 197; Taylor 2006: 22).

The US fears the Chinese challenge to its interests and 'values'. According to the US, these values include the US concern for issues of governance, human rights and economic policy. China has been attempting to bolster its standing and influence in a region beyond the Asia-Pacific (Tow 1994: 150). Arms proliferation to the Third World played a key role for that. Ethiopia, though a key ally of the US, declared China 'its most reliable partner'; it indicated its interest in closer military cooperation. President Mwai Kibaki of Kenya has been seeking Chinese investment and aid. Angola has successfully resisted IMF and other aid donors' recommendations for economic reforms after receiving more than $2 billion in loans from China (Lake et al. 2006: 44–5).

China recognizes that 'soft power' can be used to enhance its role as a rising power in regional and world affairs (Guo and Hua 2007: 2). China has been moving from engagement 'in class struggle to building a harmonious society; from denunciation of Confucianism to resurrection of Confucianism; from world revolution to peaceful rise; China has been continuing to search for the road of development to realize national power and prosperity' (Men 2007: 7). Some suggest that 'the application of traditional Confucian values emphasizing social harmony to international society in settling disputes between big and small countries may create a good model that may help reduce international conflicts'

(Zhang 2007: 125). Through stressing the peculiarity of its cultural values, China attempts to promote 'Chineseness' as uniqueness or Chinese essence (Kirby 2006: 15). 'China's culture and economic success are considered as powerful assets in creating a positive image abroad.' China's economic growth legitimizes the Communist Party, and China's development model appeals to the developing world. Its economic growth has raised its appetite for oil and raw materials. China has chosen an unprecedented development road different from the one pursued by other countries that rose in the last few centuries (Xing 2007: 145–7). Although Marxism still legitimizes the existence of the CPC in China, Chinese leadership seeks help from patriotism centred on the traditional Chinese culture (Men 2007: 33).

China's aim in the international system is to prevent or limit the development of hegemony, which would limit its space in this system as its economic growth continuously increased (Taylor 2006: 1). The Chinese aim is to counter Western hegemony through Africa, similar to the competition between China, the USSR and the West during the Cold War (Alden 2007: 16). Therefore, US–Chinese relations are a mixture of confrontation and cordiality in the face of the US military presence in Asia, closer US–Japan military ties, and US relations with and support to Taiwan. At the same time China knows that it needs the US for its technological and economic progress (Sutter 2005: 79). China seeks to outbalance these confrontations, and at the same time to demonstrate to the US that China is not a weak competitor that seeks cordiality at any cost, but is instead an equal player in the international system. China, in its Africa policy, not only pursues oil and minerals in Africa to boost its economy and its military power, but also to underline itself as global player in its economic and diplomatic activities in Africa. Therefore, the Chinese Africa policy cannot be limited to the relations between Africa and China; instead US–Chinese relations are being structurally manifested in Chinese–African relations.

Indeed, China is interested in cooperative strategies; however, China has been attempting to modernize itself by developing its economy and promoting nationalism as the leading ideological current, preserving national security and national sovereignty. Economic development is critical to domestic stability, to strengthening China's international role and preserving the country's historic boundaries (preventing Taiwan's independence). China's aim is to make the country a leading global power (Alden 2007: 197–9). China's economic interest is playing an increasingly important role in its approach to its foreign policy.

China has become the world's second-largest oil importer, and in 2004 it alone accounted for 31 per cent of global growth in oil demand; this had serious implications for its foreign policy, economic growth, social stability and for the survival of the Chinese Communist Party (Zweig and Jianhai 2005: 25–6). While some states (developing states) of the world are 'profiting' from the growing Chinese economy and its investment abroad, other states, such as the US and Japan, are concerned because China could challenge the US military dominance in East Asia (Zweig and Jianhai 2005: 26). By 2004, 11 per cent of China's oil imports were from Iran; in October 2004 China signed a natural gas agreement with Tehran that could be worth as much as $70 billion. In the same year, 28.7 per cent of China's total crude oil was supplied by Africa; China's trade and investment treaties with Africa are increasing. Since the creation of China–Africa Cooperation in 2000, China has forgiven more than $1 billion in debt (Zweig and Jianhai 2005: 26–30). Instead of traditional dependence on its Asian neighbours, China is diversifying sources of supply and expanding to regions like Africa in its focus on resource acquisition and commercial opportunism.

At least so far, China is interested in business in Africa without caring much about the political situations such as democracy and human rights (Reporters Without Borders 2008). Africa allowed China to expand its economic presence without interfering in the domestic affairs agenda, sovereignty and national independence (Taylor 2006: 65–7). China considers the Western concept of interventions, such as in Sudan, as interference in the internal affairs of a sovereign state. For China an intervention must be authorized by the UN; it must take place at the invitation of the target state, and respect sovereignty; and force is only to be used when all other options have proven ineffective (Carlson 2006: 217). For example, in Sudan, where China's National Petroleum Corporation owns 40 per cent of the oil consortium, China has been arguing that violence taking place in Darfur is entirely an internal affair.

Since 2002 Angola supplies 15 per cent of China's oil imports and in 2006 Angola passed Saudi Arabia as supplier of oil to China (Alden 2007: 8, 67). As Rocha argues, the increasingly influential role of China could undermine the good governance and democratization processes, cornerstones of the NEPAD, because African states will increasingly be reliant on China and defiant to the democratic principles (Rocha 2007: 26) in its foreign policy of 'no political strings' (Alden 2007: 8). The Chinese stance is feared to undermine the emergence and establishment of African civil societies by offering another world, which suggests that bread comes before the freedom to vote (Obiorah 2007: 38). The Chinese 'non-interference-into-internal-affairs' model is appealing to Africa's more repressive regimes (Marks 2007: 11; Alden 2007: 60). China argues, rather, that in its investment it is raising standards of living in the developing world (Zweig and Jianhai 2005: 31–4).

In Zambia, where the Chinese focus is on minerals such as copper, China is interested not in democratization but in mining. This became obvious when China's ambassador to Lusaka warned that Beijing might cut off diplomatic relations with that country if its voters elected a Taipei-leaning opposition candidate in the presidential elections of 2006. Senegal had to cut its ties with Taiwan in October 2005 and Chad in August 2006 (Alden 2007: 21; Robinson 2006: 15). China has been warning various African states against any diplomatic relations with Taiwan. On 29 January 1998 the Central African Republic (with which China broke off relations when, in 1991, it established diplomatic relations with Taiwan) bowed to such threats and dropped its relations with Taiwan (ARB 1998, 35(1): 12986). China and Guinea-Bissau resumed their diplomatic ties in April 1998, which

China had suspended for the same reasons on 31 May 1990. So far China has been successful in its attempts to displace Taiwan's official relations with African countries. Through its rapidly increasing economic power China has been gaining diplomatic ground and isolating Taiwan in Africa.

Factors that attract Africa to China's global conflict system are: the crisis in China's international relations after the Tiananmen Square incident in June 1989; the expansion of Chinese trade in the 1990s; oil; and the desire to gain the votes of the African states in United Nations decisions that affect Chinese interests, including the Taiwan issue. Moreover, African voters played a key role when China was awarded the 2008 Olympic Games. China rewards Africans by allocating the largest percentage of its development assistance (Alden 2007: 22).

China has discovered market and commercial opportunities in Africa, including African consumers receptive to the type of inexpensive products that China typically produces, and Africa's rich natural resources (oil and minerals) (Broadman 2007: 10–11; Taylor 2006: 70–1; Rocha 2007: 19). Thanks to Chinese investment in the Sudanese oil sector (about $15 billion between 1996 and 2006), Sudan became a leading oil exporter, providing about 10 per cent of Chinese oil demands. Moreover, Sudan is China's second-largest trading partner in Africa; bilateral trade was $2.9 billion in 2006, mainly because China buys 65 per cent of Sudan's oil and is the leading supplier of arms to the Sudanese government (Prunier 2007a). Chinese investments in oil and gas have been taking place in Angola, Nigeria, Algeria and Gabon (Alden 2007: 12–13, 61). Nigeria, as the most populous African state, has huge market potential for China, and it is an oil-producing country, which China desperately needs. Chinese Minister of Construction Wang Guangtao promised in February 2008 to develop Ethiopia's rich hydroelectric resources with the potential of more than 45,000 megawatts, out of which the current exploitation rate is only about 3 per cent.

The volume of African exports to Asia has grown by 20 per cent between 2000 and 2005 compared with 15 per cent between 1990 and 1995. Africa's exports to China grew by 48 per cent annually between 1999 and 2004, compared with 14 per cent for India, while 10 per cent of sub-Saharan exports, especially petroleum and metals, are now to China and some 3 per cent are to India. Sixty-two per cent of total African exports to China consist of oil, increasing at an annual compounded rate of 30 per cent. Angola supplies 47 per cent of Africa's oil exports to China, followed by Sudan (25 per cent), the DRC (13 per cent), Equatorial Guinea (9 per cent) and Nigeria (3 per cent), whereas 87 per cent of all imports from China and India to Africa comprise textile, apparel, electric machinery, equipment, medicine, cosmetic products and batteries (Broadman 2007: 82–3). China's foreign direct investment (FDI), largely concentrated in the extractive industries such as oil and mining, apparel, agro-processing, power generation, road construction, tourism, telecommunication, etc., is rapidly increasing (Broadman 2007: 11–12). For example, in July 2005, China and Nigeria signed an $800 million crude oil sale agreement through which China purchases 30,000 barrels a day for five years.

Economic reforms and defence modernization have become a necessity in order for China to keep up with international standards (Lal 2006: 5). Chinese economic strategy includes turning the country's arms industry into a top global player by 2020, resulting in arms delivery war-torn countries like Sudan, Ethiopia, Eritrea, DRC and Sierra Leone. Various African countries got arms worth $1.3 billion in 2003 from China, which is providing Africa with 6–7 per cent of its arms demands for its bloodiest conflicts, such as in Sudan, Eritrea, Ethiopia, DR Congo, Chad, Zimbabwe and Sierra Leone. Chinese corporations transferred a substantial share of US$1 billion in weapons despatched to Ethiopia and Eritrea between 1998 and 2000; various covert shipments of weapons in violation of regional or international arms embargoes with the final destination mislabelled and the weapons disguised as agricultural equipment (Taylor 2004: 89–96). China has set up factories, such as in Sudan, for small arms, rocket launchers and anti-tank weapons (Alden 2007: 62). Thanks to oil, Sudan has able to spend adequately in weapons since it began to export oil in 1999, and through the help of China, Sudan built new factories to produce weapons in the late 1990s (Kasfir 2005: 201).

In July 2008 the BBC found evidence that China has been helping Sudan's government militarily in Darfur by training fighter pilots and supplying Chinese Dong Feng army lorries, anti-aircraft guns, Chinese Fantan jets as well as Russian-made Antanovs and helicopter gunships despite the embargo of imposed by UN Resolution 1591, which requires foreign nations to take measures to ensure they do not militarily assist anyone in the conflict in Darfur.

But China is also expanding its influence by increasingly participating in UN peacekeeping in Africa. China is the largest contributor of peacekeepers among the P5; China allocated about $200,000 to fight drought in the Horn of Africa in 1999, $610,000 for humanitarian assistance in Darfur in 2004, and in the middle of 2006 it announced it would allocate $3.5 million to the AU peacekeeping in Darfur (Alden 2007: 25–6). The combination of these factors delivers China from its relatively weak and vulnerable position in the international system (Taylor 2004: 83–92). In the context of China's position in the international conflict system, China is ready to grant loans to its client African states without any conditions of human rights or good governance (Perry 2007: 25–6). Sudan has been protected by the Chinese veto sanctions in the UN on the Darfur issue. Robert Mugabe of Zimbabwe has expressed his relief that he looks to the East where the sun rises, not anymore to the West where it sets (Alden 2007: 64).

As the Forum on China-Africa Cooperation ministerial conference in October 2000 showed, the objective of China–Africa relations is to strengthen the South–South relationship on the basis of South–South 'solidarity' against the globalization led by the hegemonic US, increasingly considered as detrimental to the autonomy and sovereignty of China and Africa. The separation of Taiwan, Tibet, or Xinjiang, could mean the loss of legitimacy of the Party and of the state of China (Lal 2006: 63). It is obvious that China is attempting to advance itself as a global player beyond its immediate geographical confines (Taylor 2006: 197). China's Africa expansion is also about demonstrating the 'superiority' of China's economic policies and encouraging African countries to reform (using the Chinese model) to pursue their own development strategy, and to prevent Western domination in Africa (Taylor 2006: 72–3). However, Chinese development assistance to Africa in this decade falls around $1–2 billion per year, whereas the European Union (EU) and member countries annually contribute about $18 billion (including debt relief), multilateral institutions $9 billion, and the United States about $5 billion.

China's economic trade with Africa is forecast to surpass $100 billion by 2010. The rapid increase of Chinese–African bilateral trade from $10 billion in 2000 to $70 billion in 2007 now make China Africa's second largest trading partner after the United States. Moreover, China's direct investment in Africa increased from $491 million in 2003 to over $2.5 billion in 2006. In October 2007 the Industrial and Commercial Bank of China purchased a 20 per cent stake in South Africa's Standard Bank. Similarly in 2007 China granted a $5 billion loan and investment package for Congo that will be repaid in cobalt and copper (Reuters 2007).

China has contributed military personnel and commanders to almost all UN peacekeeping missions in Africa. In Liberia, for example, China has been cooperating with the US to rebuild the national army and military bases around Monrovia. China has been providing Liberia with vehicles, computers and some specialty training and is rebuilding at least one base. By sending 140 of the pledged 300 military engineers, China became the first non-African troop-contributing country to deploy in Darfur.

However, antipathy against the Chinese working in Africa is also increasing. During Zambia's election in 2006, opposition candidate Michael Sata campaigned with racist slogans against the Chinese: 'Zambia for Zambians' (Alden 2007: 85). In early March 2008, Chinese were beaten up, a kitchen for Chinese workers and a guard's house were set alight and hostel windows of Chinese workers were smashed at a copper smelter in northern Zambia by workers demanding better conditions. Such racist attacks had already happened against China in the 1990s in Lesotho. This suggests that many ordinary Africans are considering China not only as a development partner, but also as an economic competitor and probably also as a colonizer. If Africa does not profit equally from its ties with China, if African economies will not be diversified and if the political situation gets worse, China will be seen by Africans not as a development partner, but rather as an exploiter and economic rival (Alden 2007: 127–8). Nevertheless, China has shown the path to many other global players in Africa. Russia is one of them.

Russia and Africa

After two decades of inaction, Russia is starting to retake its role and position of the Cold War period. The foreign policy concepts of former president Putin underlines that a strong Russia can be achieved only through strong economic performance and a robust position in the international system. This is leading to competition between Russia and other global players in Africa. Moscow, a superpower in Africa during the Cold War era, has increasingly lost its influence to China in Africa. But Putin has been trying to change this.

For Russia, arms exports have been crucial to keeping munitions and weapons factories in business and to secure hard currency from abroad (Donaldson and Nogee 2005: 229–36). This factor has cemented Russia's indirect presence in various conflict zones, such as in Africa, in spite of the end of the Cold War. However, the end of the Cold War led to a dramatic reduction of Soviet military and non-military support to sub-Saharan Africa, from 12.5 billion rubles in 1989 to 400 million rubles in 1991 (Donaldson and Nogee 2005: 345). Though, after the collapse of the USSR, Russia's main foreign policy objective outside of the West was to ensure the security of the territories of former Soviet lands; the emphasis on the role of foreign policy as a contributor to economic development encouraged Russia to increase its economic presence, for instance in Africa, in order to export raw or semi-processed products that could not be easily sold in Western markets.

When Vladimir Putin visited Africa in early September 2006, he expressed Russia's intention to build new bridges with Africa through expanding trade, promoting education and supporting the fight against HIV/AIDS, thus opening new markets for Russian businesses, nuclear energy, and companies such as Alrosa, Gazprom, Lukoil, Rusal, Renova, Gammakhim, Technopromexport, and VEB and VTB banks (Saltanov 2008). Trade between Russia and the sub-Saharan African countries in the first 10 months of 2008 amounted to $1.9 billion, and surpassed the relevant period of 2007 by 30 per cent. Putin met with former South African President Thabo Mbeki on expanding economic and political cooperation between the two countries. Early in 2006 Russia concluded a $7.5 billion arms deal with Algeria. Foreign Minister Sergey Lavrov visited Angola to open a representation office of Russia's Vneshtorgbank. Lavrov then visited Addis Ababa. This shows increasing Russian interest in Africa (Klomegah 2006).

Russian aid for Africa was increased from $50 million in 2003 to $210 million in 2007, not including the written-off debts amounting to $20 billion. Russia intends to cancel further debts in Benin, Guinea, Zambia, Madagascar, Mozambique, Tanzania, and Ethiopia, amounting to more than $500 million, to provide more than $1 billion in aid to the poorest countries, including in Africa, during the period to 2010–11, and food and humanitarian aid to Guinea, DR Congo, Zimbabwe, Kenya, Somalia and Ethiopia. Russia is interested in developing cooperation with African countries, which it considers as promising partners to promote strategic stability, a multipolar world, strengthen the UN role, and counter terrorism. Russia has joined international efforts to combat piracy of the Somali coast. Current and future principal African partners for Russia include Angola, Guinea, Namibia, Nigeria, Ethiopia, South Africa, DR Congo, Gabon, Zimbabwe, Cameroon, Ivory Coast, Kenya, Madagascar and Mali (Saltanov 2008).

The new 12 July 2008 version of the Russian Federation Foreign Policy Concept emphasizes the increasing importance of Russian–African relations. Various African ministers and agencies of foreign affairs from South Africa, Kenya, Cameroon, Gabon and Madagascar visited Moscow in 2008. Parliamentarians from Gabon, DR Congo, Zambia, Madagascar and South Africa also visited Moscow. A visit of AU Commission Chairman Jean Ping to Moscow is scheduled for 2009. Russia wants to help build peace and security in Africa through the UN. About 230 Russian troops and policemen are involved in all UN peacekeeping operations in Africa, such as in the Democratic Republic of Congo, Western Sahara, Sierra Leone, Ivory Coast, Liberia and Sudan; and in 2008 Russia joined the European Union's peacekeeping operation in Chad and CAR by transferring a Russian military contingent (120 troops with four military transport helicopters) to Ndjamena (Saltanov 2008). There is already Russian–African cooperation in areas such as minerals and raw materials, infrastructure, energy and other spheres, in addition to the steps that are being taken to develop cooperation with African countries in the realm of high technologies (nuclear energy, astrophysics, exploration and development of outer space for peaceful purposes).

Putin made military modernization one of the forefronts of his government because Russia's role in the future of international affairs requires military modernization. As Putin expressed it in his Foreign Policy Concept, Russia has increasingly been alarmed by US unilateralism and has attempted to become a leading voice of multilateralism as an alternative (Caldwell 2007: 323–7). Wallander concludes that:

> Russia's is not a foreign policy driven by economic growth for economic growth's sake. It is driven by economic growth for the sake of power, autonomy, and global position … Russian interests in expanding its energy exports explain its relations with Europe, its increasing interests in the Commonwealth of Independent States, its growing attention to Japan and China, its commercial relations with Iran, and its concerted efforts to increase foreign arms sales (which amounted to more than $6.1 billion in 2005). (Wallander 2007: 458–9)

Arms trade is a key component of Russia's economic relations with Africa through equipping and advising armed forces (in Algeria, Angola and Ethiopia, about 90 per cent of the arms are Russian-made).

There are fewer Russian ideological or strategic interests nowadays; instead there are more practical and pragmatic economic and trade interests in Africa – 'with the splitting up of the former Soviet Union, Russia has found itself deprived of many of the supplies of minerals vital for its economy and which came from sites within the USSR but now outside its borders. This has stimulated the search for resources from other locations, and there has been a rise in the importance of imports from Africa of manganese, chrome, nickel, zinc, lead, etc.' (Shubin 2004: 107). Oil is also becoming increasingly important. According to the Ethiopian Trade and Industry Minister Girma Birru's statement on 14 February 2008 Russian oil company Lukoil is planning to develop deposits in Ethiopia. In the 1980s the Soviet Petroleum Exploration Expedition (SPEE) drilled nine deep gas wells (between 1982 and 1993) in the region of Calub in Ethiopia. The Calub gas reserves are estimated at 2.7 trillion cubic feet (TCF), while the Hilala gas reserves are estimated at 1.3 TCF. The Gambella basin is one of the five sedimentary basins found in Ethiopia, which are expected to be oil-prospected. Different countries, including China, India, Malaysia and the British White Nile Company, have shown interest in Ethiopia's oil exploration (Ethiopian News Agency 2008).

In addition to Russia, Japan and India are emerging in Africa. Through defence modernization and economic development, India attempts to protect its sovereignty and sustain national integrity. India advances its nuclear and non-nuclear capabilities to address threats from Pakistan and China. India is increasingly focusing on external issues, such as integration into the international system wherein the most powerful reigns and increased competitiveness lead to improved leverage (Lal 2006: 7, 83). India is modernizing its defence at a cost of US$50 billion. India's purchase of fighter jets for US$10 billion, its robust missile attack and defence programme, and its considerable progress as a sea power (including the delivery of sea aircraft carriers from Russia, such as the *Admiral Gorskov*, in addition to the ones being indigenously developed) all indicate that India's ambitions will have considerable global implications, not least because there is ongoing competition between India and China to control the waters of the Indian Ocean, while India, through its strategic friendship with the US, attempts to send a tough message to China.

Economic reforms and defence modernization are key to China as well as India. Maintaining a nuclear deterrence is critical in defending political and economic rights by increasing comprehensive national strength (Lal 2006: 41). One of the tactically helpful strategies of the emerging global powers, such as China and India, is the realization that much of the developing world is dissatisfied with the international economic order. China exploits these grievances in order to further its prestige and position as the 'spokesman' of the less developed part of the world (Taylor 2006: 200). The India-Africa summit on 9 April 2008 cemented the increasing role of India as a global player in Africa. India will play, according to the summit declaration, an increasing role in the education sector, science and technology, agricultural productivity, food security, industrial growth, infrastructure and the health sector.

Japan, though an economic global heavyweight, has not shown great influence and presence in Africa. But the Japan-Africa conference in May 2008 hinted that this could change. Japan has vowed to double its aid to Africa by 2012, which is currently $1.7 billion. Japan will attempt to increase it gradually year by year, though it is less than what the United States, Britain, France and Germany provide. This suggests that Japan is competing with other powers like China and India for influence in Africa and to have access to Africa's markets and to its natural resources. Moreover, Japan encourages Japanese firms to invest in Africa. The following chapter shows how this international system of competition becomes part of the conflict systems in Africa. Therefore, conflicts are products of states' inability to regulate the intra-state systemic interaction and to resist the negative impacts of the international security system.

Chapter 3
State and Security as Systems of Conflict and Violence

The State–Security–Conflict Interrelation

This chapter shows that conflicts emerge from states' behaviours internally and internationally. State systems, in their attempts to act as independent actors in the international system, are founded on systems of conflicts and wars that sustain them (Barash and Webel 2002: 192). The structures of the modern state are based on the capacity to organize violence through hierarchical state structures, organization of material resources directed towards the destruction of enemy countries that are seen as a threat to national security, threatening to inflict injury or actually inflicting injury to the enemy (Jeong 2005: 55; Tilly 1985, 2003). Bernard suggests that 'all social life consists of interaction within and between social systems' (Bernard 1977: 124). For Campbell security can become 'absence of movement' or death via stasis (Campbell 1998: 12), or for Deutsch a 'violent stability and stagnation' (Deutsch 1979: 5).

The conflict system is a system wherein the value of anything changes: neither the human beings nor the natural resources have constant values. A conflict system reproduces itself autopoietically (by reproducing itself) in the sense that values of the factors and actors that constitute the conflict system change continuously due to intensification of conflict. That is, the self-reproduction takes place because the role of the elements changes continuously by changing the status and type of relations between elements. There is nothing which is 'there'. In violent conflicts human beings are dehumanized (as sex slaves, raped, maimed, killed, displaced, etc.); natural resources acquire more value than the human beings, or they are destroyed if they cannot be possessed (like bridges, harvest, oil refineries, etc., which are deliberately destroyed during wars).

All three states discussed in this book are affected by civil wars. According to Waldmann 'civil war is a highly armed conflict of a long duration, carried out between two or more groups (state, ethnic or religious) in a state, in order to achieve a division or allocation of power and resources' (Waldmann 2002: 368). Waldmann suggests that civil wars happen in weak or collapsing states; where there are internal tensions and social clefts (with religious, social, ethnic and political character); when there were fights among these groups already in the past (collective memory effect); and when social and political change takes place due to demographic shift of balance or dominance, state collapse, military coup, refugees from third countries, or external intervention (Waldmann 2002: 377–9).

Crocker suggests that state failure is characterized by victory of guns over normal politics (Crocker 2003: 35). The intensity of a conflict, the quality of a regime change and the propensity to state failure are essentially interdependent. This means, the higher the intensity of a conflict, the lower the quality of the regime change resulting in regime change by military force, and the higher the propensity to state failure. Hence, according to Crocker, 'when state failure sets in, the balance of power shifts ominously against ordinary civilians and in favour of armed entities operating outside the law' (Crocker 2003: 36). But these authors focus too much on the internal factors, whereas regional and global dimensions are not given due consideration.

As the case study chapters will show, ethnicity and territory play a key role in the conflicts in DR Congo, Somalia and Sudan. Boundaries are not only sources of dispute but are also symptoms of conflict. Anthropologists suggest that territoriality has two dimensions: delimitation of boundaries and behaviour within those states. Territory becomes something symbolic and sacred that has to be defended against internal as well as external threats, even through blood sacrifice. As Goemans points out, 'we' live 'here' is a particularly powerful way to coordinate group members to provide a collective defence (Goemans 2006: 26–9). Because of the symbolic stakes that are often invested in international conflicts, territories that are devoid of resources may still become the site of violent disputes. Domestic political dynamics drive territorial conflict as much as the strategic value of the territory in dispute, and those political dynamics are often rooted in the symbolism of territory rather than in its measurable value, such as economic or demographic significance (Kahler 2006: 3–7).

Territory constitutes an important component of individual, group and national identities and determines daily behaviour and practices. The symbolic dimension determines the attachment and affiliation of the citizens to particular spaces and places because they are the location of historical and mythical events imbued with exclusive significance. Territory is more important than mere natural resources; it is the symbol to define and defend one's social, spiritual and communal world (Walter 2006: 288–9). The most important focal principles for defending territory are natural frontiers based on the common knowledge, common culture, common language and religion of the group (Goemans 2006: 27). Even the 'de-territorialized' Diaspora maintains the strongest attachment to homeland territory; it can contribute to the exacerbation of conflict back at home through lobbying and/or the raising of finance. This de-territorialized commitment of the Diaspora originates in the intangible and symbolic factors that constitute the national territorial identity, which are the most difficult to resolve between belligerents. Newman suggests that territory-related conflict resolution should move beyond the traditional discourse of demarcation, proximity, size and shape of territories to a more functional definition of the territorial phenomenon (Newman 2006: 108).

As the dynamic of various conflicts suggests, however, the territorial factor cannot be dismissed as simply 'primitive', i.e. not relevant for the modern world. Besides territoriality, the intangible and symbolic factors remain the most difficult to resolve between belligerents. 'Measuring the value of land strictly in terms of just tangible assets and making these assets the basis of a settlement is unlikely to bring long-term peace. To date, policymakers have tended to focus disproportionately on issues related to border length and placement, and the distribution of natural resources, rather than dealing with the deep psychological bonds that may exist among individuals at the local level' (Walter 2006: 294).

Natural resources, especially in Sudan and DRC, have double roles in rendering states as conflict and violence systems. They are on the one hand the economic means needed for the national economy; on the other hand the warring groups need them to make war (to finance the war). The fact that these natural resources are needed by the adversaries to wage war makes them ambivalent: both warring groups need them to finance the war; at the same time they are potential targets of destruction by either side in case they are possessed by an adversary. For Collier, conflicts in Africa are caused mainly by economic polarization and low economic growth, atypical opportunities or economic inequality rather than ethnic diversity (Collier et al. 2005: 3–8).

Conflict and Violence: Causes and Dynamics

There are various definitions of 'conflict' in international relations, psychology, sociology, and so on. It is defined as: incompatibility of interests (Bartos and Wehr 2002: 7; Van der Dennen and Falger 1990: 2); negatively correlated goal attainment (Deutsch 2006); any social situation or process in which two or more social entities are linked by at least one form of antagonistic interaction (Fink 1968); a manifestation of a belief of two or more persons or groups in incompatibility of objectives (Kriesberg 1998: 2). Georg Simmel designated conflict as 'sociation' because if we consider interaction as *sociation*, we have to consider conflict as *a-sociation* (Simmel 1964: 13). Bartos and Wehr suggest that there are three fundamental causes of conflict: first, the desire to redistribute scarce resources; second, the desire to enact incompatible roles; third, the desire to pursue incompatible values (Bartos and Wehr 2002: 7). Conflict as a malignant interaction process includes an anarchic social situation; a win–lose or competitive orientation; cognitive rigidity; misjudgements and misperceptions; unwitting commitments; self-fulfilling prophecies; vicious escalating spirals; gamesmanship orientation that turns the conflict away from issues of what in real life is being won or lost to an abstract conflict over images of power (Deutsch 1986: 131). These attributes of conflicts and violence suggest that they are a kind of organized system or human institution in which individuals have specific roles (Hinde 1990: 176–7).

Nicholson (1995) suggests that we cannot answer the question, 'How can war be stopped?' without answering the question, 'Why are wars fought?' Waltz maintains that 'war begins in the minds and emotions of men, as all acts do' (Waltz 2001: 9). Further, he argues that 'if violence among states is caused by the evilness of man, to aim at the internal reform of states will not do much good. And if violence among states is the product of international anarchy, to aim at the conversion of the individuals can accomplish little' (Waltz 2001: 14). But in spite of this Waltz concludes that 'through reform of states wars can be reduced or forever eliminated' (Waltz 2001: 83). According to Nicholson, a conflict exists when two people wish to carry out acts which are mutually inconsistent, while war is any situation of large-scale deadly violence between anonymous or political groups (Nicholson 1995: 11, 16).

Fink's definition, that social conflict is a 'social situation or process in which two or more social entities are linked by at least one form of antagonistic interaction' (Fink 1968), points to the systemic character of conflict. As shown in Chapter 1, Luhmann describes system as the totality of its parts (Luhmann 1999). If systems are interactions, they entail actual or potential conflicting interactions. The perception and definition of conflict determines the process and the method of how to deal with conflict. Conflict is usually conceived as the outcome of irreconcilable interests competing for scarce material or immaterial resources; or conflict is seen as negation of communication. Niklas Luhmann defines conflict as communication of contradictions (Luhmann 1999: 388). According to Luhmann, conflict is a parasite social system that inclines to absorb the resources of the social system in which it develops (Luhmann 1999: 390).

I argue that conflict not only absorbs the resources of the social system in which it grows, but also generates new resources that seem more precious than the resources it absorbs. These resources are not only territorial gain or natural resources, but also cohesive collective identity and national unity for which the resources of the social and economic system can be sacrificed. Therefore, conflict is the beginning and the process of the fundamental perception of the Self, the collective identity and the enemy. In the state of conflict all these factors will be defined anew and incessantly, and will constitute conflict dynamic. The changing perception and redefinition of the Self, collective identity and enemy renders identification of the main causes of a conflict almost impossible, since the issues and interests vary through this continuously changing self-definition. If a conflict analysis perceives the issues and interests as fixed and objectively identifiable realities, it will face difficulties in understanding the nature and dynamic of that conflict. The bulk of conflict theory regards the issues, actors and interests as given, and on that basis makes efforts to find a solution to mitigate or eliminate contradictions between them. Yet the issues, actors and interests change over time as a consequence of the social, economic and political dynamics of societies.

Conflict system has to be seen from two points of views: as a mode of interaction within a system as well as between systems. In the intersystem conflict a group loyalty is not constant over time but varies as a function of the threat to the group. The appearance of the hostile group serves as a trigger for a display of loyalty in the form of nationalism or patriotism symbolized through flags, anthems, constitutions, etc. (Peres and Hopp 1990: 128). 'In times of external threat, when an internal conflict would be most damaging, their symbolic universe turns to highlight unifying themes, and focuses in particular on the familial motifs. All members become brothers; the territory takes on the entity of the motherland, the leader father figure, etc.' (Peres and Hopp 1990: 130). These developments have intrasystemic as well as intersystemic consequences.

Kenneth Waltz argues that wars are 'akin to earthquakes in being natural occurrences whose control or elimination is beyond the wit of man'. Therefore, 'attempts to eliminate war, however nobly inspired and assiduously pursued, have brought little more than fleeting moments of peace among states' (Waltz 2001: 1). Further he maintains that 'our miseries are ineluctably the product of our natures. The root of all evil is man, and thus he is himself the root of the specific evil, war' (Waltz 2001: 1). The major causes of war are to be found within man, within the structure of the separate states, within the state system (Waltz 2001: 12).

Bartos and Wehr (2002), in their studies of conflicts, have identified various factors that directly or indirectly lead to the rise and escalation of conflicts: the growth of science and technology and its application to weaponry; the growth of the nation-state and its capacity to mobilize resources for control and violence; expanding population (additional human and conflict potential); large-scale civil unrest and strikes; demographic dislocation because of industrialization by the mid-nineteenth century; extreme poverty; colonial divide and rule (manipulation of group against group); gulf between worker and owner; economic depression; ethnicity (ethnic nationalism, racial superiority); education and self-consciousness and struggle for independence from colonialism; introduction of nuclear weapons; resistance to racial segregation and discrimination; civil rights movements; women's rights; environmentalists; civil conflict among ethnic and tribal groups; superpower rivalry and military support; ideological allegiances; economic hegemony; strategic and military influence stimulating wars in the developing world; international intervention (peacekeeping, peacemaking and peace-building); empowerment movements (Bartos and Wehr 2002: 2–6). Hence, for them three fundamental causes of conflict are: the desire to redistribute scarce resources; the desire to enact incompatible roles; and the desire to pursue incompatible values (Bartos and Wehr 2002: 7). Moreover, according to them, conflict behaviour can occur for six main reasons: the parties believe they have incompatible goals; they each may have achieved high solidarity; they may have organized for conflict; they can mobilize their conflict resources; they may be hostile towards their opponents; and they may have sufficient material resources (Bartos and Wehr 2002: 9). Further they argue that 'if the level of conflict *solidarity* within group increases the chances that it will engage in conflict *behaviour* increases as well'. The outcome of a conflict will be determined by the ability to mobilize both material and symbolic resources (Jeong 2005: 31).

Sometimes it is argued that scarcity causes conflicts. In many conflicts it is rather the perceptions *not* the reality of scarcity that incites hostility, hate and war towards the outside, while such perceptions of scarcity help harness feelings of solidarity, sympathy, mutual identification, shared sacrifices and sense of community towards the inside (Deutsch 1979: 9). 'Rational choice models typically assume that preferences are fixed and given; in other words, preferences are modelled to be exogenous and unchanging. What witnesses observe as changes in behaviour on the part of the decision maker do not result from changes in preferences but rather emerge from shifts in expectations and actions due to external constraints' (McDermott 2006: 52). According to the rationalist-realist model in international relations, all leaders are rational actors and are interchangeable and share the exact same preference, which is to maximize their state's power and security (McDermott 2006: 52).

But interests are not always fixed. As Kriesberg neatly suggests:

> Each side in a conflict usually blames the other's aggressive behaviour for the fight. An observer, however, might conclude that the adversaries are mistaken: each is actually seeking to defend what it has and its defensive efforts are incorrectly perceived by the other as threatening. Or the observer may believe that the basis of the conflict is a natural consequence of the imbalance of power among the adversaries and the lack of agreed-upon procedures for managing issues in dispute between them. (Kriesberg 1998: 30)

Hence, there are three different perceptions of a conflict: the *one* of the outsider (mediator), and *two* contradictory conceptions of the adversaries. As any conflict shows no conflict can be solved if the mediation process focuses mainly on a 'just' distribution of the contested material resources. The more intense a conflict, the more the immaterial conflict resources play a role than the material ones. Unfortunately, most of the time mediators of interstate conflicts focus too much on the contested material resources while forgetting the social-psychological aspect of escalation and de-escalation (Kriesberg 1998: 182). For example, the Ethiopian–Eritrean conflict cannot be solved by the positivistic approach of the Ethiopian–Eritrean Border Commission. The conflict between these states has a lot of psychological factors, like superiority complexes of both peoples going back to the colonial period as will be discussed later. Therefore, success of a mediation depends on how psychological or non-material causes are given due consideration, not only material or positivistic factors.

Cottam et al. suggest that the fundamental problem here is the *attribution error*: our tendency to attribute another person's behaviour or actions to their dispositional qualities (their personalities, motivations, etc.), rather than to situational factors in the environment that may have caused that behaviour (Cottam et al. 2004: 261). Threat and deterrence happen often by attribution error. But 'whether deterrence would fail or succeed depended not upon how many weapons each side possessed, but upon the perceptions each side possessed regarding the willingness of their opponents to really push the button' (Cottam et al. 2004: 262). The other is always perceived as aggressive and offensive (Kriesberg 1998: 30). In addition to attribution error, which leads to threat and deterrence, analogical reasoning plays an important role in gaining knowledge and a perception of the other through which an event or a situation will be designated as 'similar to (or sharing many of the same characteristics) a previous historical event' (Cottam et al. 2004: 270).

According to *prospect* theory 'states will be less likely to go to war to obtain gains from other states when potential risks are high (be risk averse in the domain of gains), and they will be far more likely to go to war with other states when their own resources are threatened (be more risk acceptant in the domain of losses)' (Cottam et al. 2004: 272). *Accountability* theory maintains that 'political leaders will take greater risks, and will be more likely to engage in conflict the more they lack accountability to a higher power (i.e., a ruling coalition, a voting public …)' (Cottam et al. 2004: 272).

Whereas Kriesberg suggests that oppressive domination and destructive conflicts are not inherent in the human condition (Kriesberg 1998: xiii), Waltz, as discussed above, concludes that 'our miseries are ineluctably the product of our natures. The root of all evil is man, and thus he is himself the root of the specific evil, war' (Waltz 2001: 3). For Waltz wars are like natural occurrences whose control or elimination is beyond the wit of man, and their major causes are to be found within man, within the structure of the separate states, within the state system (Waltz 2001: 12). This conclusion of Waltz is contentious because, if wars are part of human nature, war becomes our human destiny, which we cannot change and for which in most cases we are not responsible. This suggests that neither the individuals nor the states can do anything against that warring nature. The problem with Waltz's assertion is that individuals as well as states can argue that they are incapable of avoiding wars because they see it as part of their nature. This is mainly because the system they are in is perceived to be responsible for that. In the face of such a system, individuals pretend to be defenceless components of system machinery, and they scapegoat the system.

Conflict serves to establish and maintain the identity and boundary lines of societies and groups because it contributes to the establishment and reaffirmation of the identity of the groups, and through it the group maintains its boundaries against the surrounding social world. But at the same time conflict always denotes social interaction or communication between groups as well as within groups. Through conflicts *between* groups patterned enmities, reciprocal antagonisms, social divisions and systems of stratification will be conserved. In the case of a low degree of bridging and linking between groups, a group conflict can emerge. Because of this lack of 'bridge capital' (contrary to social capital between and within groups) conflicts are less regulated and they escalate (Karstedt 2004: 287–8). The lack of 'bridge capital' emerges because the adversaries come to view themselves as being in a zero-sum relationship (Kriesberg 1998: 39). This lack of bridge capital can be termed as intolerance, which Deutsch defines as the 'recurrent denial of social and economic opportunities to individuals because of their classification as members of a certain group' (Deutsch 1979: 50). Through conflicts *within* groups, social systems provide for specific institutions that serve to drain off hostile and aggressive sentiments towards outside through hostile stereotypes, prejudice, threat perception, general hostility and aggression. The communication between inside and outside is mostly characterized by misinterpretation (because of fear and distrust): in ethnic violence each group attributes to the adversary group the worst intention of atrocity. Radicalization of ethnic stereotypes and of symbolic hierarchization: own culture will be idealized, and past real or perceived oppression or glory magnified (Wimmer and Schetter 2002: 324). But at the same time, groups engaged in continued struggle with the outside tend to be intolerant within the group as well. Carl Schmitt, in his *The Concept of the Political* (1996), has demonstrated in detail this phenomenon of intolerance towards inside as well as outside in the enmity–amity constellation. According to Tilly, this intolerance becomes a ritual of violence in which credibility and reliability of the main agents in the conflict will be undermined if they cannot mobilize enough supporters (Tilly 2002: 563). Whether the negative qualities attributed to the 'others' are the result of conflict or its cause is a disputed issue, however.

At any rate, conflict presupposes some kind of relationship. As Simmel suggests, antagonism is usually involved as an element in intimate relationships. A conflict is more passionate and more radical when it arises out of close relationships. That is, deepest hatred grows out of broken love that results in denial of one's own past and of change of feeling. It will be recognized that the past was an error or a failure of intuition, the past splits the security and unity of our self-conception, our current intolerable feelings mean that we pay for our failures (Simmel 1964: 45–8). 'The recall of earlier agreement has such a strong effect that the new contrast is infinitely sharper and bitterer than if no relation at all had existed in the past' (Simmel 1964: 47–8).

As all conflict case studies dealt with in this book show, social changes are often disruptive to political elites. Political elites therefore generally favour the status quo and oppose changes that would reduce their power, whereas changes in a society occur as the result of actions by persons who stand to benefit from these changes. The Darfur rebels against the Sudanese government; the southern Sudanese against the Khartoum regime; the Oromo against the Amhara and Tigraians in Ethiopia; various armed and unarmed rebellions against Mobutu in Zaire; and various armed clan-based rebels since 1979 in Somalia, have been trying to change the society or state to benefit from these changes. This can happen through *scramble* competition and *contest* competition: scramble competition occurs when each participant attempts to accumulate and/or utilize as much of the critical resource as it can, without regard to any particular social interaction with its competitors; whereas in contest competition, competing individuals interact directly with each other and use the outcome of such interactions to determine access to resources (Van der Dennen and Falger 1990: 9). Conflict and competition do not lead necessarily to destruction. But sometimes they can become destructive, which is violence.

Violence

Gurr suggests that violence is 'a specific kind of response to specific conditions of social existence' (Gurr 1970: 317) characterized by explicit or implicit conflicts. According to Galtung, violence is the influence that keeps the actual somatic and mental realizations below potential realizations (Galtung 1975: 110–11). Galtung divided violence into *physical*, *cultural* and *structural*. Whereas physical violence is a violence against somatic integrity (killing, torture, rape, etc.), structural violence is about a gradual death of the poor, who are victims of a socioeconomic system. Cultural violence means discrimination and marginalization of a cultural group because of its cultural, religious or ethnic background. But here I discuss briefly only physical and structural violence.

The phenomena of violence has led various researchers to explore the causes, characteristics and manifestations of violence. Violence is a monopolistic behaviour that consciously or unconsciously attempts to prevent change, that is, it attempts to materialize stability and stagnation (Deutsch 1979: 5), sometimes known as *security*. Various scholars of state and nationalism such as Benedict Anderson (2000), Michael Billig (1995), Charles Tilly (1985, 2003), Smith (1994, 2000, 2001) etc. have contributed a lot of valuable studies on this matter. As these and other similar studies show, state- or nation-building has been accompanied by physical violence. Therefore, state, security and conflict and violence are concomitant. In the interrelation of these four phenomena even physical violence is justified for the sake of the state's national interest: material and immaterial security.

Physical Violence

The common typologies of the phenomena of violence are: individual violence (violence in the individual as well as public sphere); collective violence (political violence, group violence, civil war), and state violence (state power monopoly, dictatorship and state terrorism, violent encroachments, wars and war crimes) (Imbusch 2002: 46). For Hanagan (2002) there are four basic forms of violence: public violence (through coordinated action, asserting claims in the form of war or collective protest); symbolic violence based on symbolic consent of the citizens (the authority claims legitimacy through and from the population in order to control the unexpected and the chaotic); daily violence (routine violence: by implementing immediate means for immediate objectives through violent demonstrations, plundering, robbery…); and private violence (family violence, household violence against children, women…) (Hanagan 2002: 153–4).

Eckstein developed two approaches to violence analysis: contingency and inherency approach. Contingency of violence is about accidentality or non-continuity of violence: when violent conflict is not in the normal 'repertoire' of political competition but instead is caused by special aberrant conditions or blocked pacific dispositions. In this case collective political violence is affective and a learnt response, not calculated. Inherency of violence denotes that collective political violence is a normal response to commonplace conditions out of tactical considerations and cost–benefit calculations (Eckstein 2004: 434–5).

Heitmeyer and Hagan (2002) suggest that violence is ambivalent in its characterization, in the conception of its phenomena, in its logic of emergence, in its escalation, in its causal explanations and in its evaluation. This means that the difficult thing in violence research is to determine what violence is (Heitmeyer and Hagan 2002: 16). Imbusch differentiates between meanings of violence: *who* exerts violence; *what* happens (phenomenology) of violence; *how* (mode) of violence; *unto whom* (victims) of violence; *why* (causes) of violence; *why for* (objectives) of violence; and *wherefore* (justification) of violence (Imbusch 2002: 34–7).

Violence is also a mode of communication. In his analysis of violence Galtung goes beyond the common definition of violence as physical injury or killing, and includes the structural and cultural dimensions of violence. Hanagan defines violence as a circumstance in which physical damage is inflicted on persons or properties (Hanagan 2002: 153). For Nunner-Winkler (2004: 21) physical violence is exertion of physical coercion in order to break the resistance. For Nunner-Winkler, in addition to physical violence, there is also psychological violence, which causes psychological damage (psychological consequences) through omission (negligence), wherein damage occurs not only through infliction but also through permission (tolerating) (Nunner-Winkler 2004: 21–2).

The state as organized crime is about the elitary legitimation of violence in the building and maintaining of states (Münkler and Llanque 2002: 1219–29). Within a political community there can be collective political violence, which involves destructive attacks by groups against its regime, authorities or policies (Eckstein 2004: 432). The group members (self-)consciously engage in a power struggle on behalf of their group based on shared group membership or a common enemy while trying to control, influence or otherwise enlist for their collective interests (Simon and Klandermans 2004: 455). Accordingly, 'politicised collective identity fosters (self-)stereotyping processes at the cognitive level, prejudice processes at the affective level, and conformity and discrimination processes at the behavioural level' (Simon and Klandermans 2004: 459). These processes change the nature of the desired object.

In conflict analysis it is important to deal with how and why the fought-for resource is valued or value-changed. A combatant may value a resource more highly than does its opponent (Van der Dennen and Falger 1990: 13). The unpredictable and undeterminable value of the resource has a substantial impact on conflict dynamic. If a resource will be highly valued (which is the norm in a conflictual relationship), the 'other', the contestant, will be easily dehumanized, treated as an enemy, denied love and respect and degraded as a mere thing (Jeong 2005: 54; Barash and Webel 2002: 157–8). Girard suggests that 'it is no longer the intrinsic value of the object that inspires the struggle; rather, it is the violence itself that bestows value on the objects, which are only pretexts for a conflict' (Girard 1986a: 144). It is rivalry that continuously constitutes the value of a resource.

But 'rivalry does not arise because of the fortuitous convergence of two desires on a single object; rather, the subject desires the object because the rival desires it. In desiring an object the rival alerts the subject to the desirability of the object' (Girard 1986a: 145).

> Violence inspires counter-violence by those against whom it is directed. It consumes scarce resources that could otherwise be used to satisfy aspirations. Worst of all it consumes men, its victims physically, its practitioners mentally, by habituating them to violence as the means and end of life. The more intense and widespread the use of force the less likely are those who use it, rebels or regimes, to achieve their objectives except through total victory. (Gurr 1970: 358)

This passionate destruction applies not only to the material things, but also to the human beings. Displacement, mutilation and rape of the civilians are just the expression of this passionate violence and violent passion. The civilians are potential prey or supporters of the other party, and their 'outsideness' is an existential concussion of the rivals. Various conflicts, such as in DR Congo, Sudan and Somalia, have been strategically transforming the civilians from accidental victims into strategic targets. In the Tutsi-Hutu conflict in the eastern DR Congo, Rwanda and Burundi; the inter-clan conflict in Somalia; in the interethnic and centre-periphery conflict in Sudan, the civilians were deliberately targeted by the warring groups in order to punish the rivalling armed groups which could gain support from the civilians. Therefore, the rivals either totally monopolize the civilians or otherwise the latter will be displaced, mutilated, killed, raped, etc. Through their 'outsideness' the civilians dilute the 'truth' the rivalling rebels and the government claim. The civilians are desired 'objects' that have to be integrated into violence or destroyed through violence. This is why the desired object embodies at the same time desirability and destructibility, since it is desired by both rivals.

Hanagan argues that the emergence of the consolidated states happened simultaneously with the changes in the forms of violence, such as obligatory military service, military revolution and reorganization (Hanagan 2002: 169). The most important forms of violence (the public, the symbolic, the daily and the private), according to Hanagan, are deep-rooted in social life because they are deep-rooted in the state structure as well as in political and economic relations. The biggest changes in economic and political life lead to the biggest changes in the forms of violence (Hanagan 2002: 175). Similarly, Wimmer and Schetter argue that ethnic violence is a concomitant phenomenon of the process of the nation-state-building and democratization. Through ethnic violence the identity of an ethnic group will be set free from ambivalences and stabilized. In the violence process the belief in common identity will be elevated to purity fetishism (Wimmer and Schetter 2002: 313–15). They further argue that political modernization is a macrostructural precondition of ethnic violence: the process of the territorial separation between outside and inside creates necessarily the definition of those who are inside and outside. This process of 'internalization' and 'externalization' results in the fact that the internalized are homogenized in the process of state-building.

Violence is the process of categorically differentiating the homogenous 'internal' from hostile 'outside' (Wimmer and Schetter 2002: 318). It is a process of homogenization of a group *for* violence *through* violence against another group; or a process of unification of individual perceptions and interests into common objectives through coordination of actions and persuasion as well as coercion. Elwert concludes that violence increases, first, when institutions fail; second, when exigencies on institutions increase without that the institutions change (Elwert 2002: 363). Various studies (such as the Failed States Index of 2008) categorize DR Congo, Somalia and Sudan as institutionally failed. At the same time all states are unable to respond to the exigencies of their citizens or ethnic groups. Failed states are characterized by ethnic and/or religious conflicts against the minorities in which the victims are attacked not as individuals but instead as representatives of the group (Bjorgo 2002: 981).

As we have seen above, Heitmeyer and Hagan (2002: 16) pointed out the difficulty of determining violence. Further, Heitmeyer and Soeffner maintain that the normality of violence is its ambiguity because it cannot be definitely said when it begins (Heitmeyer and Soeffner 2004: 11). There is a widespread tendency that violent societies are the uncivilized ones. This epoch and culture (especially the Western) is seen as civilized, whereas violence is considered as pre-modern or primitive (Heitmeyer and Soeffner 2004: 12). Violence that is not legitimated by civil law, constitution or international law is considered as ignominious, unjust, disgraceful and irrational (Soeffner 2004: 62).

Girard, in his various studies, has shown that in the sacrificial act (in pre-modern societies) as well as the justification of any invasion, war and scapegoat mechanism in the modern society have something holy or sacred in their nature. The aggression in the community will be attributed to the victim that is considered responsible for the crisis in the community; and through this attribution the solidarity of the community will be repaired (this reparation is also attributed to the victim). This renders the victim enormously important for the community. Similarly, Carl Schmitt has explored this phenomenon in his *Political Theology* (2005) and *The Concept of the Political* (1996).

Ethnic-religious rebellion is a violence of expectation of salvation as a result of social loss of control or domination of one group by another (Heitmeyer 2004: 93–5). Violence is catharsis, which is the 'release or transformation of emotional pressure or impulses via activity and expression' (Heelas 1982: 56). The cathartic effect purifies by reducing internal tension and associating improvements in subjective feelings (Heelas 1982: 56). The phenomenon of violence shows us that violence is not asocial, the immoral or the anti-order as we might think (Soeffner 2004: 62). But as various studies, such as by René Girard, show, violence has its own morality, socializing function and order; it has its own system. Violence is the form of the revelation of the sacred internally. Violence is a majesty which is omnipresent and irrational (beyond rationality), creator of absolute subjugation as well as source of grace for the in-group (Soeffner 2004: 71–4). Violence represents a social order: social regulations emerge *through* and *against* violence, and violence is the lasting proof of the capability and will of the society to establish and defend its own order (Soeffner 2004: 82). Violence is on the one hand the social loss of control, but on the other hand the attempt to control violence is the very source of the violence (Heitmeyer 2004: 87).

One of the hypotheses of Gurr about forms of political violence maintains that 'the likelihood of internal war varies directly with the intensity and scope of elite and mass relative deprivation' (Gurr 1970: 335). Elite are meant here as those individuals who possess acquired or inherited personal qualities required for high-value positions (Gurr 1970: 335–6). However, Collier argues that income inequality does not necessarily lead into civil war (Collier 2007: 23). The main factor causing civil wars in Africa, according to Collier, is low economic growth. Inequality is, nevertheless, seen widely as one of the main causes of conflicts and violence. Various development aid policies of the EU, US, etc., argue that conflicts in Africa with global implications are caused by inequality. However, the role of structural injustices in initiating wars and conflicts caused by interaction of local and global factors cannot be ignored. The structural violence that kills the poor gradually and constantly is based on internal, regional and global structures.

Structural Violence

Globalization can be designated as 'rationalization' of international relations. But it is strongly disputed how rational it is. Such perception of rationality compels us to redefine *rationality*. Those who plead for globalization argue that the benefit of globalization outweighs isolationism and that in the long run it furthers the welfare of all; and the actors in this global system are considered as rational actors. According to rational choice theory, states or people are consistent and coherent about what they want (Nicholson 1995: 4). That is not only the system is rational but also the elements that constitute the system are considered as rational. They act strategically and rationally in the use of and in the face of violence. Instead of the system-level and element-level rationality, we see irrational behaviour of structural exploitation. It is irrational because structural exploitation will sooner or later undermine those who believe they are making profit rationally, accumulate their wealth or guarantee their security.

Galtung has extensively demonstrated that structural violence occurs when nobody feels personally responsible for the damage; violence is inherently part of the system and manifest in unequal power relations (Galtung 1975, 2004). Structural violence is seen as avoidable death due to misdistribution of income caused by the social structure of the society (Nunner-Winkler 2004: 22–3).

Life expectancy in Africa at birth is currently only 47 years on average. Infant mortality is 106 per 1,000 births as compared with China (32), India (69) or Indonesia (41). About 50 per cent of sub-Saharan Africans live on less than $1 a day. The majority of the sub-Saharan population is affected by high rates of unemployment, especially among historically vulnerable groups (e.g. women and youth). But in the African case employment does not necessarily reflect economic self-sufficiency; instead, according to an ILO study, the working poor account for almost 45 per cent of the total number of the employed, and an estimated 110 million people considered employed earn below the poverty wage and are unable to provide their families with decent living conditions (ILO 2004). In global affairs such as trade, Africa's participation is not even 2 per cent (UNCTAD 2004). Various studies show that high rates of HIV/AIDS are decimating labour forces in Africa. At the end of 2005, 25.8 million people with HIV/AIDS were living in sub-Saharan Africa, and out of 4.9 million new infections worldwide, 3.2 million (65 per cent) occurred in sub-Saharan Africa. AIDS has become the leading cause of death for both children and adults in Africa, and it has reduced average life expectancy from 62 years to 47 years in Africa. A typical characteristic of structural violence is that urban areas have better access to treatment, and vulnerable groups are inadequately targeted, fuelling the epidemic (WHO 2006: 2).

Caused by internal as well as global structures (fighting terrorism, securing economic and strategic interests), basic political rights are absent, and basic civil liberties are widely and systematically denied (Cf. Freedom House 2007). Bribery and the embezzlement of state assets have been the symbol of Zaire/Congo. It is estimated that the former president of Zaire (Democratic Republic of Congo), Mobutu, plundered some $5 billion, an amount equal to the country's external debt at the time. As recent studies of the World Bank show, the current external debt of Congo makes 149.2 per cent of its GDP, whereas the average external debt of sub-Saharan Africa is 34.2 per cent of GDP. In spite of the oil production of 235,900 bbl/day and exports of diamonds, copper, crude oil, coffee, cobalt, etc., life expectancy in Congo is just 44 years, the infant mortality rate 129 per 1,000 live births, under-five mortality is 205 per 1,000; 31 per cent of under-fives suffer from malnutrition and are underweight. There is no physician per 1,000 people; DRC is characterized by discrepancy between rural and urban areas: 82 per cent of urban population has access to an improved water source, whereas it is only 29 per cent in rural areas. The brutal colonial period by Belgium, the alliance of various global actors such as France and the US with Mobutu during the Cold War, and exploitation of Congolese natural resources by regional and global actors made Congo one of the highly corrupt states of the world. A World Bank study shows that in the scale between –2.5 (worst) and 2.5 (best), Congo's control of corruption scores –1.43, voice and accountability –1.62, political stability –2.31, government effectiveness –1.61, regulatory law –1.51, and rule of law –1.68 (World Bank 2008: 25).

Ethiopia's total external debt consists of 55 per cent of its GDP; this is 66.2 per cent in Sudan. While the gross primary school enrolment makes 93.4 per cent in Ethiopia and 60.4 per cent in Sudan, in Somalia it is just 16.9 per cent whereas in sub-Saharan Africa it is 97.6 per cent of the relevant age group. According to the World Bank studies, the gross secondary school enrolment is 31.3 per cent in Ethiopia and 34.2 per cent in Sudan of the relevant age group (World Bank

2008: 38, 88, 92). The main causes of this decrease are that parents are unable to pay the school fee, buy exercise books and clothes. Moreover, because many poor Ethiopian farmers need their children to help them during the preparation of soil, sawing, weeding, harvesting, looking after cows, etc., for many children it is not easy to attend their classes. In the Somali case, state collapse and continuous violence have dramatically affected the lives of children, who are carrying weapons rather than exercise books.

These poor countries are further affected by changes in the global structure. Rapidly rising food prices are leading to violence in various African countries. In February 2008 protests about such price rises resulted in looting stores and burning government buildings, such as in Burkina Faso. Similar protests days later in Cameroon resulted in the deaths of 20 people. Senegal and Mauritania experienced similar protests in late 2007. Booming economies such as China and India have the cash to buy ever-greater quantities of food, and have been growing less of their own food (Walt 2008: 42–3). Another common problem is the increasing consumption of meat, which has driven up demand for grain and its price. High oil prices have caused high fertilizer prices. The unpredictability of weather for those who are completely dependent on rain for their agriculture is hugely affecting the poor farmers who produce their grains without an irrigation system. Droughts, erosion and floods are frequently affecting poor African farmers. In predominantly agricultural societies such as the DR Congo, irrigated land consists of zero per cent of the cropland, in Ethiopia only 2 per cent, 11 per cent in Sudan, and 19 per cent in Somalia. Food protests will threaten the survival of not only authoritarian and corrupt, but also democratic, governments. In spite of the fact that the majority of Africans depends on agriculture, Africa still spends only 4–5 per cent of its total national budget on agriculture; instead it invests 22 per cent of its GDP in order to achieve the MDG, which requires sustaining an economic growth rate of 6 per cent over the next 10–12 years, from a baseline of data from 1994–98 (Fan et al. 2008). Studies suggest that at least 70 per cent of the African workforce is at least partially engaged in agriculture, which is a source of $0.33–0.80 per day. Therefore, increased investment in agriculture by 20 per cent, or $32–39 billion per year, is a development pathway that Africa cannot bypass (Fan et al. 2008). Instead, instable and undemocratic states spend more in arms than agriculture. During the war between Eritrea and Ethiopia in 1998, 1999 and 2000, Ethiopia spent 6.7 per cent, 10.7 per cent and 9.6 per cent of its GDP in the military sector respectively. The governments of low-income countries (to which all three countries studied in this book belong) devote an average of nearly 16 per cent of their budgets to the military (Cohen 2007: 20).

The global structure has been considerably contributing to killing, displacement, rape, destruction of economic infrastructure, hunger and famine. As the Sudanese case shows, the humanitarian tragedy in Darfur cannot be explained by internal factors only. The global structure of incoherent donor-country policies (Cohen 2007: 19) driven by contradicting interests plays a decisive role in Congolese, Sudanese and Somali conflicts. Various studies suggest that the P5 provide over

60 per cent of the legal arms transfers to Africa, contributing to conflict, instability and food insecurity (Cohen 2007: 19). Whereas the European Community imposed an arms embargo on Sudan in March 1994, the USA imposed sanctions on Sudan only in 1997. When the UN in 2004 (Resolution 1556) imposed an arms embargo, China ignored it. Moreover, the UN first embargo was imposed only on transfers to non-governmental armed forces and individuals operating in Darfur in spite of reports accusing the Sudanese government of supplying weapons and military support to the Janajawiid. Russia chose to care for its economic interests in Sudan by selling MiG-29S combat aircraft in 2004 and military helicopters in 2005. China and Russia undermined the arms embargo when they abstained from the vote on UNSC Resolution 1591 in March 2005, which attempted to expand the scope of the UN arms embargo to cover all parties to the 2004 Humanitarian Ceasefire Agreement – the Sudanese government, the Sudan Liberation Movement/Army, and the Justice and Equality Movement – and any other belligerents in Darfur, including Sudanese government forces, that were active in the region (Fruchart et al. 2007: 16–17). Similarly, the Congolese government, as well as rebels, have been supplied by various arms suppliers. For example, according to an Amnesty International report in 2005, though no state had reported to the UN an authorized export of arms to the DRC, there had been no shortage of arms and ammunition arriving in the DRC since 2003. According to a UN Panel report in 2003, between 10 and 15 containers of small arms and light weapons were imported from Jordan. Or according to the UN Expert Group on the DRC, a Libyan aircraft nominally registered in Equatorial Guinea but based in Sharjah (United Arab Emirates), with a Romanian crew on board, transported tons of weapons to Congo. Moreover, the DRC government's agreements with arms suppliers in Eastern Europe (Czech Republic, Bulgaria, Georgia and Ukraine), China, Zimbabwe; large quantities of small arms from Western Europe such as 30,000 pistols and revolvers from Germany in 2001, $570,000 of small arms ammunition from Italy in 2001–02 and $250,000 of munitions from France in 2000–01 (Amnesty International 2005: 18) are decisive global structures that exacerbate physical violence as well as destroy economic infrastructures exposing the poor and displaced to malnutrition and hunger. Research suggests that in Sudan and the Democratic Republic of the Congo (DRC), more than 90 per cent of the conflict deaths are attributed to indirect or structural causes such as public health and epidemiological factors in conflict zones because armed violence imposes huge socioeconomic costs on states and societies, including direct medical costs, direct non-medical costs (policing and incarceration, legal services, and post-conflict reconstruction costs), tangible indirect costs (productivity losses, protection and security costs, insurance, and lost investment), and intangible indirect costs (quality of life) (Krause 2007: 3). Two important conclusions can be drawn from the conflict-related structural violence: an disproportionately high amount of money is being invested in the military sector; and conflicts not only kill directly those living in the conflict areas but they also destroy the fundamentals of existence of the poor. As a consequence of such conflicts and economic destruction, more than 60 per cent of the 87.8 million

people assisted by the World Food Programme in 2006 were in Africa (WFP 2007: 2). Moreover, donors require recipients to spend aid receipts on products from the companies of the donor countries. The US, in its 'tied aid' policy, requires the recipients to spend about three-quarters of the aid money buying US products (Easterly 2006: 192).

In the complexity of the global system the non-identifiability of the actors is a key component of structural violence. As William Easterly suggests, 'all the rich governments and international aid agencies are supposed to work together to achieve the Millennium Development Goals. So when the goals are not attained, no one can be held accountable. This weakens the incentive of any one agent to break its neck to reach the goals … [It] is understandable that aid agencies want to share the blame with as many other agencies as possible if something goes wrong' (Easterly 2006: 172). This complexity of actors involved makes difficult not only the identification of main actors, but also exacerbates the structural violence against the poor, who are made dependent upon these structures.

The increasing global demand for oil is one of the leading factors of structural violence. By 2015 Africa is expected to provide 25 per cent of US oil imports. New deposits are being developed offshore in deep water all along the African west coast as far north as Togo and Ghana. Many US oil and natural gas firms, like Exxon-Mobil and Chevron-Texaco and their affiliates, are expected to spend about $60 billion in the coming years on exploration and development in countries such as Nigeria, Mozambique and South Africa. In eastern Africa, in addition to Sudan, Ethiopia will be a producer of oil. Agreements between Russia and Ethiopia in February 2008 have confirmed this. Such new discoveries usually do not benefit the poor or further broad-based institutional reforms. Angola is a good example.

In addition to oil, the development of modern technology has become an integral component of global structural violence. A lot of sub-Saharan African countries dispose of a lot of natural resources like gold, diamonds, uranium, copper, coltan. The demand for the latter has been increasing over the last few years because it is used for the production of cell phones, computer chips, missiles, etc. Nevertheless, most of the African countries that dispose of natural resources are characterized by poverty, corruption, environmental damage, human rights abuses, violence and war, resulting in the 'paradox of plenty'. The Transparency International Corruption Index shows that those African countries that dispose of a lot of natural resources have mostly poor results. In 2005, for example, it was Chad that was declared to be the most corrupt country in the world. However, countries such as France and China are strong allies of Chad. The extraction of raw materials very often leads to human rights abuses like, for example, poor working conditions, poor health conditions for those employed in the natural resources extraction sector, limited rights to organize themselves, environmental destruction displacement, etc.

Regions rich in raw materials can contribute to the dissolving of social structures, such as migration (into cities), prostitution, conflicts of communities about the new resources and increasing illegal business. A lot of wars and conflicts in Africa are based on natural resources facilitating structural violence. Structural violence on the national or global level results in slavery and abductions, debt slavery, forced domestic labour, commercial sexual exploitation, forced overtime labour under threat of dismissal, unpaid compulsory labour for public servants, and trafficking in persons (ILO 2005). The following chapters will show that the causes of conflict in Africa are systemic and complex. The dimensions of interactions of actors and factors in DR Congo, Somalia and Sudan at different levels will be discussed.

Chapter 4
Conflict Systems in the Democratic Republic of Congo

Right after its independence, Congo was marred by conflicts and violence caused by competing political elites. This competition resulted in the independence of Katanga and Kasai, in the blockade of the government when President Kasavubu and Prime Minister Lumumba suspended each other, and in the murder of Prime Minister Lumumba on 17 January 1961. Mobutu assumed power through a military coup to end this political impasse. After having destroyed the country for 32 years, he was deposed by a rebel alliance in 1997. Violence continued even in the post-Mobutu era. Why?

According to Ottaway et al. (2004), African big states such as DRC, Sudan or Ethiopia fail simply because of internal political, ethnic and religious differences and because of their inability to establish institutions to moderate these differences. Prunier (1995, 2007b) sees wars such as the one in Darfur more from an intrastate point of view and less from external influences and repercussions. Collier et al. (2005) and Collier (2007) underscore rather the economic factors of conflicts, such as resource abundance, deprivation and low economic growth. This chapter attempts to show that the different systemic levels have not been adequately addressed in various analyses of Congo's conflicts.

Intrastate Conflict System

Most of the conflicts in Africa had already begun in the pre-independence period. In preparation for independence from colonial power, national interests were defined and arranged not only to fight the colonial power, but also to administer the state when the colonial powers left. The colonists supported one or the other group during the so-called preparation of the local political elites to overtake the task of state-building after the departure of the colonists. The local elites attempted to repair the destruction that happened during the colonial period (Tetzlaff and Jakobeit 2005: 59–79).

This political background structurally supported the emergence of key components of the Congolese internal conflict system: the politics of regime survival, 'ethnicized' political violence, the state's inability to arbitrate land disputes, state officials' manipulation of such disputes, the weakness of the central state, and disputes over citizenship, which were also manipulated by state officials and political leaders (Tetzlaff and Jakobeit 2005: 138–41). As will be shown in the regional dimension, efforts by neighbouring governments, especially from Rwanda and Uganda, later tried to force the DRC government to adopt their policy priorities as its own (Nest et al. 2006: 12–13). These regional actors exerted considerable impact on the politics, economy and society of Congo. Hence I will discuss the internal conflict system of Congo divided into three subsections: *politics*, *economy* and *society*.

Society

Though ethnic heterogeneity in itself is not an essential cause of conflicts, its role in the conflict dynamic is considerable. Congo consists of some 200 ethnic groups, of which the majority are Bantu (80 per cent), Sudanic-speaking groups in the north and northeast, and about 100,000 Pygmies in the cuvette area. The Bantu group consists of Kongo in Lower Congo; the Luba in East Kasai and Shaba; the Mongo and others in the cuvette area; and the Lunda and Chokwe in Bandundu and West Kasai; the Bemba and Hemba in Shaba; and the Kwango and Kasai in Bandundu. This ethnic heterogeneity often results in social fragmentation.

Social fragmentation in Congo has been playing a key role in the violence dynamic since its independence. The unilateral declarations of independence of Katanga on 11 July 1960 as well as the independence of Kasai were caused by social fragmentation that was exploited by political opportunists. Such fragmentations are manifested very often through ethnic tensions. In the independent Congo, whereas Joseph Kasavubu's ABAKO party was based mainly among the Kongo people, Patrice Lumumba's MNC (Congolese National Movement) tried to overcome party divisions along ethnic lines. Nevertheless, ethno-nationalism emerged and the threat of regional separatism in Katanga became imminent. CONAKAT (*Confédération des Associations du Katanga*), headed by Moïse Tshombe, was created. CONAKAT supporters described themselves as 'authentic Katangese'. They were essentially drawn from the Lunda and Yeke peoples of southern Katanga, and were most resentful of the presence of Luba immigrants from Kasai. On the other hand, BALUBAKAT (Association of the Luba People of Katanga – *Association des Baluba du Katanga*) was created. At the same time, a rift happened between the MNC-Lumumba (from Lulua group) and the MNC-Kalonji (from Luba elements) in Kasai. This fragmentation resulted in the independence of Katanga as well as interethnic conflicts. Michael Mann asserts that Africa's weak states are caused by multi-ethnicity, low ideological intensity and rare demand for territorial revisionism (Mann 2005: 428). As we shall see later, especially after Mobutu was deposed, ethnic conflicts occurred mainly in the eastern part of the country with catastrophic consequences.

In the case of such fragmentations, politicians usually pursue at least three important strategies to extricate themselves from difficult political demands. *One* possibility is that they ignite conflicts among the opposing groups in a fragmented society so that they engage each other mutually. *Second*, they divert the attention of the rioting population from internal dissent towards an outside enemy in an attempt to unite the citizens. In the Congolese case, this objective was pursued when the nationality law was 'reformed' against the Tutsi in 1981. According to the Citizenship Decree of 1981, the citizens' ancestry had to be traceable to 1885. The citizenship of those who could not fulfil these requirements was revoked. The Nande and Hunde politicians in North Kivu demanded these strict citizenship regulations to undercut political and economic influences of the immigrated Rwandese and Burundians (Lemarchand 2000: 334; Mamdani 2002: 244). The distinction between 'indigenous' and 'non-indigenous' became salient. Moreover, the Hutu-Tutsi distinction became an important factor leading to conflicts (Mamdani 2002: 251). Banyarwanda and even some Banyamulenge were affected by this decree, which was reaffirmed by the Sovereign National Conference in 1991 (Mamdani 2002: 246–7; Nest et al. 2006: 21). Further, the incoming Hutu refugees – as a consequence of the 1994 Rwandese genocide – and the 'autochthones Congolese' felt imperilled by the Tutsis in Congo supported by the Tutsi-dominated government of Rwanda, as will be shown in the regional dimension. Three things paved the way for the conflict: the long-standing tensions in the North Kivu and South Kivu over land, regional political dominance and access to state-distributed resources; the citizenship disputes between autochthones or 'indigenous' Congolese and Kinyarwanda speakers; and the relationships of these groups to the DRC state. This situation was further exacerbated by the fact that Kinyarwanda speakers are divided into Banyamulenge and Banyarwanda, and they are further identified as Hutu and Tutsi (Nest et al. 2006: 20–21; Nzongola-Ntalaja 2004b).

Third, they 'reform' the society. Mobutu attempted on 27 October 1972 to 'authenticate' Congo by changing 'Congo' to 'Zaire', suggesting that 'Congo' was a Belgian construction. Through his 'authenticity' ideology, Mobutu constructed his 'idea of the state' (Clapham 1996: 45). Mobutu created 'authenticity' as a unifying philosophy for fragile Zaire. Wrong calls it 'the most intellectually ambitious project of his career'. Mobutu believed, because of the size and richness of the country, that Zaire deserved a great ideology on the continent (Wrong 2006: 93–4).

As Clapham correctly observed, 'the rhetoric of unity papered over the extreme reluctance of any but a very small minority of African rulers to sacrifice any of their power in the interests of any continental grouping' (Clapham 1996: 106). Ali Mazrui's 'On the concept of, we are all Africans' is a very helpful analysis here (Mazrui 1967).

Adelman suggests that 'authenticity' is based on the negritude that created a type of mythical being. For Senghor the African is an emotional, natural human being in harmony with the rhythm of the universe. Negritude has also been criticized for presuming a uniformity, or at least a great similarity, of values within the black world. But according to Adelman, there is a basic difference between *authenticity* and *negritude*. The former does not make an assumption of uniformity. Adelman concludes that, on the contrary, authenticity leads to an emphasis of cultural differences rather than similarities (Adelman 1975: 136). At any rate, the poorly addressed social fragmentation based on ethnic heterogeneity has been a decisive factor for conflict and economic destruction.

Economy

Authenticity influenced Zairianization and radicalization. 'Until Zairianization, the economy had grown by an average 7 percent a year' (Wrong 2006: 98). But since Zairianization, Mobutu 'killed that economy'; corruption and kleptocracy skyrocketed. Second, in spite of his attempt to unite the state as one nation, he was paradoxically pursuing the divide-and-rule strategy of clientelism, the system of 'everyone-suspects-everyone else' (Wrong 2006: 103–4).

Economy has been a melting point of many actors competing to exploit the abundant natural resources of Congo. These natural resources furthered corruption, patrimonialism and brutal war after the demise of Mobutu (Tetzlaff and Jakobeit 2005: 138–40). Between the period of 1920 and 1990 copper was the dominant source of government revenue and most foreign exchange. Resource dependence established under Belgian colonial rule, absence of a democratically accountable regime and patrimonial profit-making robbed Congo of its chance to be a stable and prosperous state. Because of the fragmentation of the state and decline of the formal economy, and the informal and illegal economy of the huge natural resources, Congo became the symbol of Congo's economic and political chaos. To increase its political independence from the foreign-owned corporations that dominated the mineral sector, Mobutu nationalized plantations and the mineral industry (the copper-mining company Gécamines in 1967 and the diamond-mining company Miba in 1973) and expropriated many businesses owned by non-Congolese. Through nationalization Mobutu attempted to build the Zairian nation by creating an indigenous bourgeoisie and 'Africanizing' the workforce (Nest et al. 2006: 17–18). The economy was characterized by prodigality and corruption by the mid-1970s. Mobutu often held 20 per cent of the national budget for himself. Despite the mismanagement and misgovernment, Mobutu's regime survived until 1997 due to the support of the Western world (ICG 2006a: 2). To stabilize his power, Mobutu created a patrimonial political network by satisfying the private interests of politically corrupt entrepreneurs within the state bureaucracy and distributing resources to selected local leaders and communities in order to reduce opposition to the state while simultaneously deepening competition between communities to prevent the emergence of a united local opposition, and to easily manipulate them. Instead of erecting an integral state, Mobutu set up an informal system of rule through patronage, intermediary institutions and clandestine markets (Tull 2007: 114). Warlords who derived power from connections to Mobutu and from their control of coercive forces emerged and levied taxes from producers and the public and established near monopolies on the buying, selling and illegal export of commodities, especially coffee, gold and diamonds (Nzongola-Ntalaja 2004a: 16; Nest et al. 2006: 18–21).

Those who deposed Mobutu inherited an economically destroyed Zaire. There was no state budget between 1998 and 2001. ADFL had only military capacity but few cash resources at the start of its campaign, and it set about selling state-controlled assets – assets it did not yet control – in exchange for cash (Nest et al. 2006: 23). In 2000 Congo had an inflation rate of 511 per cent inflation and $100 per capita income, while 40 years earlier at independence it was $259. After the death of Kabila in 2001, a period of relative growth followed and donors started engaging in Congo again. In 2005 its GDP grew by 7 per cent and national revenues tripled between 2000 and 2005 (ICG 2006a: 3).

Cobalt, copper and coltan in Katanga, especially copper, were the pillar of the Congolese economy when the Belgians conquered the Katanga region in 1891 and 1892. Congo has been the world's largest producer of cobalt, and one of the main suppliers of industrial diamonds and copper. It is also rich in gold, tin, uranium, zinc and silver. While mining made up 80 per cent of GDP at independence, by 1990 the sector had collapsed due to mismanagement and the decline in world copper prices. In 2004, it accounted for only 9 per cent of GDP and less than 1 per cent of revenue.

In order to raise money for war, all belligerents sold minerals to and signed mining contracts with multinational corporations and smaller mining entrepreneurs. Further, resources such as diamonds (between $80 million and $100 million every year) and gold have been smuggled out of the country. According to the Congolese government, gold exports in 2004 were only $7.4 million, in stark contrast to the $58 million in Congolese gold that the UN estimated went through Uganda that year. Uganda itself has traditionally produced less than $50,000 a year in gold (ICG 2006a: 6–7).

The so-called 'Lutundula Commission' was established by the peace deal in 2002 to review all contracts during the two wars (1996–97, 1998–2002) to make sure that they did not contravene Congolese law. The commission's report, submitted to parliament in June 2005, suggested that the management of natural resources had worsened during the transition period (2003–06) and recommended that many contracts be reviewed. The report stresses that unless the terms of existing contracts are altered, the mining sector will bring little profit to the state and it therefore recommended that all contracts signed since the beginning of the transition be cancelled. However, if the government had tried to renegotiate or cancel existing contracts, it could have been sued for violating international investment conventions, leading to international arbitration (ICG 2006a: 8–9; Lutundula Report 2005).

A structural problem in Congo was that neither Mobutu nor Kabila hired administrative staff base on their quality, but instead hired them because of their connections. In 2003, two-thirds of the budget was spent outside normal procedures with only retrospective justification. By 2004, this had declined to 27 per cent, but the presidency overspent by $18 million. In November 2004 a World Bank report evaluated that the Congo only fulfils 3 out of 16 good governance criteria. Civil servants have been steadily impoverished by graft. In 2006 there were an estimated 90,000–120,000 soldiers and 600,000 civil servants. Their monthly base salaries ranged from $50 for a janitor to $77 for a doctor and $111 for a judge. In contrast, ministers received monthly wages of $4,000, while some directors of state-run companies earn $25,000. Low salaries have also led to brain drain, as many qualified officials take jobs with NGOs or the UN. A driver for the UN Development Program (UNDP), for example, made between $450 and $750 monthly (ICG 2006a: 4–5).

The state-run Office of Customs and Excises (OFIDA) generates the most income of all fiscal services – $348 million in 2005, 36 per cent of total revenue. However, private consultants and the UN estimated that between 60 and 80 per cent of custom duties – $870 million to $1.7 billion – are not collected. The reason was that a variety of duties were embezzled. For example: It was estimated that $4 million was embezzled or not collected by OFIDA in Ituri in the first 10 months of 2005. Kasumbalesa in Katanga, the only export route to Zambia, was the site of large-scale smuggling of copper, cobalt and other minerals. A study by Global Witness concluded that the 13,720 tons of cobalt and copper concentrate exported legally through Kasumbalesa each month was likely only a fraction of actual exports. With cobalt prices around $35,000 per ton and the export tax on cobalt 1 per cent, the estimated monthly loss was in the millions of dollars. The World Bank helped to reform the procedures at the port of Matadi by replacing at least 12 different payments with one, the result being that the revenues increased by 280 per cent between 2002 and 2005 (ICG 2006a: 5–6).

In December 2005, a foreign military expert close to the defence ministry suggested that $3.5 million to $5.8 million was being embezzled monthly from the $8 million set aside to feed and pay soldiers. The consequence was that the troops, who lived in terrible conditions, in turn harassed and extorted the local population. The army is itself the largest single threat to civilians in the country. The Congo contains around half of Africa's rain and humid forests. Smuggling is a major problem: it is estimated that timber export is seven times higher than shown by official figures. Nevertheless, the EU and the Ministry of Environment reviewed the forestry sector in 2004 and concluded that the Congo was producing only one-sixtieth of its potential. Due to the very low concession fees, companies occupy much more land than they actively use, hoping to sublease to other investors once there is economic and political stability (ICG 2006a: 10). In Congo 34 per cent of children under five are affected by malnutrition, and infant mortality per 1,000 live births is 129. Destroyed by a decade-long war, only 46 per cent of Congolese have access to an improved water source. Life expectancy at birth is about 46 years. Gross primary enrollment makes only 61 per cent of school-age population, and the literacy of population age above 15 is 67 per cent. To address these problems it is important how the economy grows in the coming years. In 2007 the industry sector grew by 11 per cent, manufacturing by 8 per cent, services by 8.5 per cent and agriculture by 3 per cent. The GDP per capita is expected to grow between 2007 and 2011 by 4.3 per cent (World Bank 2008). But for this to happen, political stability is decisive.

Politics

> Ethno-political conflicts and tensions have been kept simmering for a long time by Mobutu regime's supposed rationalization of politics and progress through the major projects of the nation-state and one-party set-up from 1965–97. The democratization of the 1990s was meant to bring about 'authentic nationalism' and "national unity" and do away with tribalism and regionalism, but in fact it led to social implosion and gave free rein to power struggles which brought about crisis of identity. (Bwenge 2005: 90)

The rebellion of the Banyamulenges in October 1996 confronted the Zairian government with huge tasks. Mobutu's idea of the state of overcoming intrastate divisions was not yet successful after 32 years in power. First of all, Mobutu's decade-long wrong policy created a structure for distrust among the Banyamulenge and the 'indigenous' Zairians: Mobutu stripped the Banyamulenge of their Zairian nationality in 1981. The Banyamulenge complained that their land was wrongly curved into the Zairian territory by the reckless European colonizers, leading to continuous insecurity (O'Ballance 2000: 163–4; Lemarchand 2000: 327).

Security

Right after the independence of Congo on 30 June 1960, the fledgling process of democratization was rapidly reduced to the level of ethnic competition before spilling over into rebellions in the west and eastern regions, ending up as 'national tribalism', which makes the paradox between nationalism and ethnicity. In spite of all the slogans proclaiming nationalism as the philosophy behind the political management framework of the new one-party system (*Movement Populaire de la Révolution*), ethnicity and regionalism were still active. National groups from the province of eastern and western Kasai under the leadership of Etienne Tshisekedi on the one hand and those of Mobutu from Equateur confronted each other (Bwenge 2005: 92). The 'expressions of violence have occurred since 1992 in the conflicts between the Lendu-Ngiti and Hema, and those in the provinces of North and South Kivu. It was these conflicts, crystallized by the conflagrations in Rwanda and Burundi, which led to the October 1996 war of the Alliance of Democratic Forces for the Liberation (AFDL) of Congo, and the August 1998 war of the Congolese Assembly for Democracy (RCD) in the DRC' (Bwenge 2005: 92–3).

After Congo became independent on 30 June 1960, about 25,000 Congolese soldiers mutinied on 5 July 1960 against the Belgian commanders, who were making the time before and after independence factually the same. This mutiny not only demonstrated the frustration of the soldiers expecting higher pay and positions of authority, but also resulted in various riots in Kinshasa and other areas of the country. Belgians and Belgian interests were attacked. The breakdown of chains of command in the police and military following independence led to a unilateral decision by the Belgian government to return troops to protect Belgians and their property. Belgium supported the independence of Katanga under Tshombe, which was declared on 11

July 1960, and similar moves in the diamond-rich Kasai in August of the same year to protect its economic interests in the regions (Schraeder 1994: 53).

The Tshombe-Kasavubu struggle for presidency, the independence of Katanga, and the arrest and killing of Lumumba on 17 January 1961 all characterized years of violent struggles after Congo's independence. Between 1960 and 1963 Congo's territorial control and coercive monopoly was seriously challenged by the Katanga secessionist movement, a movement enhanced by an alliance around mining interests, including the Belgian government (Nest et al. 2006: 17). Through the help of the UN troops' intervention in August 1961, the Katangese soldiers were disarmed and Katanga's secession ended in 1963; Tshombe was driven from power. However, in 1964 President Kasavubu appointed Tshombe as prime minister, but a year later he was dismissed by President Kasavubu from the government and replaced by Kimbe. For his part Tshombe arranged a vote of no-confidence against the Kasavubu-Kimbe government. In the face of this political chaos, on 25 November 1965 Mobutu intervened militarily and dismissed all politicians; he suspended the constitution and political parties. However, soon after, his regime was characterized by a culture of exhibitionism, presidential cult, deception and political unanimity (Ikambana 2007: 30).

In the 'Pentecost Plot' of 2 June 1966, Kimbe and three ex-ministers attempted to overthrow Mobutu. A mutiny took place in Kisangani on 26 July 1966 led by Colonel Tshipulsa (a Katangan), and former Katangan gendarmes attacked newly trained Congolese National Army (CNA) soldiers. Tshombe and Tshipulsa were condemned to death on 12 March 1967.

To consolidate his power, on 17 April 1967 Mobutu created a one-party system – *Movement Populaire de la Revolution* – and established the Second Republic, replacing the First Republic of 30 June 1960. On 5 July 1967, because of Mobutu's failure to provide timely payment of salaries and because of the kidnapping of Tshombe to Algiers under Mobutu's orders, over 1,500 Katangan troops loyal to Tshombe mutinied and seized several cities, such as Kisangani and Bukavu; and on 6 July 1967 European mercenaries from Belgium, Spain and England arrived at Kisangani to overthrow Mobutu, and the fight on this front lasted until 4 November 1967. In addition, on 3 November 1967 another small European mercenary force entered Katanga from Angola (O'Ballance 2000: 100–104). Despite Washington's provision of over $850 million in economic and military aid from 1960 to 1976, the Mobutu regime was incapable of stemming the advance of some 1,500 FLNC insurgents.

Mobutu's regime had been threatened by rebels infiltrating, especially from Angola. In early 1977, led by Nathaniel Mbumba (a former Katangan police officer), about 2,000 men from there invaded Katanga, which had been renamed Shaba by Mobutu. But Mbumba later withdrew to Angola. Early in 1978 Mbumba returned with twice as many men, but French and Belgians drove Mbumba's men out of the city (Edgerton 2002: 214–15). In the so-called Shaba Crisis II in January 1978, anti-Mobutu groups seized a number of hostages (government officials, Europeans and missionaries). A sabotage campaign began on 18 February 1978 with the primary objective of gaining control around Kolwezi, the main copper area (O'Ballance 2000: 124–6). In the same year, about 67 army officers made a serious coup attempt; later 19 of them were executed. Fights between the group and forces of Mobutu began on 11 May 1978. The French and Belgians arrived at Shaba on 19 May 1978, secured Kolwezi Airport and rescued numerous Europeans. In September 1983 an unsuccessful coup attempt took place led by Kengo Wa Dondo (the first state commissioner). In November 1984 rebels (a mixture of CFRD – Congolese Front for the Restoration of Democracy, based in Brussels – and some government soldiers) attacked Moba, near Lake Tanganika in the Shaba province; and the second invasion of Moba took place in June 1985 by the CFRD from Tanzania (O'Ballance 2000: 127–49).

To consolidate his power Mobutu changed the constitution, prohibited political parties other than his, made membership in his party (Popular Revolutionary Movement – MPR) obligatory, and controlled the key trade union, the student association and security services positions. The political instability was further exacerbated by the collapse of the Zairian economy due to corruption in the government and the price decline of minerals such as copper. As a consequence, Mobutu was forced to turn to foreign powers such as the US for greater levels of economic aid. Nevertheless in 1980, when 13 members of parliament, including Etienne Tshisekedi wa Mulumba, wrote a 52-page letter to Mobutu, the demand and struggle for institutional and political reform started to emerge. In 1982 *Union pour la Démocratie et le Progrès Social* (UDPS) was created. However, it was not until his 'Third Republic' speech on 24 April 1990 that Mobutu announced political reforms such as revising the constitution, the multi-party system and a one-year transition period ending in April 1991 (O'Ballance 2000: 141; Edgerton 2002: 219).

A series of interrelated social, political and economic turbulences took place in the early 1990s: in addition to various political instabilities, copper production collapsed in September 1990 when the roof of the Kamoto mine collapsed, causing Gécamine's total production to drop 90 per cent by 1993; former Cold War patrons withdrew financial and diplomatic support in the early 1990s; in response to domestic and foreign pressure to reform the political system in various African countries, demands for democracy intensified. Moreover, from 1990 to 1993, unpaid soldiers engaged in widespread looting of private property (Nest et al. 2006: 19). The genocide of 1994 exacerbated the situation when Hutu refugees and perpetrators of massive violence from Rwanda flooded into Zaire. AFDL was created in 1996. AFDL consisted of the People's Democratic Alliance (*Alliance Démocratique des Peuples*, ADP – a grouping of Congolese Tutsi led by Déogratias Bugera); National Council for Resistance and Democracy (*Conseil National de Résistance et pour la Démocratie* CNRD – a Lumumbist guerrilla group established in 1993 in eastern Congo by André Kisase Ngandu); Revolutionary Movement for the Liberation of Zaire (*Mouvement Révolutionnaire pour la Libération du Zaïre*, MRLZ – a South Kivu opposition group led by Anselme Masasu Nindaga), and Kabila's PRP.

Under the leadership of L.D. Kabila, AFDL entered Kisangani on 15 March 1997, Lubumbashi on 9 April 1997, and Kinshasa on 17 May 1997. On 16 May 1997 Mobutu left Zaire for Togo and then Morocco. By 1999, the AFDL itself disappeared as a political organization, to be replaced by the *Comités de Pouvoir Populaire* (CPP). It disappeared because there was disunity concerning the agenda of the alliance; some were interested in looting, vengeance or sheer violence (O'Ballance 2000: 168; Nzongola-Ntalaja 2004: 14). Laurent D. Kabila banned political parties, like Mobutu, and laws were issued by presidential decree.

When the second DR Congo war broke out in 1998, the DR Congo was challenged not only by rebels (Congolese Democratic Coalition, launched on 16 August 1998) attempting to overthrow Kabila, but also by several different guerrilla groups in the east: the Wanglima in Beni and Butembo (Nord Kivu); Mai Mai in Masisi (North Kivu); and fighters from the *Conseil de Resistance pour la Liberation Nationale* led by Asema bin Asema (South Kivu). The Mai Mai, a rural militia, have had a long-standing tradition in the Kivu with a mixed aim of self-defence and profit accumulation.

Laurent Kabila was believed to have encouraged military coordination between the Mai Mai and Interahamwe and the Burundian Hutu rebel FDD in order to counter Tutsi activities as well as the presence of Rwanda in the Congolese territory. Since the end of 1999 and throughout 2000 the Kivus was transformed into a patchwork of militia fiefdoms and local conflicts in which the Mai Mai were sometimes allied, and sometimes opposed to the Interahamwe. The conflict in South Kivu exacerbated because the Tutsi there were accused by other inhabitants to be collaborators and proxies of the RPF regime in Rwanda (ICG 2000: 32–5).

In the transition period (between 2003 and 2006) the number of rebel groups reduced considerably, partly due to improving relations between Rwanda, Uganda and DR Congo. However, rebel leader General Nkunda and his Tutsi rebels (CNDP with about 6,000 fighters) continued to engage the armed forces of DR Congo. They claim to fight to protect the Tutsi community from attacks by Rwandan Hutu rebels, FDLR, which consist of 6,000–7,000 Rwandan Hutus as well as Mai Mai pro-government militia of about 3,500. Various sources suggest that they were allying with the Congolese armed forces (90,000 nationwide). Because of this coalition of the government forces and the rebel groups, General Nkunda threatened in November 2008 to take over the whole country if President Joseph Kabila's government continued in its refusal to negotiate with him and continued to support the rebels. This confrontation between the government forces and rebel forces displaced a quarter of a million people in the strife-torn region around Goma, the capital of North Kivu province. The UN Secretary General Report of November 2008 (UN 2008a) blamed both the rebels as well as the national armed forces for looting and raping. The Nkunda army has been demanding direct talks with Kabila, integration into the national army, and security of the Tutsi community, meaning a democratic and reliable government representing the interests of all Congolese.

Democratization

Democracy deficit has been characteristic of Congo since its independence. During the night of 11 May 1990, members of Mobutu's elite presidential guard attacked a group of student protesters at Lubumbashi University and killed at least 12 (some figures go as high as 150), when the university students took Mobutu at his word after his address in April 1990 on multi-party democracy. Nevertheless, upon popular demands for a multi-party political system, Mobutu legalized opposition parties and pledged to establish a 'national conference on democracy' to refashion the constitution and to prepare for the multi-party elections to end his one-party rule and introduce reform (Schraeder 1994: 103–4). He gave up his ideology of 'authenticity'. The use of the word 'citizen' introduced during the authenticity ideology as a form of address and the wearing of traditional costume ceased to be obligatory. Mobutu resigned as leader of the MPR, announced the 'depoliticization' of the police, the civil service, the army and the parliamentary forces. His promise of political reforms was not enough to pacify the unhappy soldiers as well as the civilians. On 3 September 1991, badly paid Zairian armed forces and angry political demonstrators (complaining about delays in the constitutional reform process) protested against Mobutu (O'Ballance 2000: 141–3; Edgerton 2002: 219). In this unstable situation *Conférence Nationale Souveraine* (CNS) opened on 7 August 1991 and ended on 6 December 1992.

According to the Draft Constitution of the CNS, the Third Republic envisioned a parliamentary regime within a federalist state (Ikambana 2007: 40). However, disrupted by political disputes and by politically motivated violence and looting in September and October 1991, the CNS was suspended by Prime Minister Nguza Karl I Bond on 19 January 1992.

Instead of the CNS, Mobutu authorized the creation of a *constitutional* conference, which was rejected nationwide. Mobutu suggested a non-sovereign *national* conference, and on 7 January 1993 he declared he would ignore the HCR's (High Council of the Republic) ultimatum that he be removed from power unless he reversed his decision to dissolve the transitional government (O'Ballance 2000: 145; Ikambana 2007: 44). On 14 January 1994 Mobutu dissolved the HCR, and created the High Council of the Republic-Parliament of Transition (HCR-PT) and on 14 June 1994 Kengo wa Dondo was elected as Prime Minister. The HCR-PT adopted on 6 October 1996 a draft constitution that provided for a federal state with 26 provinces (instead of the existing 11 regions), a bicameral assembly, and the president with executive authority (O'Ballance 2000: 163). However, it was too late for Mobutu.

Mobutu was removed as president in 1997. The Sun City (South Africa) Peace Agreement of 19 April 2002 was the first major step towards peace and democracy. It involved the following groups: Joseph Kabila; the RCD-Goma (controlling the east with Rwanda's support); the MLC (controlling the north backed by Uganda); the unarmed political opposition, consisting of Tshisekedi's UPDS, the Mobutuists, and those in between; representatives of civil society organizations; the two factions of the RCD breakaway group *Rassemblement Congolais pour la Démocratie/Mouvement de Libération* (RCD-ML); the original group led by Professor Ernest Wamba-dia-Wamba and the group that was still militarily active in the northeast under warlord Mbusa Nyamwisi; the *Rassemblement Congolais pour la Démocratie/National* (RCD-N), a breakaway faction from the RCD-ML; and the Maï-Maï fighters, a grassroots resistance group with multiple factions engaged in armed struggle against the invaders in the east (Nzongola-Ntalaja 2004a: 19)

Though the transition could integrate the six main armed groups into a national army, this integrated army itself has been one of the worst human rights abusers. The UN Secretary General Report of November 2008 on DR Congo confirmed this (UN 2008a). Ill-disciplined and partisan security forces fuelling popular resentment allowed militias to prosper. During the transition period justice was sacrificed for peace, which came at the cost of impunity for human rights abuses. Corruption and patronage networks were tolerated, undermining necessary institutional reform. The division between western (opposition supporters) and eastern Congo (Kabila supporters) showed that the country was not yet genuinely unified.

Since Kabila's support in the West was very limited, in order to rally support in this part of the country and avoid being perceived as the 'President of the East', Kabila created an Alliance for Presidential majority (AMP) by making a deal with Nzanga Mobutu, leader of the Union of Mobutists' democrats (UDEMO), who was fourth in the first round with 4.77 per cent, and with Antoine Gizenga, leader of the Unified Lumubist Party (PALU), who was third with 13.06 per cent; whereas Bemba with 15 other parties and first-round losers formed a union called *Union pour la Nation* (UpN). The transition period was marred by some internal problems: Kabila's military and political dominance against RCD (Congolese Rally for Democracy) and MLC (Movement for the Liberation of Congo) in the transition; the withdrawal of RCD leadership from the transition in August 2004 (until it turned to government several weeks later) after the massacre of 160 Tutsis in the Gatumba refugee camp in Burundi; the loss of popularity of RCD in Congo due to its Rwanda links (because of which it won less than 6 per cent in legislature seats, and its leader Azarias Ruberwa less than 2 per cent of the presidential vote in 2006); the weakening of MLC due to failed confrontation, internal divisions and international pressure (ICG 2007a: 3–4).

The elections of 30 July 2006 and 29 October 2006 were the first 'real' democratic process since Congo's independence. The next elections are to be held in 2011. In the second-round elections Joseph Kabila was elected president with 58 per cent of the vote, whereas Jean-Pierre Bemba Gombo got 42 per cent. During the elections there were 70 parties in the National Assembly, 26 parties in the Senate. 338 of 500 seats of National Assembly and 55 of 108 seats in the Senate were won by AMP. Under the new constitution the president is elected by popular vote for a five-year term (eligible for a second term). The AMP is clearly dominant, whereas the UpN has only 132 seats in the National Assembly (ICG 2007a: 7). After losing the elections Bemba lost his moderate allies. François Mwamba, a leading Bemba moderate, tried to negotiate a new power-sharing government. The moderate *Mouvement pour la Libération du Congo* (MLC) Executive Secretary

Thomas Luhaka also decided to leave the country. Bemba was left surrounded by people who want a harder line against Kabila, such as Joseph Olenghakhoy.

The bicameral legislature, which consists of a National Assembly, consists of 500 seats (61 members elected by majority vote in single-member constituencies, 439 members elected by open list proportional-representation in multi-member constituencies). The National Assembly serves for five terms. The Senate (which has 108 seats) is elected by provincial assemblies, also to serve five-year terms. According to the elections of 30 July 2006, the National Assembly consists of the following seats: People's Party for Reconstruction and Democracy, or PPRD (Joseph Kabila): 111; Movement for the Liberation of the Congo, or MLC (Jean-Pierre Bemba): 64; Unified Lumumbist Party, or PALU (Antoine Gizenga): 34; Social Movement for Renewal, or MSR: 27; Forces of Renewal, or FR (Mbusa Nyamwisi): 26; Congolese Rally for Democracy, or RCD (Azarias Ruberwa): 15; independents: 63; others: 160 (includes 63 political parties that won 10 or fewer seats, such as the Christian Democrat Party, or PDC (Jose Endundo); Convention of Christian Democrats, or CDC (Pierre LUMBI); Union for Democracy and Social Progress, or UDPS (Etienne Tshisekedi); Union of Mobutuist Democrats, or UDEMO (Mobutu Nzanga)) (ICG 2006c).

Following the 2006 presidential election, media owned by Jean-Pierre Bemba as well as community radios were particularly targeted. Journalists are imprisoned frequently. Journalist and editor on local *Radio Okapi*, Serge Maheshe, was killed on 13 June 2007; and a few weeks later Patrick Kikuku Wilungula, a freelance photographer working for the *Agence congolaise de presse* (*ACP*) and privately owned weekly Kinshasa-based weekly *L'Hebdo de l'est* was killed (RWB 2008: 11). But all aspects of the societal, economic and political components of the intrastate conflict system are essentially interlinked with the regional dimensions.

Regional Conflict System in Congo

DR Congo borders nine countries that are in latent or open internal wars, except Zambia. In all the neighbouring countries politics, economy and society have mutual negative interactions. The outcome of such negative interactions of the neighbours has a huge destabilizing impact on Congo. The fall of the Mobutu regime resulted from a combination of internal weaknesses and the exploitation of these weaknesses by neighbouring countries to get rid of a bad symbol of a political figure in the region (Nzongola-Ntalaja 2004: 13). Two main reasons characterized the involvement of the neighbouring states in Congolese politics: first, economic interests; second, armed rebels of the neighbouring states operating from the Congolese territory. In the same way, Congolese rebels were also using neighbouring states as refuge. For example, already in December 1967, rebellious troops retreated to neighbouring Rwanda (Clark 2002: 27). Even before that, Kabila was organizing his rebellious project from the Tanzanian territory (Gleijeses 2003: 88). In the 1970s Congolese rebels were operating from Angola.

Rwanda and Uganda perceived the power holders in DR Congo and rebels operating from there as a threat to their security (Tull 2007: 117). In addition to pursuing the armed rebels of neighbouring Congo, another reason caused immediate neighbours and other African states to create a coalition of states in eastern and southern Africa, including Angola, Eritrea, Ethiopia, Namibia, Tanzania and Zimbabwe to get rid of Mobutu: a new wave of pan-Africanism solving the problems of Africa by Africans. On 18 October 1996 the coalition made Laurent D. Kabila the leader of AFDL at Lemera, in South Kivu, to liberate Congo from Mobutu (Nzongola-Ntalaja 2004: 13).

From the regional point of view two important factors contributed to the start of the second Congo war: first, Kabila sought greater autonomy from Rwanda and Uganda; second, there were a large number of disgruntled and powerful Congolese unhappy with the pace and extent of the political reform. As a consequence, Rwanda and Uganda took advantage of the existence of this domestic disgruntle, and the Rwandan and Ugandan governments facilitated efforts to build a political and military campaign against the DRC government (Nest et al. 2006: 24). Laurent Kabila decided to join the SADC after having taken control of the Congo. Since the militarily strong Rwandese and Ugandans made Congo dependent on them for its security, Kabila hoped to become less dependent on these two states if he joined the SADC. Especially when Uganda and Rwanda invaded Congo in August 1998, Kabila hoped for support from the SADC member states (Chimanikire 2003: 187).

The Rwandan Dimension of Congo's Regional Conflict System

The main cause of the Rwanda–Congo conflict has been the genocide in Rwanda in 1994. However, there were some preceding events that contributed to it. Though the peak of the Rwandan–DRC conflict system was in the 1990s, the period since the Rwandan 'social revolution' of 1959–62 that overthrew the Tutsi Monarchy and gave power to the Hutu (Lemarchand 2000: 331; Mamdani 2002: 103–31) had decisively paved the way for the future genocide and regional conflict system. When two attempts of incursions by some Rwandan Tutsi refugees and ex-soldiers failed in 1963 and 1967, the retaliation against Tutsis in Rwanda led to the killing of up to 20,000 (Mann 2005: 438).

President Juvenal Habyarimana of Rwanda, a Hutu, came to power in 1973 and refused to allow Tutsi victims of the 1959 pogrom. He decided to defend his power through 'majority democracy' and 'Hutu power' (Mann 2005: 435). When the leadership of the Rwandese Patriotic Front (RPF), the Tutsi Diaspora in Uganda, launched a military campaign to overthrow the Habyarimana regime in October 1990, France, Belgium and Congo prevented an RPF victory. However, on 4 August 1993, Habyarimana's government and RPF signed the Arusha Peace Agreement, which included a ceasefire agreement, protocols on the rule of law, power-sharing with a new transitional government created in April 1992 (which was supposed to consist of all five parties, including FPR), repatriation of refugees and resettlement of displaced persons, and integration of armed forces. However, all those issues remained unimplemented because the president's party, NRMD (National Republican Movement for Development), refused to cooperate. In the meantime, the FPR revolt escalated, and the situation was exacerbated when Habyarimana's plane was shot down on 6 April 1994, initiating the genocide against the Tutsis accused of shooting down the plane. As a reaction, Tutsis and moderate Hutus were indiscriminately targeted by two of the most notorious Hutu killer groups: NRMD's *Interahamwe* ('those who stand together') and CDR's (Coalition for the Defence of the Republic) *Impuga Mugambi* ('the single-minded ones') (O'Ballance 2000: 154; Mamdani 2002: 218). Adding to the already existing instable border situation in eastern Zaire, more than a million Rwandan refugees brought the trauma of post-genocide Rwanda to Zaire around mid-1994. 'As the tension between Hutu and Tutsi increased in Rwanda, it also did in North Kivu' (Mamdani 2002: 251).

The refugee-generating violence was transformed into violence-generating refugee flows (Lemarchand 2000: 331). In 1972 Mobutu, through a new Citizenship Law, gave them the right to control and administer the Kivu region of eastern Congo, and they were recognized as full citizens of Zaire. In 1981, because of the grievance of the 'autochthonous' population, Mobutu revoked the Citizenship Law of 1972. The new law limited full rights only to those who entered Congo in the nineteenth century. In 1991 the Sovereign National Council reaffirmed the 1981 Citizenship Law. Therefore, those with Rwandan origins faced triple threats: first, they were stripped of their citizenship (there was no law to protect them); second, they were hated by the autochthonous as 'strangers'; third, Hutu refugees and perpetrators of Rwandan genocide of 1994 came into the Congolese area where the Tutsis were living.

Therefore, the Congo–Rwanda conflict consists of two main interrelated regional phenomena: the social crisis of post-genocide Rwanda and the citizenship crisis in the region (Mamdani 2002: 261). *First*, because of the genocide Rwanda became a diasporic state, or political community that is politically and morally obliged to protect the Tutsis wherever they are. *Second*, the survival of the Tutsis depended on their solidarity against hatred of Tutsis. They were persecuted at least three times: in 1959 when Tutsi power was routed and dismantled at the local administration level as well as transferred government power from Tutsi to Hutu (Mamdani 2002: 104); in 1973 when President Juvenal Habyarimana of Rwanda, a Hutu, came to power and refused to allow Tutsi victims of the 1959 pogrom and purged the Tutsi from the public sector (Mann 2005: 438); and in 1994 when the genocide against the Tutsi took place.

The Tutsi citizenship was a key aspect of the regional conflict system. Some radical Hutu suggested that the Tutsi are non-indigenous and should be excluded altogether from citizenship, whereas Habyarimana suggested to tolerate the Tutsi as long as Hutus controlled the public realm and the migrated Tutsis do not return (Mann 2005: 438). During the genocide, Tutsis were dehumanized as 'cockroaches', 'snakes', 'rats', 'scum', 'hybrids', 'beings with two heads', 'parasites', which had to be 'destroyed', which should not be allowed to 'get away', to be 'wiped out'; the nation had to be 'cleared' and 'cleansed' by the 'democrats', the 'majority people' or the 'great majority' (Mann 2005: 444, 445). Such murderous ethnic cleansing pointed to a dark side of democracy (Mann 2005: 473) that constructs the identity of the dehumanizer and of the dehumanized. When the Rwandans and Ugandans organized AFDL against Mobutu in 1996, the Congolese Tutsis willingly joined the alliance of revenge against Mobutu for the new Citizenship Law of 1981 and 1991 and for its consequences.

When the Tutsi Rwanda Patriotic Front entered Rwanda in reaction to the ongoing genocide, the Armed Forces of Rwanda (FAR) and their Interahamwe militia escaped with virtually all of the weapons at their disposal into the Congo. The RPF's pursuit since 6 October 1996 to destroy the UNHCR refugee camps in Kivu, where the ex-FAR (Armed forces of Rwanda) got refuge and received help from the Mobutu regime, led to Mobutu's demise (Nzongola-Ntalaja 2004: 8). This Interahamwe militia (extremist Hutu militias) group that had been chased into exile by the RPA in 1994 has been managing to escape and reorganize itself. The militia group could increase its number by recruiting its members that infiltrate from Katanga, Tanzania, Zambia, Congo-Brazzaville, Gabon and the Central African Republic. Interahamwe was fighting not only against the Rwandan government, but also against the Banyamulenge in Zaire. Just one example of the tactical cross-border alliances was that Burundi's Tutsis were supporting the Banyamulenge in Congo, whereas the Mai Mai were supporting the Ex-FAR and Interahamwe (Lemarchand 2000: 344). Towards the end of 1996 the Interahamwe, taking advantage of the UN-administered refugee camps in eastern Zaire, killed and displaced thousands of Banyamulenge from the area, from where they launched attacks against the Rwandan troops (O'Ballance 2000: 164). The persecution of the Tutsis by the Interahamwe further provoked the invasion of Rwanda; Zaire was not able to protect itself from the invasion of Rwanda. Since Interahamwe was acting against the Congolese Tutsis from the refugee camps in Congo, Congolese Tutsi fled into Rwanda. The consequence was that the RPF trained and armed Congolese Tutsi, whereas the Interahamwe and Ex-FAR formed FDLR (*Force Democratic de Liberation du Rwanda* created in 2000) and began to collaborate with the Zairian army, and the native authorities in Kivu created their own militia called Mayi Mayi (Mamdani 2002: 256, 258).

Even after the overthrow of Mobutu in May 1997, the style of mobilization against the Tutsis in DR Congo was similar to that against the Tutsis during the genocide in Rwanda. Radio stations such as *Radio des Patriotes* in DR Congo were broadcasting an appeal to all Mai Mai, ex-Zairian Armed Forces, and ex-FAR and all Hutus to fight the Tutsis. Because of the escalating conflict in DR Congo, Tutsis from DR Congo had to flee to Mudende in Rwanda, where they were sheltered in a camp. Towards the end of 1997, especially between October and December 1997, groups such as Palir (Armed People for the Liberation of Rwanda – emerged in 1995) in the north-west of Rwanda called on all Tutsis to leave Rwanda 'before it is too late' (*Africa Research Bulletin*, Vol 34, No 12, 1997: 12942).

Such genocidal mobilizations forced Kigali to go on the counteroffensive. The Tutsis who had been displaced by the massacre of the Tutsis in DR Congo and had fled to north-west Rwanda faced further massacre there. These developments have been making it impossible to heal the rift between the Rwandan Tutsi-dominated government and the Hutu majority, rebuild the economy and move the nation towards democracy. In 1998 militant Rwandan Hutu were spreading throughout eastern Congo, Burundi and western Tanzania and frequently moving in and out of Rwanda because the states in the region were not able to control border areas. When Rwanda, Uganda and DR Congo were still allies, the Congolese population backing Laurent Kabila was being attacked by Hutu forces in the DR Congo.

Rwanda was rewarded by the DRC for its role in the demise of Mobutu's power by naming James Kabarebe, a Rwandan military officer, as chief of staff of the newly established Congolese army, the *Forces Armées Congolaises* (FAC). Further positions, such as foreign ministry, personal secretary to the president, and secretary general of the AFDL, were held by Rwandese (Nzongola-Ntalaja 2004: 14).

On 16 August 1998 a Congolese Tutsi, Bizima Karaha, announced that the Congolese Democratic Movement (later Congolese Rally for Democracy – allied mainly with Rwanda) was created (Lemarchand 2000: 339; O'Ballance 2000: 181). RCD (*Rassemblement Congolais pour la Démocratie*, or the Congolese Rally for Democracy), led by Professor Ernest Wamba dia Wamba (the long-time Mobutu opponent), provided a Congolese political face to Rwanda's war in Congo. From the outset it was plagued by internal divisions. It was a coalition of opportunistic politicians who shared little more than a common antipathy for Kabila.

Besides Rwanda, General Nkunda and his fighters were claiming to protect the Tutsi community in the Democratic Republic of Congo against attacks from ethnic Hutu Interahamwe militias who fled there after the Rwandan genocide of 1994. Between June and August 2008 some 250,000 people fled their homes in DR Congo's North Kivu province, when General Nkunda began an offensive on the regional capital, Goma. In January 2009 General Nkunda was arrested in Rwanda within the framework of cooperation between Rwanda and DRC to address the destabilization caused by Nkunda's CNDP, Mai Mai militias and the FDLR. As a consequence of this inter-rebels conflict, Rwanda hosts more than 50,000 Congolese refugees.

Hutu rebels (FDLR) were driven from their bases by a joint force of Congolese and Rwandan troops but later they managed to retake their Congo bases. All rebel groups have been killing civilians, burning houses and raping women in North Kivu province. Since late 2006 the fighting involving the FDLR, CNDP, Rwanda, Mai Mai militia, and the Congolese army and other militias has displaced more than 1 million people (UNHCR 2009a). When towards the end of May 2008 the Rally for Unity and Democracy (RUD), a breakaway faction from FDLR, agreed to disarm, the FDLR stated that its mission remained 'liberation' of Rwanda.

The Rwandan factor is tantamount because it also affected Congolese–British relations. The UK, a key ally of Rwanda and its extended arm RCD, had limited leverage to influence Congolese politics, whereas France, the US, South Africa and Angola were playing a significant role. However, the Rwanda–Congo conflict systemically embraces the Ugandan–Congo conflict system. Rwandan and Ugandan forces fought to control mineral resources and, in the Kisangani area, Rwanda used hydropower without any financial compensation. It became increasingly evident that, besides protecting the Tutsis in Congo, the Rwandan army was involved in the Democratic Republic of Congo for the exploitation of natural resources (UN 2001: 8). The military involvement of Rwanda and Uganda in the Democratic Republic of Congo would have been an expensive adventure without the active exploitation of the Congo's natural resources (UN 2001: 5). Rwanda maintained it spent 4.3 per cent of GDP for defence in 1998, whereas according to the IMF the real figure was closer to 8 per cent. At any rate, cooperation between Rwanda, DRC and Uganda is indispensable for peace in Rwanda as well as in the region.

The Ugandan Dimension of Congo's Regional Conflict System

Uganda occupied Ituri from 1998 to 2003. The objective of the Ugandan in the DRC was to destroy the ADF's (Allied Democratic Forces) Congolese rear bases. Besides the WNBF and the LRA, the ADF that emerged in 1995 was supported by Sudan, and later joined by some former Mobutu soldiers. ADF has not only been fighting the Ugandan army but it has also been terrorizing the inhabitants of southwest Uganda with its viciously brutal attacks. The ADF could not be so easily defeated, as they demonstrated in early December 1999 with a series of deadly attacks (ICG 2000: 38–40). Though the movement is not strong enough to pose a military threat to the regime, it has however given a good pretext for Uganda to get involved in the Congolese, both to fight the ADF and to exploit the natural resources of DR Congo (Lemarchand 2000: 325). Uganda too was involved in the exploitation of the natural resources of the Democratic Republic of Congo (UN 2001: 5).

In 1996, the UPDF (Ugandan People's Defence Force – Armed Force of Uganda) controlled regions of north-eastern Congo and profited an estimated $60 million in gold exports for Uganda. Official figures by the Bank of Uganda, which controls all gold exports, show that Uganda's gold exports shot up from $12.4 million in 1994–95 to $110 million in 1996, which corrected the country's $600 million trade deficit. This facilitated Uganda's defence expenditures and attracted UPDF officers to get increasingly involved in Congolese profits of war (ICG 2000: 40–41). When Kabila ordered on 27 July 1998 all foreign military to leave, he provoked simultaneous invasion by Rwandan and Ugandan forces and uprisings by military units in Goma and Kinshasa whose commanders were part of the ongoing plot and allies of Rwanda and Uganda (Nzongola-Ntalaja 2004: 16).

The UPDF presence in the Ituri region worsened the conflict between the Hema and Lendu. The Ugandan army exacerbated the situation since it protected the interests of the rich Hema. The reason for the dispute between the Hema and Lendu is mainly land. Belgian colonial authorities favoured the Hema and permitted them to amass large landholdings before they left Congo. The two groups lived alongside one another peaceably until the closing days of the Mobutu regime. After the demise of Mobutu and when the state administrative structures collapsed, the rich Hema landowners upset the status quo by seeking to acquire additional land-holdings from the Lendu. Powerless to respond legally because they lacked documentary proof of ownership, the Lendu turned to violent means of resistance. It was estimated that in the fighting that followed the demise of the Mobutu era, more than 10,000 people died in a period of 18 months, especially in two areas – Djungu and Fataki.

The complexity of the conflict system was exacerbated when Uganda and Rwanda began to fight against each other. The disintegration of another important coalition – Uganda and Rwanda – brought a new dynamic to the conflict system in the region. The coltan exploitation (destined for countries such as Germany, the US, China and Russia) played an important role in the conflict dynamic (Todd 2006: 6–10; Montague 2002: 105). Erstwhile allies Uganda and Rwanda fought each other on three separate occasions – in August 1999 and May and August 2000 – to control the resource-rich areas of Kisangani (Nzongola-Ntalaja 2004: 16, 18). Further, in 1999, disagreement between the UPDF and RPA over who should lead the RCD resulted in the group splitting into two factions: RCD-Goma (supported by Rwanda) and RCD-ML (Liberation Movement – aligned with Kampala and relocated to Kisangani). Following clashes in the first half of 2000 between UPDF and RPA over control of the local diamond industry in Kisangani, the RCD-ML moved its headquarters once again – this time to Bunia, closer to the Ugandan border (Human Rights Watch 2003: 5). The RCD-ML also briefly merged with the MLC to form the *Front de Libération du Congo* (FLC) headed by Jean-Pierre Bemba. However, the FLC soon split along the lines of the original parties following acrimony over Bemba's attempt to extract high taxes from areas under the RCD-ML's control and bring these areas under the direct military control of the MLC. These coalitions resulted in three broad zones of military influence in territory occupied by antigovernment forces: an MLC/Ugandan zone in north and western Congo, an RCD-ML/Ugandan zone in the northeast, and an RCD-Goma/Rwanda zone in central-eastern DRC (Nest et al. 2006: 26–7). The peace agreement between Uganda-backed MLC and RCD-ML, and the Congolese government in April 2002 was an important step towards peace in Congo. The peace accord between Rwanda and Congo in July 2002 did not bear much fruit, though Rwanda agreed to withdraw its troops.

Rwandan, Angolan, Namibian and Zimbabwean troops withdrew officially between October and December 2002 from the DRC, and Ugandan troops pulled out in May 2003 based on the Ugandan-DRC bilateral accords of September 2002. An all-inclusive power-sharing agreement was signed in Pretoria on 17 December 2002. Rwandan soldiers and military advisers remained integrated with RCD forces in eastern DRC. The Pretoria Accord was formally ratified by all parties on 2 April 2003 in Sun City, South Africa. The 'official' conflict – the one addressed by the peace process – concluded in June 2003, when the Government of National

Unity and Transition, comprising different belligerent groups and civil society representatives, was established.

Though Uganda stopped its support for MLC early in the transition, it continued to destabilize the Ituri district by facilitating the creation of an alliance of Ituri armed groups, Congolese Revolutionary Movement (MRC), on its territory in July 2005. Moreover it has entered the Congolese territory several times in pursuit of LRA rebels. The Ugandan, Congolese and the South Sudanese joint military operation, known as 'Lightning Thunder', against the Lord's Resistance Army launched in December 2008 caused attacks on innocent communities. As a reaction to the operation the LRA attacked villages in DR Congo and South Sudan and massacred more than 900 people. However, there is no alternative to regional cooperation. The Burundi–Congo rapprochement has shown this.

The Burundian Dimension of Congo's Regional Conflict System

Already in the early 1970s, the internal conflict in Burundi paved the way for the 1990s Burundian dimension of the Congo's regional conflict system. After the massacre of about 200,000 Hutu in Burundi in 1972, many refugees flooded into Zaire, which caused tension between immigrants/refugees and the 'nationals' (Mamdani 2002: 243)

A Hutu, President Bagaza, came to power in a 1972 military coup and was overthrown by a Tutsi, Pierre Buyoya, in 1987. After the uprising and violence of 23 November 1991, between the Tutsi and Hutu, more than 50,000 people, mainly Hutus, fled to Rwanda and Zaire. Rwanda, Burundi and Zaire agreed on 4 January 1992 to repatriate them and effectively to control borders, but nothing substantial took place (O'Ballance 2000: 144). Buyoya and his party (UPRONA – *Union pour le Progress National*) lost power after the presidential elections of June 1993, which were won by a Hutu, Melchior Ndadaye, from an opposition party (FRODEBU). In the military coup on 21 October 1993, President Ndadaye was killed by the Tutsi-dominated army. At least 50,000 people lost their lives (Lemarchand 2000: 336). As a result, as many as 270,000 Burundian refuges left for a refugee camp in Tanzania, which was controlled by PALIPEHUTU and CNDD. His assassination was decisive for the future Burundian civil war with regional implications. After the assassination of Ndadaye, Léonard Nyangoma, a founding member of Ndadaye's FRODEBU party, refused to accept a power-sharing deal with those who had killed the president. In order to re-establish the institutions of 1993, the most important Burundian rebel group was created (the CNDD – *Conseil National Pour la Défense de la Démocratie*) in 1994. Its military wing, FDD, headquartered in Lubumbashi (Capital of Katanga), recruited troops in the refugee camps of Tanzania, and trained and organized them in the Congo from where they launched their attacks against south-western Burundi. In February 1994, Cyprien Ntaryamira, a Hutu, was installed as president, and was killed, together with President Habyarimana of Rwanda, in an air crash on 6 April 1994.

When Ntaryamira was installed as president, Tutsis at Bujumbura were building barricades for fear of Hutu attacks and they even protested against the presence of the OAU troops, accusing them as protectors of the Hutu regime (O'Ballance 2000: 148–50). Retired South African military officers and *Executive Outcomes* supported CNDD and facilitated weapons import to Hutu groups based in eastern DR Congo. CNDD-FDD had been receiving weapons from UNITA.

In mid-1996 Mobutu's Zaire was providing CNDD-FDD with rifles and ammunitions. The fact that FDD had its training camps in Zaire until 1996 caused Burundi to send its troops there. Later the FDD was forced to flee to Tanzania when the AFDL (*Alliance des Forces Démocratiques pour la Libération du Congo*), Rwandan and Burundian troops entered Zaire. After that, some Burundian troops stayed in the Congo to protect its commercial interests. The main reason was that Burundi was under regional trade embargo after the July 1996 coup through which Buyoya came to power. The uncontrolled territory of Congo gave them a good opportunity to bypass the restrictions of the embargo imposed by the regional states (DR Congo, Kenya, Tanzania, Rwanda, Uganda and Zambia). In spite of the embargos, arms were flowing into Burundi from eastern European states such as Bulgaria through the territories of regional states such as Angola, Kenya, Rwanda, Tanzania and DR Congo (ARB 1998a: 13078).

When Kabila came to power, he supported the FDD. The reason was that Kabila needed them for his military service. Various sources suggest that many of the Congolese troops in the Katanga offensive in October 2000 were FDD. On 17 November 2000, President Buyoya emphasized that Burundi had no territorial intentions or interests to exploit natural resources in DRC (UN 2001: 9). But Burundi has had vital security interests in DRC. When Burundi was involved in the DR Congo, its armed forces were fighting the FDD/FNL (*Forces de la Défense de la Démocratie/Front de Libération Nationale*). Around 3,000 Hutu rebel fighters left the Tanzanian refugee camps to join the FDD in Congo. Kabila made FDD into millionaires and permitted them to control a territory three to four times bigger than their own country. Because of the close cooperation between Kabila and the FDD, Kabila could rely on their readiness to fight for him and recruit new FDD troops for Kabila from the Kigoma camps in Tanzania. Moreover, the FDD were able to move easily between Tanzania, Burundi, Zambia and the DRC. The Burundian Hutu militias fighting the Tutsi-dominated Burundian army not only operated from the Congolese territory, but they also supported the DR Congo army during the Kabila era.

After two years of negotiations to end their civil war, 19 political parties signed a peace agreement in Arusha, Tanzania, on 28 August 2000. But the war started again because the FDD (armed and equipped by the DRC, protected by Tanzania) had no interest in joining the peace process (ICG 2000: 20). Since Kinshasa was supporting CNDD-FDD rebels, and when these rebels came to power in the 2005 elections in Burundi, ties between DR Congo and Burundi improved easily (ICG 2007a).

In May 2008 Burundi's last active rebel group, National Liberation Forces (FNL), and Burundi's government under ex-rebel Pierre Nkurunziza, elected president in 2005, signed a ceasefire. This has been a significant move to implement a 2006 peace agreement (which broke down after the government rejected rebel demands for power-sharing) and to end violence which has been devastating this small central African country. Moreover, this peace agreement is a non-negligible factor for sustainable peace in DR Congo.

The Angolan Dimension of Congo's Regional Conflict System

The Angola–Congo conflict goes back to the period of the Cold War. At various times Angolan rebels were operating from the Congolese territory, and Congolese rebels were operating from the Angolan territory. Between 1975 and 2002 Angola was marred by a civil war between the National Front for the Liberation of Angola (FNLA), the Popular Movement for Liberation of Angola (MPLA) and the National Union for Total Independence of Angola (UNITA), which had regional implications as well as global involvement. After the FNLA fell apart, the US switched to supporting the UNITA. The US refused to support the *de jure* MPLA government because it was 'too communist'. At the peak of US (clandestine) involvement, President Ronald Reagan labelled the brutal UNITA's guerrilla leader Jonas Savimbi a 'combatant for liberty' (Hentz 2004: 28). Mobutu served as an important conduit to the FNLA in Angola for the CIA. In spite of some tensions with the US from the early 1970s, Zaire remained important to dictate the political make-up of the soon-to-be independent Angolan government, where the Soviet-backed MPLA would dominate politically. UNITA, operating from Congo, was supported by South Africa and China (Taylor 2006: 46).

Angola contributed significantly by sending around 1,000 elite units armed with tanks and planes to halt the Congolese rebels attempting to enter Kinshasa. Similarly, in October 1997, Angola contributed to the victory of Sassou Nguesso over his rival Pascal Lissouba in the Republic of Congo. However, both in the Congo Republic as well as in the DR Congo, the main motivation of Angola's intervention was (besides securing its share of Congo's oil and support Dos Santos' ally Kabila): to destroy the UNITA rebellion and to protect the Angolan enclave of Cabinda against separatists. Weapons flowing from Russia were decisive for the operations. For such operation Angola was buying weapons from Russia. Such conflicts are especially profitable for Russia in its attempt to modernize and expand its defence industry, show its military power and boost its economy by exporting arms into conflict areas of the world.

The intra-Angolan conflict also expanded to the Republic of Congo. The existence of the former president Pascal Lissouba's 'Zulu' militias and former prime minister Bernard Kolelas' 'ninja' militia in Congo-Brazzaville complicated the regional situation. UNITA was supporting not only the Cabindan pro-independence groups, but also intensified its collaboration in 1997 with the leader of the Congolese Movement for Democracy and Integral Development (MCDDI) and its 'ninja' militias of Congo Republic. The ninja militias fought with the deposed Congolese president Pascal Lissouba not only against the troops of President Denis Sassou Nguesso, but also against the Angolan troops operating from the territory of the Republic of Congo to fight the UNITA operating from there against Angola. Moreover, three Cabindan movements (Flec-FAC – Armed Forces of Cabinda; Flec-Renovada; and Cabindan Democratic Front) had been equipped by UNITA and trying to mobilize the support of Bakongo ethnic groups in the Republic of Congo, DR Congo, Cabinda and Angola in order to eject the Angolan troops from Congo-Brazzaville, defeat Nguesso, reinstall Lissouba, and achieve Cabinda's independence. Moreover, President Denis Sassou Nguesso became dependent upon Angolan military help, because of which Angola was increasingly involved in the Congo Republic's war. UNITA's Savimbi had contact with Kolelas, who had about 1,000 militiamen in the Lower Congo region of the DR Congo in early 1998. Later the rebel group was led by Frédéric Bintsamou, alias Pasteur Ntoumi.

On the other side of the DRC, the porous Zambian border between Zambia and DR Congo also benefited the UNITA to operate against Angola. An intricate web of logistic routes, brokering agents and transport companies in the central and southern African region of South Africa, Zambia, Tanzania, Uganda and the DR Congo had been facilitating arms transfer to the UNITA.

Though during most of 1998 the Armed Forces of Angola could cut UNITA's supply lines and access to the DRC's ports and airfields, in December 1998 UNITA resumed its offensive. On 26 March 1999 UNITA captured the important frontier city of Maquelo do Zombo supported by Congolese rebels as well as directly or indirectly by Rwanda and Uganda, because these two states were against the Angolan support for Kabila. To recapture the town, for their part the heads of state from Angola, Namibia, Zimbabwe and the DRC agreed to mount a joint operation not only against UNITA but also against the Congolese rebels and against Rwanda and Uganda on the Congolese–Angolan borders. This resulted on 28 April 1999 in UNITA's expulsion from the DR Congo. Between September 1999 and September 2000, the Angolan government controlled 92 per cent of the 157 districts of the country; and the last provincial capital in UNITA hands, Cazombo on the Zambian border, fell on 19 September 2000 along with its strategically important three-kilometre airstrip (ICG 2000: 65–6).

The involvement of Angola in the Congo conflict is caused not only by the intent to defeat the UNITA rebels. There has been an economic reason behind it. Dos Santos owned a share of the DRC's petroleum distribution and production networks as a reward for Angolan armed forces' assistance to DR Congo (ICG 2000: 57). Moreover, thousands of Angolans were involved in illegal diamond mining and trafficking along the border (ARB 1998c: 12996).

DR Congo is a strategically important ally of Angola in its fight against the Liberation Front of the State of Cabinda (FLEC), as well as economically important. DR Congo geographically surrounds Cabinda, which produces 60 per cent of Angola's oil. As a result, Angola has been training thousands of Congolese soldiers, as well as Kabila's Presidential Guard, and police. Moreover, Angola is competing with South Africa to become a regional player, and Congo offers a good opportunity for this.

Starting with the January 2003 exploratory talks between the government and the FLEC-FAC, the Angolan government expressed its willingness to grant a measure of autonomy and the economic security of FLEC troops, but not independence to Cabinda. As a result an agreement on a ceasefire and general amnesty, *Memorandum of Understanding for Peace and Reconciliation in Cabinda Province*, was signed between the government and a coalition of FLEC factions, Cabindan Forum for Dialogue, on 1 August 2006. Though not all FLEC factions were party to the deal and low-intensity fighting continued for some time, a promising framework for enduring peace could be established (EISA 2008). UNITA no longer plays a destabilizing role in Congo since the death of Savimbi in 2002. This has been a positive development not only for Angola but also for DR Congo as well as for the Republic of Congo. The ninja militias signed a peace agreement with the Congo government; FLEC no longer strives for independence. Besides Angola, Zimbabwe has been one of the significant competing countries in the region to play a dominant role in the DR Congo as well as in the region as a whole.

The Zimbabwean Dimension of Congo's Regional Conflict System

Zimbabwe does not share a border with DR Congo, but ambitions for regional precedence (because Mandela's prestige eclipsed Mugabe in the region) and economic interests in the Congo's natural resources pushed Mugabe to be militarily present in the DR Congo. However, the Foreign Ministry of Zimbabwe argued that Zimbabwe was liberating Congo from Uganda, Rwanda and Burundi, not exploiting its resources (UN 2001: 9). Mandela's conciliatory attitude towards the rebels was because he was disappointed by Laurent Kabila, who was incapable of stabilizing the DR Congo, whereas Mugabe espoused Kabila's cause, together with the desire to counterbalance the Ugandan influence and to compensate for the loss of Mozambique to South African intrusion. For South Africa, DRC is a huge reservoir of raw materials and an enormous potential market, and South Africa has been one of the major investors in the DR Congo. On the other hand, Namibia has been a traditional ally of Zimbabwe and Angola in the fight against apartheid and colonialism. The impacts of the Cold War and colonization enjoined this southern axis of Namibia–Angola–Zimbabwe, whose byproduct has been concerted aid for DR Congo. Moreover, Angola, Namibia and Zimbabwe have been trying to emerge as major players in SADC.

According to the SADC's 1995 Inter-State Defence and Security Committee (ISDSC), an aggression against a member state can be considered as aggression against SADC as such. Therefore, for the Angolan–Namibian–Zimbabwean axis, SADC had some justifications to intervene in Congo. Chimanikire argues that the SADC intervention in the DRC is consistent with Article 51 of the UN Charter regarding collective security by member states, upholds the principles of international law relating to aggression, and consistent with the OAU Harare Declaration of 1997 regarding the removal of the government by force (Chimanikire 2003: 189).

But the Zimbabwean motivation for the intervention was primarily based on the internal political situation of Zimbabwe and Mugabe's interest in intervening there. Mugabe's August 1998 decision to intervene with 600–800 men in the DR Congo was mainly pushed by his desire to assert his leadership as an African statesman (ICG 2000: 60); and to set himself free from the overwhelming shadow of Mandela in the region and in all of Africa (Lemarchand 2000: 347; Meredith 2003: 147; Chimanikire 2003: 188). Moreover, the economic interests of the ruling elite pushed Zimbabwe to enter Congo on 20 August 1998 to support the government and defeat the Congolese rebels and Rwandese forces advancing on Kinshasa. Kabila had already promised Zimbabwe a great deal in return for its military support (ICG 2000: 60). Accordingly, on 4 September 1998, Kabila and Mugabe signed a deal providing for a 'self-financing' intervention by the Zimbabwean National Defence Force (ZNDF). The number of Zimbabwean troops increased to 11,000 from the initial 3,000, costing Zimbabwe $1 million a day (Lemarchand 2000: 346; Meredith 2003: 148), or according to other sources, between $3 million and $25 million per month (Chimanikire 2003: 191).

According to the Congo–Zimbabwe agreement, Zimbabwe Defence Industries (ZDI) would provide arms and munitions to the DRC; the Zimbabwean mining company, Ridgepointe, would take over the management of Gecamines and receive a 37.5 per cent share of the Congo's state mining company. Moreover, between 20 and 30 per cent of the Congo government's 62.5 per cent of the firm's profits was to be used for financing the Zimbabwean war effort (Chimanikire 2003: 190–91). To the disappointment of Mugabe, due to bad management and illegal exploitation of the mining in Gecamines, the state mining company was producing at the time only between 10 and 20 per cent of its 1980s production of 475,000 tonnes of copper and 17,000 tonnes of cobalt. Even after Zimbabwe took over the management, contrary to the initial provisions of the contract with Gecamines to supply 2,500 tonnes of copper monthly to a reprocessing plant in Zimbabwe, it produced only 3,000 tonnes over 18 months (ICG 2000: 60–61). Because of the badly calculated business in Congo, Mugabe got stuck there. He could not pull out without achieving a return on what he had already foolishly invested in the war. He was under powerful domestic political pressure. Because of the economic crisis it did not seem reasonable to fight in the Congo with uncertain outcomes and benefits for Zimbabwe (ICG 2000: 69–70).

The internal crisis of Zimbabwe played some role in pushing Mugabe to get involved in the Congo. After 20 years in power, the president as well as ZANU-PF came under serious pressure from the opposition Movement for Democratic Change (MDC) to show some successes in the politics and economy of Zimbabwe. Public dissatisfaction due to the bad economic situation of the country resulted in riots, an incomparable inflation rate of over 62 per cent and unemployment rate of over 50 per cent at that time. All foreign aid programmes were under review or cancelled, and both the World Bank and the IMF had suspended loans; Zimbabwe was listed among the world's poorest countries in a World Bank report (ICG 2000: 63). All these gloomy records at home pushed Mugabe to look for some opportunities abroad.

Mugabe miscalculated not only in his expectations for economic gains from the intervention in the Congo. Military setbacks of Kabila's troops as well as the capture and killing of Zimbabwean troops in October 1998, defections and mutinies of senior Zimbabwean officers dashed the expectations of Mugabe (Meredith 2003: 149). Mugabe was motivated to be militarilyy present in Congo by the rich natural resources and a desire to increase his own power and prestige in Africa. But the overall developments in Zimbabwe proved not propitious for that for almost a decade. In July 2008, Zimbabwe's annual inflation was 431 million per cent, unemployment 80 per cent, life expectancy 37 for male and 34 for female; 45 per cent of the population is malnourished and affected by broken-down sanitation systems and the spread of cholera in Harare.

Besides Zimbabwe, other states of the region have been trying to be stakeholders on the side of the DRC. Several family members of Namibian President Sam Nujoma are deeply involved in Congolese mining. Chad was encouraged by France to join those supporting Congo as a means of regaining influence in a region where the French had retreated in disgrace after the 1994 genocide of Rwanda. Libya too was involved. Gaddafi wanted to play an increasing role in African politics to break out of the international isolation following the Lockerbie Bombing of 1988. Though without a significant military presence inside the Congo, Sudanese government forces were fighting rebels in Orientale of Congo. This Sudanese involvement was mainly meant against Uganda.

However, ultimately neighbouring actors like Uganda and Rwanda have come to realize that cooperation and coordination with the DRC are indispensable to solve their security problems. For Uganda the DRC's role will be important in resolving the more than 20-year war with LRA because the LRA rebels have been operating within the territory of the DRC. Similarly, Rwanda hopes that the Hutu militias along its border are kept at bay by the DRC. Uganda is undergoing an internal security crisis. Besides the various armed rebels, the biggest ethnic institution in Uganda, the Buganda Kingdom that helped Museveni, is in conflict with Museveni. However, Congo wants Uganda to pay up the $10 billion in damages for pillage and plunder awarded against Uganda to the DRC by the International Court of Justice. Angola needs DRC's cooperation against the secessionists in the enclave of Cabinda. This means the DRC is increasingly becoming a crucial player in the systemic security apparatus of the Great Lakes region and the geopolitics. However, the question is whether these neighbours wish a powerful, rich and stable Congo, which could tilt the balance of power in the region and possibly revenge for past sufferings of Congo.

Ultimately, as Nzongola-Ntalaja suggests, any chance for peace in Congo is dependent on successful political culture consisting of rule of law, a democratically elected and inclusive government and national reconciliation. In the long run, stability of the Great Lakes region cannot be sustained without a stable and effective government in the DRC. Nevertheless, there can be no stability in the DRC without stability in Rwanda, Uganda, etc. A peaceful coexistence between Hutu and Tutsi in Rwanda and Burundi is crucial for Rwanda as well as Burundi (Nzongola-Ntalaja 2004: 21–2). In her 1998 tour of Africa, Madeleine Albright acknowledged that events in DR Congo would always have a direct bearing on what goes on in the central, eastern or southern regions of Africa. However, neither the intra-Congolese nor the regional conflict system can be addressed if the global dimension is not taken into account. The global dimension is as significant as the internal and regional levels for a peaceful Congo.

Congo's Global Conflict System

The involvement of the global actors in Congolese politics and economy has been obvious for decades. But Lemarchand claims that the criminalization of the state, undermining of civil society and violence in the Great Lakes region is caused by 'global disengagement' (Lemarchand 2000: 324). According to Lemarchand, international actors are responsible 'for either doing too little too late, or too much at the wrong time on behalf of the wrong party' (Lemarchand 2000: 343). But I doubt if there has ever been global disengagement in the Congo. In this section I argue that there has always been enough global engagement in Congo in different forms. Global actors contributed their part to the complexity of the conflict by pursuing their own interests rather than peace. The fluidity of domestic and international alliances have made the conflict in Congo and in the Great Lakes region intractable since the Belgian king and Belgian state set their feet on Congo soil.

The Belgian Dimension of Congo's Global Conflict System

Congo became part of the global system mainly since the second half of the nineteenth century. King Leopold II of Belgium created the African International Association in 1876 to found a chain of commercial and scientific stations running across central Africa from Zanzibar to the Atlantic to attack the slave trade and to protect Christian missions. Stanley in 1877 completed his coast-to-coast journey by descending the Congo River, took service under King Leopold in 1879 and established a practical land and water transport system from the head of the Congo estuary to Stanley Falls. Leopold persuaded the European powers that it would be preferable to have the Congo basin as a free trade area under his 'international' regime than to let it fall to any national rivals (Oliver and Fage 1966: 184). Belgium inherited Congo from King Leopold in 1908, but it did not prepare 'competent' Zairian military forces and administrative officials for independence. Belgium supported the independence of Katanga and the assassination of Prime Minister Lumumba.

For example, Belgium favoured the pastoralists Hema over the agriculturalists Lendu in the eastern Congo. Even after independence this social structure was maintained that the Hema continued to be administrative, landowning, and business elite, whereas the Lendu remained excluded from administrative positions in the farming, mining or local administrative sectors. The first major conflict between Hema and Lendu emerged in June 1999 when the Hema systematically tried to evict the Lendu from the land through false documents. Uganda intervened militarily because there was no strong Zairian authority in the region, and the Ugandan 'administration' altered administrative boundaries in this area of Ituri. This gave the Lendu the impression that the Ugandan army was siding with the Hema landholders (Human Rights Watch 2003: 18). The Hema–Lendu conflict in the Ituri region and around Bunia led to the military intervention of EU troops in 2003 in response to the killing of about 60,000, whereas tens of thousands were displaced. Since Belgium has been perceived by the Congolese as a negative power, it could play only a very limited role in helping stabilize the post-Mobutu Congo. Moreover, the violence in the Ituri region or elsewhere in Congo would not have been so dramatic had global players like the US not contributed to it. The US was acting in Congo through Belgium after the independence of Congo.

The US Dimension of Congo's Global Conflict System

The reckless US Africa policy has shown that good governance, human rights and democracy have never been the genuine priority of the US Africa policy, not even in the Carter era. In spite of the Apartheid in South Africa and devastating corruption and kleptocracy in Zaire, the US developed special relationships with Zaire and South Africa. To deny the spread of Communism in this region of Africa, the US tolerated Mobutu as well as Apartheid's racism and discrimination (Keller 2006: 4), and the 25 November 1965 coup of Mobutu was encouraged by the CIA and US government as a necessary military solution to the country's political chaos (Edgerton 2002: 206). The USSR wanted to use Congo as its stepping stone into Africa (Wrong 2006: 66). Therefore, Zaire was the 'anticommunist bulwark in the region' (Smith 2005: 232).

Initially the US had a rather low-key approach towards Congo for various reasons. It favoured the Euro-centric approach of indirectly supporting the European responsibility to influence their newly independent African regions and to put the former colonial power 'out front' when confronted with a crisis situation, while remaining attentive to European preferences in Africa. Though the US was watching the situation attentively, the low-key engagement was partly because, for Washington, the levels of communist influence were negligible and the levels of US economic interests in the region were relatively modest (Schraeder 1994: 52). However, when a mutiny at the prison on 13 January 1961 resulted in Lumumba's liberation for a few hours, the US got nervous. As a consequence, the concerned CIA station cabled Washington that the refusal 'to take drastic steps [assassination] at this time' would eventually lead to defeat of US policy in Zaire. With the principle of 'plausible deniability', the covert efforts authorized by Eisenhower and the CIA effected the murder of Lumumba on 17 January 1961 (Schraeder 1994: 56–9).

During the Katanga crisis after Congo's independence, mindful of Belgium's desire to maintain Tshombe as a buffer to instability in other portions of Zaire, conservative Congress members and the State Department for European Affairs opposed any build-up of UN forces that could possibly lead to military conflict in the province of Katanga. Moreover, Tshombe was 'an ardent friend of America'. As a consequence, since the US government was supporting Tshombe as well as the military presence of Belgium and opposing the UN troop deployment in Congo on 17 July 1960, Lumumba and Kasavubu wrote an ultimatum that if Belgians won't leave, they would ask for USSR intervention.

After the Eisenhower era, a coalition government between moderate nationalists (Kasavubu and Adoula) and radical nationalists (Gizenga – proponent of Lumumba) was established, and a 'moderate' US policy started to emerge. Nevertheless, for the Kennedy administration only 'moderate' Congolese forces were acceptable, and the 'radicals' with ties to USSR had to be excluded from exerting a major influence, like winning the position of prime minister (Schraeder 1994: 60–61; Nest et al. 2006: 18). In order to tighten the US–Congo relations, Kennedy decided to sign bilateral military agreements with Zaire. The decision was important for Washington for three reasons: first, to assume an important responsibility for the continued 'stability' of Congo; second, to strengthen a bilateral military relationship in which the US Defense Department became more intimately involved in maintaining the integrity of any future Congo regime; third, to move away from the State Department's Africanist coalition's ideas of fostering a moderate and democratic civilian regime towards supporting, as the CIA suggested, a military strongman capable of ensuring stability, meaning Mobutu. On 31 May 1963, Kennedy congratulated Mobutu, 'if it had not been for you, the whole thing would have collapsed and the communists would have taken over' (Schraeder 1994: 66–7). When the Lumumbist insurgents controlled Kisangani and about 50 per cent of the Congolese territory, on 4 August 1964 the Johnson administration accepted the suggestion of the CIA and Pentagon to dispatch mercenaries. In cooperation with Congolese security services, the CIA spent more than $30 million in transferring weapons and other *matériel*, and it dispatched 83 CIA officers to CIA posts in Lusaka, Kinshasa, Pretoria and Luanda (Clark 2002: 27).

For Washington through the military intervention of Mobutu, Congo's future looked brighter 'due to careful and skilful exercise of power' by Mobutu. On 25 November 1965, General Mobutu dismissed all politicians, with the support of the CIA, and announced that he would stay in power for a transition period of five years. To quell the 5 July 1967 uprising of the Tshombe loyal rebels, Washington gave Mobutu three C-130 aircraft and accompanying pilots and 150 US soldiers (Schraeder 1994: 76–7).

When the rebellion between July and December 1967 concluded, three important developments came to light: first, in the superpower competition, the US demonstrated its supremacy in Africa; second, Mobutu became *the* man of stability as the US wished; third, the US acted unilaterally where the former colonial power was not able to fulfil its responsibility. According to Washington, chaos (territorial disintegration, regional instability, communist expansion) was the only alternative to Mobutu. An annual $3–4 million was budgeted for training military personnel through the International Military Education and Training (IMET). Kissinger gave directives in 1975 to grant Zaire $60 million in Security Supporting Assistance for 1976. This was a significant rise in military aid from $3.8 million in 1975 to $30.4 million in 1977. The White House decided to equip Zaire with much more sophisticated weaponry (including 150 armoured cars and tanks) to counter possible Soviet threats from Angola.

In February 1977 President Neto of Angola accused Zaire of planning to invade northern Angola and the oil-rich Cabinda. On 8 March 1977 an armed force of about 2,000 men (some of them former Katangan gendarmes or FLNC (Congolese National Liberation Front)) invaded Katanga from Angola to topple Mobutu. While Mobutu accused President Agostino Neto of Angola and his government for conspiring to oust him from power (O'Ballance 2000: 112–14), Angola accused Mobutu of making Zaire FNLA's rear base. Between 1987 and the mid-1990s, annual outlays were between $1 million and $2 million for joint US–Zairian military manoeuvres and selected military construction projects at Kamina Airport for the use of the Pentagon (O'Ballance 2000: 141; Edgerton 2002: 217). This was officially meant as a project to stabilize Congo and the region as a whole. When Zaire had militarily been supported by the US, it could maintain the unity of the state through militarization. At Kamina air base in southern Shaba for several years the US had carried out secret joint military exercises with Zairian troops. This base was used by the CIA to provide UNITA against the Communist Angolan government with weapons to arm Angolan rebel FNLA forces under Holden Roberto opposing the country's communist regime.

Whereas for the Carter administration initially human rights in Zaire was a sensitive issue, for the State Department Mobutu was a 'necessary evil'. Washington could not do anything to make Mobutu meet the human rights demands. For the Carter administration prior to and during the Katanga I crisis of 7 March 1977 (less than seven weeks after Carter's inauguration), human rights and basic human needs seemed to be more important than containment of the Soviet Union and communism in Zaire. About 1,500 FLNC entered Congo's Shaba province from neighbouring Angola and threatened to capture the mining centre of Kolwezi in Shaba.

Carter's human rights policy started to change in May 1978 when the FLNC (Front for the National Liberation of Congo) invaded Zaire from Zambia and captured Kolwezi, the economic heart of Katanga province, and took 2,500 Europeans and 88 US citizens hostage. When in May 1978 the Angola-based Shaba (Katanga) insurgents invaded Congo, resulting in Shaba Crisis II, the US moved with alacrity to support the French and Belgian militarily in the aid of Mobutu (Legum 1979: 636). Further, the Carter administration accused Cuba and the USSR of attacking Zaire from the Angolan territory, where more than 20,000 Cuban soldiers were stationed.

Reagan assumed power with a highly ideological world view to offer greater support to anti-communist, strategic allies, such as Zaire. The Reagan doctrine was an ideologically based programme for arming insurgencies intent on overthrowing self-proclaimed Marxist regimes in the Third World. Mobutu was hailed as a distinguished 'peace-broker' when he presided over a diplomatic effort that resulted in a ceasefire between the MPLA and UNITA on 22 June 1989, and recognized as an 'old and strategically important friend' (Schraeder 1994: 88–9). George H. Bush's speech on 29 June 1989, during Mobutu's visit to the White House, was remarkable:

> Zaire is among America's oldest friends and its President – President Mobutu – one of our most valued friends on the entire continent of Africa … One of Africa's most experienced statesmen, President Mobutu has worked with six US presidents. And together, they and we have sought to bring to Zaire, and to all of Africa, real economic and social progress, and to pursue Africa's true independence, security, stability as the basis for that development. (Schraeder 1994: 51)

The African Affairs Bureau in the State Department stressed the political benefits accruing to Washington as a result of US–Zairian special relationship. Washington praised the pro-Western, 'moderate' foreign policy and 'voice of reason' of Mobutu. Until the student massacre at Lubumbashi University on 11 May 1990, the US State Department was strongly supporting US–Mobutu relations, in spite of the corruption and violation of human rights, on the grounds that aid cut-off could only serve to undercut the process of reform. 'If we push Mobutu too hard, we will only have ourselves to blame for political chaos and instability that surely will ensue' (Schraeder 1994: 105).

From 1985 Zaire again became an important regional component for the US for intervention in favour of anti-communist insurgents in Angola. Since Congo for the US without Mobutu could entail a Zaire engulfed by chaos, *incrementalism* instead of rapid change was preferred (Schraeder 1994). The changing political situation in Angola led to the changing of US policy towards Congo. When Angola's MPLA expressed its willingness to conduct internationally monitored multi-party elections in 1992, for the US there was no longer an urgent necessity for a regional ally that served US paramilitary interests against the anti-communist insurgents.

However, even after the Cold War, the US's negative influence, at least indirectly, was not negligible. For example, Rwanda, acting through its proxy (RCD), started to reduce its destabilization in the Kivus in the transition period only under pressure from the US, the UK and South Africa, who had invested politically and economically in the peace process. The US played a key role in facilitating and sponsoring the Pact on Security, Stability and Development in the

Great Lakes region signed by the heads of state of Congo, Rwanda, Uganda and Burundi (ICG 2007a: 4). The US facilitated the Nairobi Communiqué (9 November 2007) and Goma Accords (23 January 2008); played a leading role in the Tripartite Plus mechanism (Burundi, DR Congo, Rwanda and Uganda); and it supported UN efforts to create a Joint Verification Mechanism to monitor the border between the DRC and Rwanda. According to State Department information, the US has provided DR Congo with more than $700 million in aid in 2008 through both bilateral and multilateral programmes.

Does this show that the US has corrected its negative role during the Mobutu era and is now going to play a more stabilizing role in Congo? We have to wait and see what the Obama administration is going to do. At any rate, France has been watching carefully what the Anglophone states have been doing in Congo and in the region in general. Therefore, France's role in Congo and in the region is important for Congo and the whole region.

The French Dimension of Congo's Global Conflict System

On 22 June 1994 the UN Security Council approved a resolution that allowed France to intervene in Rwanda, but this was opposed by FPR (which would announce its victory in the civil war on 19 July 1994) on the grounds that France was a supporter of the Habyarimana regime, which hindered the implementation of the Arusha Accords as well as undermined the transitional government created in April 1992 (O'Ballance 2000: 155–6). France continued through its *Operation Turquoise* to protect those responsible for genocide and to arm the Interahamwe even when they were operating from the refugee camps in Congo. However Mobutu and France pretended to be protectors of the Hutu refugees (Mamdani 2002: 254–5). Former French Prime Minister Edouard Balladur acknowledged Operation Turquoise was 'purely and simply to provide cover to the genocidal forces'. French military support of $8 million worth of heavy machine-gun and mortar ammunition and further tons of weapons and equipment were flown to Kigali or Zaire for the Hutu extremists. Some of these weapons arrived on 17 July 1994 after the UN arms embargo (UN Resolution 918 of 17 May 1994), which banned the supply of any arms and related *matériel* of all types, including weapons and ammunition, military vehicles and equipment, paramilitary police equipment and spare parts. In 1997 some newspapers in France claimed that Francois Mitterand said that genocide in countries like Rwanda is not very important if it is about curbing American influence and interests in the region; defeating English-speaking Tutsi rebels, long exiled in Uganda, whereas the Hutus were seen as a vital bulwark against the US influence and the English language in South Central Africa (*Africa Research Bulletin* 1998b: 12979).

Call for change of French–African politics began gradually by the mid-1990s because France acknowledged that, so far, French–African politics had failed mainly because of the crisis in the Great Lakes region (because of and) despite the French alliance with Habyarimana of Rwanda (with whom France concluded a military cooperation agreement in 1975) and Mobutu. France had supported and equipped a corrupt, undemocratic government of Rwanda; had sent troops to defend it from rebellion; and had established safe havens for Hutus – including perpetrators of genocide to protect friends of France from public scrutiny and accountability. France supported Mobutu militarily and diplomatically, insisting that his government remained Zaire's only legitimate authority. Moreover, France became increasingly isolated within the international community, abandoned by Washington and European allies who began supporting Kabila. Moreover, African countries who were most strongly implicated in the crisis became increasingly hostile to Mobutu and hence to the French stance (Utley 2002: 132–3).

France has had huge economic and strategic interests in DR Congo. Though French oil company *Elf* ties France with President Dos Santos, the growing Angolan role in the region has been worrying France: Angola could replace France as regional gendarme. One strategy of France to prevent Angola from being regional gendarme was to assist the regime in Congo-Brazzaville to be strong and self-sufficient instead of depending on Angola for its security against the rebels such as the MCDDI (Movement for Democracy and Integral Development) led by Kolelas. Moreover, since Angola was accusing Gabon for supporting UNITA, France was worried that Angola could attempt to overthrow Bongo's regime, which would undermine the French gendarme role in the region. However, France has been trying to avoid any confrontation with Angola. The oil abundance in Angola created competition between France, the US and the UK in the 1990s. Now China has joined the club, with a huge impact on the competition. Global players such as the US, Canada, South Africa and Belgium have entered into a competition with China seeking to control strategic reserves of minerals such as copper and cobalt. Corruption and impunity are disregarded in this global competition (ICG 2007a: 3). Such competition among the global actors in Congo not only exacerbates conflicts, it has also been undermining the role of the UN in Congo. And the UN, in spite of its massive presence in the DRC, is unable to put pressure on any of the significant global actors.

The UN Dimension of Congo's Global Conflict System

The global conflict system consists of the political hegemony and economic interests of the global players as well as regional actors in the conflict areas of the world, with negative implications on the UN itself. Congo had already had a bad experience with the UN right after its independence. It was the weakness of the UN that contributed to the murder of Lumumba, when he turned to the USSR. The UN was not willing and able to stop the Belgian support for the Katagan secessionists. The UN was there but with insufficient mandate and lack of genuine will to help end the internal rivalries between Kasavubu and Lumumba, and the secession of Katanga.

When the Katangan rebels attacked Zaire in 1977–78 from the Angolan territory, Belgians and French intervened militarily in favour of the Congolese government.

To replace those 'non-Africans' or 'neo-colonizers', it was suggested that African states contribute to stabilize the situation with the mandate of the UN. Morocco, Egypt, Senegal, Ivory Coast, Central African Empire, Togo, Gabon and Sudan supported the Zairian government and contributed about 2,700 troops to replace the Belgians and French. Carter suggested that pan-African peacekeepers supported by Western powers would be preferable to the military presence of the Western states. France supported it because it would be commanded by Francophone Africans, whereas the UK complained because of under-representation of Anglophone Africans (Legum 1979: 636).

During the discussion of how to deploy forces in Congo, a division emerged not only between East and West, but also within Africa. The question was under whose influence should the African peacekeeping force be: under East or West? Under the anti-Western and anti-Chinese African states (Angola, Mozambique, Ethiopia, Guinea-Bissau, Cape Verde, Sao Tome and Principe and Congo, Madagascar, Equatorial Guinea and Benin) or under the pro-Western states (Egypt, Sudan, Somalia, Ivory Coast, Senegal, Togo, Malawi, Gabon, Morocco, Tunisia, Mauritius, the Central African Empire, Swaziland, Gambia, Zaire, Chad, and the Comoros) (Legum 1979: 638–40).

In the post-Mobutu era after many peace agreements were signed, a UN mission (MONUC) took the task to help rebuild Congo. The Lusaka Ceasefire Agreement was signed on 10 July 1999 by the DRC, Angola, Namibia, Rwanda, Uganda, Zimbabwe and, later, by the *Mouvement pour la Liberation du Congo* and the *Rassemblement Congolais pour la Democratie*. Whereas on 30 July 2002 in Pretoria DRC and Rwanda signed their peace agreements, on 6 September 2002 in Luanda, DRC and Uganda signed their agreements. Inter-Congolese Dialogue was signed as an Agreement on Transition on 17 December 2002 in Pretoria. On 6 March 2003 the Agreement on Constitution of the Transition was signed in Pretoria, and on 4 April 2003 the Transitional Constitution of the Democratic Republic of Congo came into force. On 30 June 2003 the Government of National Unity and Transition was announced. The Memorandum of 29 June 2003 on Security and the Army paved the way for the establishment of an Integrated Policy Unit (IPU) out of the Congolese National Police. These developments were decisive for the UN mission.

Since its establishment in 1999 the success of MONUC has been limited. Fighting continued through all those years despite the 17,500-strong UN peacekeeping force, costing $1.15 billion a year. The UN does not have the power to control or enforce policies on those states or multinationals that undermine Congo by exploiting its natural resources. These multinationals have been contributing to corruption and state weakness. A variety of (unsuccessful) initiatives, such as the Extractive Industries Transparency Initiative, have been created and were signed in November 2005 to urge those economic actors to publish their revenues from companies in the natural resources sector, but no concrete steps have been undertaken so far due to the lack of a coherent policy by the global actors – mainly the permanent members of the UN Security Council.

The acts of barbarism taking place in Eastern Congo are nurtured greatly by global economic structure. Due to demand for coltan (columbite-tantalite) needed for high-tech consumer goods such as cell phones, laptops and PlayStations, Motorola has contributed to and profited from the Congolese conflict (Todd 2006: 3, 12; Kern 2007: 94; Montague 2002: 105). Some 80 per cent of the so far known coltan (important for military and non-military technology) is in Congo. The demand for gold, diamonds, copper, zinc, uranium, cobalt, cadmium and timber has contributed to the brutal and reckless involvements in Congo (Kern 2007: 94). Rwanda and Uganda not only financed their wars by looting these minerals from Congo, but they also became leading shippers to address the increasing demands for coltan for goods such as PlayStations (Kern 2007: 95–6; Todd 2006: 10–12; Montague 2002: 103–4). The Pentagon classified it as a strategic mineral (Todd 2006: 4). The abundance of diamonds led Rwanda to create diamond marketing agencies (Lemarchand 2000: 345).

If the interest of the major global players that constitute the P5 of the UN Security Council were international peace and security, how can we understand that they themselves contribute to the perpetuation of the Congolese conflict? China began to show its interest in the DR Congo immediately after Laurent Kabila took power. In mid-December 1997 Kabila visited China and announced that the DR Congo government had chosen China as a model for Congo because the Chinese economy is growing rapidly, and it was transforming itself through its own efforts. In 2007 China agreed to lend the Democratic Republic of Congo $5 billion to modernize its infrastructure and mining sector. Accordingly, $3 billion were earmarked for strategic highway and railroad projects, whereas $2 billion was earmarked to revive Congo's mining sector of copper, cobalt, gold, nickel, uranium and diamonds (Reuters 2007). But at the same time China is one of the leading arms exporters to Congo (Control Arms Campaign 2006).

Various business enterprises violate the OECD guidelines for multinational corporations. Of 85 business enterprises that have violated the principles according to a report in 2002, 21 companies are Belgian, 12 South African, 10 British, 8 American, 5 Canadian and 4 each for Germany and Zimbabwe. Various individuals with close ties to Presidents Joseph Kabila, Paul Kagame, Robert Mugabe and Yoweri Museveni of DRC, Rwanda, Zimbabwe and Uganda; Victor Bout (the notorious former Soviet military major who retired from the service in 1993 and became an international arms dealer); and a number of Antwerp-based diamond dealers were reportedly connected to illegal business in Congo (Nzongola-Ntalaja 2004: 17)

The US gave Rwanda $75 million in military aid right after Kagame took power. Military training including counterinsurgency was given by the US military for Rwanda after the genocide. Before the Kabila government was installed and stabilized, various US and other corporations began negotiating with Kabila for minerals in various parts of the DR Congo (Kern 2007: 99; Montague 2002: 106). On the one hand the IMF and World Bank urged Rwanda and Uganda to pursue peace relentlessly, but on the other hand they praised these two states – enriching themselves with the plunder from Congo – for their GDP, such as containing inflation to single-digit levels (Kern 2007: 100).

The Rwandan government profited from its military aid from the US and the UK in order to finance its military intervention in the Eastern Congo or to support its rebel allies, Congolese Rally for Democracy (RCD-Goma) (Kern 2007: 98). Both the UK and the US wanted to undermine the role of France in the region by taking advantage of the bad reputation France had in the region, and to present themselves as forerunners of the values of the 'international community' (Blair 1999), or they fight for 'global values' (Blair 2007). Not until June 2000 did the UN Security Council criticize Rwanda and Uganda's invasion of Congo. Even worse, Washington was providing them with counterinsurgency training (Adebajo 2004: 207).

As Adebajo suggests, the UN's peacekeeping successes and failures are contingent on the domestic, regional and external dynamics of conflict situations. According to Adebajo, there are six factors that often contribute to success in UN peacekeeping missions in Africa:

- 'the willingness of internal parties to disarm and accept electoral results;
- the development of effective strategy to deal with potential "spoilers";
- the absence of conflict-fuelling economic resources in war zones;
- the cooperation of regional players in peace processes;
- the cessation of military and financial support to local clients by external actors and their provision of financial and diplomatic support to peace process;
- and the leadership of peacekeeping missions by capable UN envoys' (Adebajo 2004: 196).

The Security Council has not taken tough decisions and measures against that because none of the UN P5 are immediately or strategically affected by the conflict.

Secret shipments of arms from countries such as South Africa, Bulgaria, France and Belgium to DR Congo were pouring to Congo and into the region (Lemarchand 2000: 345). Belgium was accused of becoming 'a hub of international trafficking in arms to Burundi'. Romania, Israel, Belgium, the USA, etc., were transferring arms to Rwanda and through Rwanda to Laurent D. Kabila in the 1990s before and after the genocide in Rwanda and during the civil and interstate war in the DR Congo.

Some UN officers were involved in gold and other minerals smuggling in 2007. Since 2005 there have been a lot of reports that some peacekeeping troops of MONUC were abusing children sexually (Citizens for Global Solutions 2005). Civilians have been complaining that MONUC is not protecting them. For example, in Kiwanja when the November 2008 massacre took place, a contingent of about 100 UN peacekeepers was just a few minutes' away (*Africa Research Bulletin*

2009: 17790). Such dire civilian situations pushed the EU to be active militarily in Congo. But is the EU better than the UN?

The EU Dimension of Congo's Global Conflict System

The EU started to be active in Congo, mainly since the mid-1990s, to address the causes of conflicts, prevent conflicts and for post-conflict reconstruction. The European Security Strategy of 2003 underlined this. As a consequence, the European Strategy for Africa of 2005 and the Joint EU-AU Strategic Partnership of 2007 outlined the policies of the EU towards Africa, and also of the EU cooperation with sub-regional organizations of the African Union. The EU allocated hundreds of millions of euros for the period 2008–10 in order to strengthen African capabilities for the prevention, management and resolution of conflicts, security sector reform and disarmament, demobilization and reintegration, assistance to the counterterrorism centre and the African Centre for Study and Research on Terrorism (CAERT) based in Algiers, and stemming the illicit flow of small arms and light weapons; strengthening the AU's African Standby Force; strengthening fragile states; conducting in Africa EU's own crisis management missions, including potential deployment of EU battle groups. As a consequence, as the military interventions of the EU in the DRC have shown, the EU is not only financially supporting the African peace and security initiatives, but it is also militarily present in Africa.

In July 2003 the RCD-ML was fighting an alliance of small Mai Mai factions in the east (which was usually allied with the Congolese government against the RCD-Goma), claiming RCD-Goma was arming them. The existence of varieties of ethnic groups favoured the formation of such alliances in the region: Lendu, Hema (Hema-Gegere from the north and Hema-Nyoro from the south), Alur, Bira, Nyari, Mambisa, Ndo-Okebo, Lugbwara, Kakwa, Logo, Lese and Ngiti (also known as Lendu-Bindi).

As a reaction to the ongoing violence, the European Union decided to conduct a European Union military operation in the Democratic Republic of Congo, named *Artemis* in accordance with the mandate set out in UNSCR 1484 (2003). Based on Article 15 of the Treaty of the European Union the Council adopted a Common Position 2003/319/CFSP on 8 May 2003 to support the implementation of the peace agreements through rapid implementation of the process of disarmament, demobilization, repatriation, reintegration and resettlement (DDRRR). According to the plan of action, the forces were deployed to operate in accordance with the objectives set out in the 'Framework for EU action in response to the crisis in Bunia' adopted by the Council on 5 June 2003 (Council Joint Action 2003/423/CFSP) and launched on 12 June 2003. It lasted from 12 June 2003 to 1 September 2003.

The objective was to contribute to the stabilization of the security conditions and the improvement of the humanitarian situation in Bunia, to ensure the protection of the airport, the internally displaced persons in the camps in Bunia, and to contribute to protect the civilian population, UN personnel and the humanitarian agencies in the area, and to secure the airport of the north-eastern provincial capital Bunia and to protect the local population from further massacres exacerbated by the abundance of coltan in the region (Todd 2006: 6).

The Joint Declaration on UN-EU Cooperation in Crisis Management of 29 September 2003 underpinned the mission's international support. Upon request of 20 October 2003 by the government of the DRC addressed to the High Representative for the CFSP for European Union assistance in setting up the IPU, on 15 December 2003 the Political and Security Committee of the EU (PSC) agreed to support the establishment of the IPU. The EU showed its readiness to rehabilitate and refurbish a training centre and provide with basic equipment; train the IPU, composed of 1,008 police officers from the whole country, with a target of training 38,000 police; and monitor and follow up the implementation of the IPU mandate following the initial training (Loeser 2007). On 17 May 2004, the Council adopted Joint Action 2004/494/CFSP to support the process of the consolidation of internal security in the DRC. Its tasks were the protection of the state institutions and transitional government authorities, maintenance of law and order.

On 9 December 2004, the Council adopted a decision setting up a European Union Police Mission – 'EUPOL-Kinshasa' – which monitored, mentored and advised on the setting-up and the initial running of an Integrated Police Unit (IPU) in Kinshasa, to ensuring the protection of the state institutions and reinforce the internal security apparatus. The activities of the mission have been to: support the Integrated Police Unit (IPU), train police at the Kasangulu centre, reform and reorganize the Congolese National Police (CNP), and establish an element of coordination of the Congolese forces maintaining order during the election period.

EUPOL Kinshasa, in close liaison with MONUC, contributed *inter alia* to (i) the identification and training of IPU members, (ii) the inspection and verification of equipment and assistance in managing it, (iii) the definition of organizational and functional standards, and (iv) the drafting of relevant rules and regulations for the IPU' (Löser 2007). Three essential missions of EUPOL-Kinshasa were: to provide escorts to certain actors of the transition as well as personalities during official missions, as well as guarding the institutions; to contribute to the security of the elections by organizing prevention and intervention patrols; to constitute a tactical reserve force of 350 men for the purposes of maintaining order (Consilium 2006).

Starting from 1 July 2007 EUPOL RD Congo succeeded EUPOL-Kinshasa, which was deployed in Kinshasa between February 2005 and June 2007 to secure the parliamentary and presidential elections. The mandate of the mission is extended until 30 June 2009. The objective of the EUPOL RD Congo has been to support the Congolese authorities SSR in the domain of police and its interaction with justice. EUPOL RD Congo consisted of 30 international agents from police, justice, security experts, and human rights experts.

Among EU member states, France has trained and equipped 2,000 rapid intervention police at a cost of €2 million and provided a senior officer to assist the rapid intervention police commandant for two years (ICG, Africa Report N°104, 13 February 2006). The UK earmarked up to £8 million, in addition to £4 million that has already been provided, for training and equipping the police to provide election security and supplied £253,000 in radio equipment. In 2007–08, British humanitarian allocation totalled £39,435,956 (DFID 2008).

In June 2005 the EU created a civilian mission for Congo called EUSEC RDC whose establishment was based on UNSCR 1592 (2005) for the security sector reform (SSR) in Congo (Löser 2007). The assistance programme of SSR encompasses the development of norms of good practice in the security sector; the control, collection and destruction of small arms; enhancing civilian control over the military; and community-based policing and justice reform; poverty reduction; conflict prevention; post-conflict reconstruction; promotion of human rights and democratization; facilitate the integration of the Congolese army; support good governance in the field of security; provide logistical support to the modernization of payment of DRC armed forces and its financial administration (Löser 2007; Helly 2006). The €7 million project was intended to run for 12 months, and EUSEC would provide 40–45 military advisers to check the payment flow as well as information technology and equipment, as well as support for the administration reform.

Former colonial power Belgium was highly ambitious to play a significant role in the DRC's army reform. Within the framework of EUSEC SSR Belgium spent about €30 million annually on security sector reform out of its total Congo funding of about €125 million (ICG 2006b). It led a series of workshops that developed practical guidelines for army integration; trained the First Integrated Brigade in Kisangani between January and June 2004; trained 285 Congolese instructors in Belgium and another 250 in Kinshasa; engaged in a joint training programme with the South Africans for the Third Integrated Brigade in Kamina. Belgium agreed to equip three brigades fully and three partially; it offered €500,000 to equip additional brigades; it provided 30 trainers to work together with three French trainers; it provided military engineering training to promote reconstruction of infrastructure (ICG 2006b).

The EU Commission provided €20 million through the World Bank Multi-Country Demobilization and Reintegration Program (MDRP). The UK provided some funding to the South African-led integration process, registration support and census process. Moreover, to the World Bank's MDRP programme the UK pledged $25 million over five years; it provided short training courses in the UK for some Congolese officers, and pledged an additional $5 million of non-combat aid for army integration, conditional upon implementation of the EUSEC plan. Whereas the Netherlands provided €5 million in funding to help the South Africans refurbish the integration centres in North Kivu, France provided support for officer training (ICG 2006b).

The United Nations Security Council Resolution 1671 (2006) of 25 April 2006 authorized the EU, acting under Chapter VII of the Charter of the United Nations, to deploy forces (EUFOR RD Congo) in support of MONUC during the election process, with initial financial reference amounting to €16,700,000 for the common costs of the EU military operation for a four-month period. EUFOR RD Congo consisted of 2,400 soldiers from 21 EU member states and Turkey. The highest contributions were from France (1,090) and Germany (780). The objective was to support the 17,000-strong UN peacekeepers while preparing and securing the parliamentary and presidential elections of 2006 and to secure the airport of Kinshasa.

Many European citizens expected an explanation from the EU about why it has intervened in the DRC because they do not see an immediate threat to the Europeans coming from Congo. An important motivation of the EU for its intervention has been its Security Strategy Paper of 2003. But increasingly the EU migration policy has been playing a role: to stop the ‘wave’ of African refugees into Europe. Therefore, the EU’s intervention in the Congo cannot be reduced only to the emergency situation in Congo. Moreover, the EU attempts to show that it is capable of conducting military operations independently outside of the NATO framework as a global hard (military) power. But whether such calculations would help achieve its objective of making genuine peace in Congo is still unclear.

The capture of General Nkunda on 22 January 2009 is a decisive development in the chance for peace in DR Congo. Since 1994 the war, which started to include the neighbouring states, claimed the lives of 5.4 million people. But whether this is the beginning of the end of the suffering of the Congolese will depend on whether intrastate, regional and global aspects are systemically addressed.

This chapter has shown that there are contradicting interests on the intrastate, regional and global levels. Therefore, peace in Congo does not depend on how well equipped the international missions of the EU or UN are, but instead on how the societal, economic and political aspects of Congo’s conflicts are equally addressed regionally as well as globally. Moreover, the interaction of the three conflict systems has resulted in the integration of interregional conflicts. Through Ugandan rebels operating in northern Uganda, southern and south-western Sudan and north-eastern DR Congo, the conflict systems in Sudan and Congo in particular and between the Great Lakes and the Horn of Africa in general are practically interrelated. Therefore, conflicts in both regions have to be addressed on intrastate, regional, interregional and global levels. The Horn region will be discussed in the following chapters.

Chapter 5
The Intrastate Conflict System in Somalia

In 2008, 32,000 Somalis made the perilous crossing to Yemen, and 365 have gone missing. Within the first two months of 2009 more than 10,000 Somalis made their journey to Europe. About 1.3 million Somalis are displaced. In 2008 some 100,000 Somalis sought refuge in Kenya, Ethiopia, Djibouti and Yemen. The number of Somali refugees in asylum countries now stands at 438,000 (UNHCR 2009b). The number of people in need of assistance has increased to nearly 3.2 million people, or 43 per cent of the Somali population. This is a 77 per cent increase since January 2008.

This chapter deals with the factors and actors of the internal conflict system that contributed to Somalia's collapse, and the factors and actors that protract this states of collapse. The first section deals with the factors and actors on the societal level. The second section deals with how different civil wars destroyed the economy of Somalia. In the third section we shall see militarization of Somali society, the stumbling blocks to democratization, and the unfavourable conditions for security. The combination of all of these constitutes the intrastate conflict system in Somalia.

Society

Clanism

Somalia's ethnic groups consist of clan families and major sub-clans: Darod 20 per cent (consisting of Dolbahante, Majertain, Marehan, Ogadeni and Warsangeli); Hawiye 25 per cent (consisting of Abgal, Ajuran, Degodia, Habr Gedir, Hawadle and Murosade); Isaaq 22 per cent (consisting of Eidagalla, Habr Awal, Habr Toljaala and Habr Yunis); Rahanwein 17 per cent; Dir 7 per cent (consisting of Gadabursi and Issa); Digil 3 per cent; and other minorities 6 per cent (CFR 2002). Darod are predominantly in the north-eastern, northern and south-western part; the Hawiye mainly live in and around the capital, Mogadishu; the Isaaq in Somaliland; the Dir in the northwest; and the Digil and Rahanwayn in the river areas of the south (*New York Times* 1992). In this section we shall see the components of the societal dimension that contributed to the state collapse as well as to the unending suffering of the Somalis.

The role of this factor had already been acknowledged before independence when the SYL (Somali Youth League) was mobilizing the Somalis across all clan groups. However, ultimately neither the SYL nor Siad Barre was successful in overcoming this clan factor. Moreover, in this clan-based societal heterogeneity, minorities such as Rahanweyn, Bantu, Ajuran, Jarso, Madhiban, Ashraf, etc., have been marginalized. Especially when the central government became weak and could not effectively enforce its power as a legitimate authority, these minorities became targets of majority clans.

Though ethnicity does not necessarily lead to violence, it can play a decisive role when other co-determining factors are given, such as economic and political marginalization. When Siad Barre failed in 1991, the clan factor combined with the colonial legacy contributed to the unilateral declaration of independence by the Isaaq of Somaliland in 1991 from the rest of Somalia. The British colonized Somaliland, inhabited by the Isaaq clan, whereas the Italians colonized the rest of Somalia, inhabited by the Hawiye, Darod, Dir and other minor clans. In spite of this clan-based heterogeneity, Somalis have one religion and more or less one culture; there are nomads as well as agriculturalists. Therefore, the main challenge for the political elites was to overcome clanism.

The Somali National Movement (SNM) created in 1981 was mainly an Isaaq-clan organization though not exclusively. Even if it originally included Hawiye and Harti clan members as well, it was the political expression of the Issaq clan family in Somaliland (Matthies 2005: 151). However, Somaliland is inhabited by heterogenous sub-clan groupings. Somaliland is populated by Gedabuursi and Isse in the Awdal region, Habar Awal (Isaaq) in the south-west region, Habar Jecllo, Habar Yonis, Isse Muse (all Isaaq) and Dhulbahante (Darod) in the Togdheer region. The Sanaag region is inhabited by Isaaq clans of Habar Yonis, Habar Jecllo and Isse Muse together with Darood clans of Dhulbahante and Warsangeli. Dhulbahante (Darood) live in the Sool region. However, urban areas like Hargeisa, Berbera and Borama are ethnically mixed. The non-Isaaq tribes of Somaliland, such as the Gedabuursi, Isse, Dhulbahante and Warsangeli, did not participate in the armed liberation struggle for independence of Somaliland; and most of these tribes were armed and mobilized by Siad Barre as local militias to wage wars against the Isaaqs and their movement (SNM). Siad Barre took advantage of this heterogeneity of the ethnic members of the SNM as well as of the military/political organizations and managed to divide Hawiye and Harti, which joined the SNM fighters (Brons 2001: 207).

The already fragile political institutions of Somalia were destroyed after the coup of 15 October 1969, and six days later the military controlled the key points in the capital resulting in the suspension of the Constitution, abolishment of the Supreme Court, and closure of the National Assembly, in order to eliminate corruption and tribal nepotism. On 1 November 1969 the membership of the Supreme Revolutionary Council was announced, which took charge of the administration through a 14-member committee of 'Secretaries' until it was disbanded in March 1982 (Lewis 2002: 207, 249). As its new national objectives the military government attempted to implement 'scientific socialism', which in this context consisted of 'togetherness' or unity, self-reliance, self-help, wealth-sharing (socialism) based on wisdom (scientific dimension). Siad Barre equated tribalism with class in a society struggling to liberate itself from lineage affiliations and allegiances (Federal Research Division 1993: 87–8). This new ideology was furthered by state control through thought control or revolutionary inculcation and indoctrination; management of the economy (state-run economy); socialism and development through a mass literacy campaign; attempt to blend socialism and Islam; attempt to overcome the persistent force of tribalism. In this process socialism became a religion and scientific Syadism (Lewis 2002: 209–23). One of the key objectives of this new political 'religion' was to overcome the dangers of clan-centred identities and oppositions that could threaten the collective national identity.

However, Siad Barre pursued paradoxical strategies to weaken any opposition. He favoured his own clan (Marehan) and fanned clan animosities, but at the same he outlawed 'tribalism' and banned clan gatherings during engagements, weddings, etc.; he destroyed independent institutions, civil organizations and political parties that had clan affiliations and challenged him (Human Rights Watch 1993: 4). Besides the politics of ethnic allegiances, the Siad Barre regime was famous for stirring up competition among sub-clans and clans, manipulating them in order to create inter- and intra-clan conflicts. Since the Ogaden war of 1977–78, Siad Barre distributed weapons freely to his clan-based allies to fight his foes. This resulted ultimately in the spread of modern weapons originated from Russia, the West and Arab world throughout the country, facilitating mutual killing and displacement of the population. Second-hand tanks were as cheap as cars (Lewis 2002: 262). The Galgala, for example, were displaced from their houses in Mogadishu and Gedi-Hir in Middle Shabelle region in early 1991 by the Abgal clan with whom they share a territory. The Siad Barre regime armed the Galgala against the Abgal (Lewis 2002: 264).

When the regime collapsed, the Galgala faced brutal Abgal reprisals that led to the displacement of thousands of Galgala from their lands. This pattern was repeated against the Bantu, who were forcibly displaced from their lands in the Gosha area, by Majerten and Habargedir militias who fought over the control of resources and territories between 1991 and 1993. It was reported by the displaced that Majerten and Habargedir militas ensured extermination of the Bantu, appropriation of the best lands, and obstruction of relief food to force the Bantu farmers to abandon their lands. As a last resort many Bantu sought protection in Kismayo and other neighbouring villages, while others migrated to Mogadishu and other areas in Puntland and Somaliland. Since the strategy of Siad Barre was to divide and rule, he armed the Galgala minority against the Abgal. Later this led to retaliation against the latter when the regime fell. The Bantu minorities, who ultimately suffered and were despised as slaves, were displaced from their lands in the Gosha area by Majerten and Habargedir militias who fought over control of resources between 1991 and 1993 (Asylumlaw 2000).

Though through unification right after independence both Somalia and Somaliland attempted to form the Republic of Somalia, the loyalty of the individual to his clan and kin remained (Lewis 2002: 166). Somali political elites knew this very well. In order to address this problem the SYL Party strongly pursued political integration and economic modernization. This approach of the SYL to the clan factor is comparable to the strategies of various African political elites at that time. In different parts of Africa there was an attempt to rally the whole population of the state behind a single nationalism and 'idea of the state' (Clapham 1996: 45). It was about instigating nationalism: in Zaire it was the *authenticity* ideology; in Ethiopia when Mengistu Haile Mariam took power he attempted to fight ethnic nationalisms, which were designated as 'narrow nationalisms', such as Eritrean, Tigrayan or Oromo independence movements. As a consequence, Ethiopia as well as Somalia tried to implement political centralism and socialism.

In Somalia there was a near prohibition of the mention of clans by declaring all Somalis as 'comrades'. SYL as well as Siad Barre attempted to fight this parochialism, which represented narrow constituencies rarely reaching across clan lines (Lewis 2002: 121–31). For example, during the 1964 elections there were 18 parties, and during the 1969 elections there were 60 parties. Such clanization of parties and politics resulted in corruption, personal aggrandizement, making the legislature a sordid marketplace of wealth accumulation (Talentino 2005: 99–100). When Siad Barre staged a coup in 1969, fighting this clanism was one of his primary objectives. Even the road-building projects advancing rapidly were meant not only to link various parts of the country but also to form a national unity (Prunier 1995: 4).

When the Siad Barre regime lost the war against Ethiopia, the significance of clan politics revived. When the regime lost its internal legitimacy, Siad Barre resorted to clan clientelism, which was contrary to the Greater Somali project. This led to the military coup attempt by a group of Majerteen officers to overthrow the now weakened Siad Barre dictatorship on 9 April 1978. This was a decisive start of clan-based distrust or trust. Patrimonialism and the failed Greater Somali project resulted in rebellion, clanism and state disintegration. When on 26 October 1978 the officers were executed after the failed coup, the regime focused mainly on the Majerteen civilians as well as officers. Out of the 17 executed officers on that day, 12 were from the Majerteen clan (Mohamed 2001). Through the patronage network most of the political positions were occupied by Barre's clans. In the course of the political era of the Siad Barre regime the three Darod sub-clans gradually dominated: Marehan (Barre's father clan), Ogaden (Barre's mother clan) and Dulbahante (sons'-in-law clan). The lost irredentist war against Ethiopia was the beginning of the end of the Siad Barre regime. Siad Barre, whose support was increasingly limited to the Darod clans, built his power by manipulating clans and playing their interests off against each other (Talentino 2005: 100). Further, the clan constituencies split into sub-clan, sub-sub-clan, etc., with clan or sub-clan members sometimes fighting against each other.

Even before the era of Siad Barre the political landscape after independence was not favourable for an integrative and inclusive state-building. The central political power formation had been inherited from the colonial period; politics concentrated in southern Somalia (Mogadishu), and the north was marginalized (Brons 2001:

162). Although the party of the post-independence era (Somali Youth League – SYL) was trying to be integrative by including the clans, such as Hawiye, Dir, Isaaq and Rewin, it was dominated by the Darod clan (Brons 2001: 162). However, it did not take long until one-party rule (SYL) was transformed into one-man rule and state coercion, repeating the politics prior to Siad Barre's era.

Looting and selling of the spoils of war in the markets of Mogadishu by Barre troops was openly practised. As a result of this state weakness, clan-based armed opposition started to emerge. For the Isaaq clan in British Somaliland the SNM (Somali National Movement) was founded in April 1981 in London to fight the south-biased regime (Brons 2001:185). Towards the end of the 1980s the Somali government began to sponsor some militia forces in order to crush the SNM (Shinn 2003: 3). One of the causes of the creation and consolidation of the SNM was Siad Barre's strategic concentration of political control and power in the capital city, ignoring the rest of the country. The regime began to persecute not only the SNM fighters, but also the Isaaq civilians (Human Rights Watch 1993: 3). The SNM disapproved of Siad Barre's Greater Somalia policy, the scientific socialism, secularization, and the superpower allegiance. Ethiopia was supporting the SNM. Towards the end of May 1988, the SNM attacked and gained Burao completely, and Hargeisa partly (Mohamed 2001). Instead of retreating after the Ethiopian support was cancelled (owing to the agreements between the regimes in Ethiopia and Somalia), the SNM intensified its military action (Cornwell 2004: 3). This was the most serious blow that caused the fall of the Somali regime (Mohamed 2001). Moreover, additional armed groups started to emerge against the Barre regime primarily on clan basis. In 1989 SPM (Somali Patriotic Movement) and USC (United Somali Congress – formed in 1989 in Rome) took up arms (Mohamed 2001). SPM was founded by the Ogadenis (Ethiopian Somalis), who were recruited into the Somali National Army (SNA) after Somalia lost the war against Ethiopia. However, towards the end of the 1980s, the Ogadenis began to dessert from the SNA after Barre began to rely rather on Marehan clan (his father's clan) while neglecting the Ogadenis (his mother's clan). In the second half of 1990 SPM joined the coalition of SNM in British Somaliland and USC (led by Aideed from the Hawiye clan). The emergence of various armed groups and coalition among them led to the ouster of Siad Barre's regime in January 1991. Since the collapse of Somalia some radical Islamists have emerged, according to whom Somalia can be rebuilt only by going back to Islamic religious roots.

Religion

The Transitional National Government (TNG) was the first significant attempt to rebuild Somalia. There were many peace processes between 1991 and 2000 for Somalia. In 2000 the TNG was established. But from the outset it was suspected by Ethiopia to have been infiltrated by terrorist Islamist organizations like Al Itihad. The Somali federal government, which was established in Nairobi in 2004, was regionally and internationally hoped to address the rising Islamic fundamentalism.

However, towards the end of 2005, three rival power centres began to emerge: the TFG led by President Yusuf; the Islamic Courts led by Hassan Dahir Aweys and their chairman, Sheikh Sharif Sheikh Ahmed; and the Mogadishu Group composed of the opposition ministers and their local allies (ICG 2006d: 7). The TFG failed to establish a functioning local administration. Islamic courts emerged to fill this gap (ICG 2006d: 9; Marchal 2004: 133).

Since the collapse of the Somali state, Islamism has started little by little to gain ground. A fight in southern Mogadishu between the militias working for Islamic courts and Mohammed Qanyareh Afarah's followers controlling the neighbourhood on the one hand, and between Ali Mahdi's forces and supporters of Yalahow in the north of the capital on the other hand, killed about 80 people between April and May 1998. Furthermore, the fighting in the lower Shabele region and the south between RRA and Islamic Tribunal, and between Gare and Gido sub-clans, left at least 28 people dead in June 2000. When a Somali national reconciliation conference scheduled for 15 May 1998 to establish a transitional government was cancelled because of lack of funds, Sheik Ahmad Qasim (Chairman of the Hizb Al-Islam – Islamic Party) said that only Islam and tougher Islamic Sharia law can end the civil war and create jobs and security.

Islamism is an ascendant ideology, and Islamic institutions like schools, hospitals, charities and local Sharia courts seemed to be the only sources of order in the collapsed state of Somalia (Menkhaus 2005: 25). In June 2006 ICU (Islamic Courts Union) demonstrated its power but in December it was driven out by the armed forces of transitional federal government of Somalia through the support of Ethiopian troops. By 2005 the ICU consisted of 11 Islamic courts from different clans operating in Mogadishu under Sharia (Human Rights Watch 2007: 19).

Many Western states as well as states in the Horn of Africa are very concerned that Islamic fundamentalism and terrorist organizations such as Al-Qaeda are using Somalia as a breeding ground and springboard for terrorist activities. It was because of his stance as anti-Islamist that Ahmed Yussuf was supported by the Ethiopian government and the US as President of the Transitional Federal Government in 2004 to replace the Transitional National Government elected in 2000 and which was allegedly infiltrated by Islamist elements (Menkhaus 2005: 24).

President Ahmed Yussuf was against Islamic ideologies in the TFG. However, the TFG has failed to make itself a genuine government of national unity and economic and political reconstruction. In this gap of credible authority to integrate the Somalis, the Islamic courts emerged. At least originally these courts were considered as a platform for opposition from large sections of the Hawiye clan, the most powerful kinship group in southern Somalia (ICG 2006d: 17). The courts have been claiming to transcend clan-based societal heterogeneity through principles of Islam (ICG 2006d: 5). Further, ICU's objective has been to eventually unite all Muslims in the Horn of Africa; to annex the Somali region; to support and use the Oromo Liberation Front (OLF) and the Ogaden National Liberation Front (ONLF) in destabilizing Ethiopia while relying on foreign jihadists. Further, the Islamic courts are heterogeneous, encompassing a range of religious traditions and political perspectives. Probably as a contrast to the ethnicized politics and warlordism during and after the regime of Siad Barre, the courts tried to make simplicity their key moral virtue for the whole Somali society.

These courts were composed of 11 separate, Sharia courts. They consist of a shura (council) which includes respected political, traditional, business and religious leaders from the clan; a chairman appointed by the shura; and a militia commander appointed by the chairman, subject to the shura's approval (ICG 2006d: 9). Right after the collapse of the Barre regime and Somali state, Islamic courts appeared during the early 1990s. Since many attempts to make peace in Somalia failed, Sheikh Hassan Dahir Aweys, who is known as former vice chairman and military commander of the jihadi Islamist organization al-Itihaad al-Islaam (AIAI), established a new brand of court after being defeated by Ethiopia and the forces of Ahmed Yussuf in 1998 (ICG 2006d: 9; Human Rights Watch 2007: 18). Later, with the support of Ethiopia and the US, ARPCT (Alliance for the Restoration of Peace and CounterTerrorism) was founded on 18 February 2006. It consisted of warlords who were members of the TFG as well: Muse Sudi Yalahow (trade minister), Mohammed Qanyare Afrah (security minister), Botan Issa Alim (disarmament minister), and Omar Mohammed Finish (religious affairs minister), as well as several business and other militia leaders (Human Rights Watch 2007: 19). However, in May and June 2006 ARPCT had already steadily lost ground. This weakness of the ARPCT encouraged the Islamic courts to advance in Somalia without a significant fight (ICG 2006d: 13).

The resignation of President Yussuf of the TFG and the withdrawal of Ethiopian troops towards the end of 2008 has paved the way for an uncertain future in Somalia. These two events have buoyed the Al-Shabab (The Youth) and other Islamist militants who have continuously imposed Sharia in the areas they controlled. They declared Sharia law in Baidoa, a day after seizing the central Somali city, a stronghold of the country's transitional government. Former Ethiopian bases in Somalia were taken over hours after Ethiopian troops pulled out. Al-Shabab promises to rule Somalis with justice and Sharia law instead of any man-made constitution; and fight anyone who opposes it. They took over the airport controlled by Ethiopians and the AU forces in Somalia, and the parliament building and president's home in Baidoa in January 2009. Though Ethiopian troops have left Somalia and Ethiopia's prime minister has welcomed the election of the new Somali president, Sheik Sharif Sheik Ahmed, who was leader of moderate Islamists on 1 February 2009, Ethiopia will certainly be securing its borders and carefully watching the Islamist developments in Somalia as well as in the region as a whole.

President Sheik Sharif Sheik Ahmed has appealed for foreign military help at an African Union (AU) summit in Ethiopia in 2009. But he is accused by the hardline Islamist militia Al-Shabab, of selling out to the West. Sheikh Mukhtar Robow, leader of the radical group, vowed to intensify its attacks on the AU mission in Somalia (Amisom). In fierce fighting on 23 February 2009 in Mogadishu, 15 people were killed and 90 injured when the Islamic Party (a coalition of four insurgent Islamic groups) insurgents fired mortars at the presidential palace and a base for

African Union and government troops. Two days later Al-Shabab, an opponent to the UN-sponsored reconciliation efforts in Somalia, overpowered the government forces. About 50 civilians were killed and 120 injured in Mogadishu.

The conflict has not only killed and displaced many Somalis, but it also has destroyed the economy. The infrastructure has been destroyed. Exacerbated by environmental factors such as drought, the conflict has been escalating, and many Somalis are dependent on continuous humanitarian aid. The pre- and post-collapse conflicts have strongly affected the Somali economy.

Economy

In the 1980s the Somali political crisis was exacerbated by economic crises. Somalia is dependent on agriculture. But its production declined in the 1980s, at least partly due to the fact that peasants were neglected in favour of building up ports. Moreover, Somalia has been heavily dependent on food imports. The problem was further exacerbated by low profits of livestock market and diminishing foreign exchange, and decline of foreign aid. Even the foreign military aid was cut off in 1988. A poor human rights record led the US Congress to cut the aid; and in 1989 it completely denied any aid to Somalia. Aid from outside made 25 per cent of Somalia's GDP and the foreign debt-to-GNP ratio of 203 per cent against a per capita GNP of $175 (Talentino 2005: 101–2).

Prior to 1991, Somalia was the largest exporter in East Africa of bananas and was supplying markets in Europe, especially Italy, and the Persian Gulf. But as a consequence of the unending civil war over 3.2 million people (43 per cent of Somalia's population) in different parts of Somalia require emergency assistance and livelihood support through June 2009 (Reliefweb 2009). According to the UNDP Report, between 2000 and 2005 the probability at birth of not surviving to age 40 was 38.9 per cent. In 2004, 71 per cent of the Somali population was not using an improved water source. Between 1996 and 2005, 26 per cent of Somali children under age five were underweight for their age. In 2000 maternal mortality ratio per 100,000 live births was 1,400 (UNDP 2008). By the end of 2008 more than 3 million people needed food aid.

Between 2001 and 2007, the population and labour force have been growing 3.0 and 2.9 per cent respectively. Infant mortality (per 1,000 live births) is 90, and 33 per cent of children under five suffers from malnutrition. Only 29 per cent of the population has access to an improved water source. Since 1987 it is not possible to acquire reliable data on the literacy rate, gross primary enrolment, gross capital formation, exports of goods and services, gross domestic savings, gross national savings, current account balance, interest payments, total debt, total debt service, exports, percentage of agriculture or industry or manufacturing (World Bank 2008: 1–2).

Already in the second half of 2006, about 1.8 million Somalis – including up to 400,000 internally displaced persons – were in need of critical assistance until the end of 2006 caused by low-level cereal production due to the war. The World Food Programme (WFP) dispatched 3,300 metric tons of food aid commodities (Report of the Secretary-General 2006: 7–8). More than 1 million Somalis are internally displaced. 20,000 fled Mogadishu each month in the first half of 2008. Somalis's life expectancy is only 47 years. High food prices and drought have heavily affected the population. Between February and early March 2008, between 200,000 and 500,000 people fled the fighting in Mogadishu between government forces and a combination of Islamist insurgents and clan opponents of the regime. They are forced to live under miserable conditions. In mid-March 2008 UN Secretary General Ban Ki-moon presented the council with a report proposing the deployment of 27,000 peacekeepers to replace 2,400 African Union peacekeepers (of a planned 8,000-strong AU force). In his report of 17 November 2008 Secretary-General Ban Ki-moon expressed that a United Nations peacekeeping operation, which envisages 22,500 troops, would take place, but only following the establishment of a credible, inclusive ceasefire agreement. This UN force would take over the task of an international stabilization force supposed to support the implementation of the Djibouti agreement on 26 October 2008 (UN 2008a: 8–10). The agreement was signed by the Somali parties among others on the cessation of armed confrontation. Such UN operation presupposes political progress and improvement in the security situation (UN 2008a: 8).

Politics

The third key component of the internal conflict system is the manner in which the quality of the political life of the citizens is organized. The objective of this sub-section is to show how the different elements of politics influenced the societal and economic dimensions. For society–politics interaction, the 'idea of the state' is significant.

Idea and Ideology of the State

The role of elites is decisive in the formation, dynamic as well as termination of conflicts and violence. In their struggle and mass mobilization, elites construct and expand certain ideas and ideologies of state in the form of national identity. In preparation for independence, some Somali elites tried to underline a common concern of all Somalis. The Somali Youth League represented certain ideas and ideology for all Somalis. First, the Somali Youth League (SYL), which was already active in the early 1940s, had expansive intentions, and it (the League) was supported by the UK, which administered Somalia after the Second World War. The nationalist Somali Youth League was nurtured by the British Military Administration between 1941 and 1948 in spite of its expansive nationalist ideology. The UK did not realize the dangers of such nationalist policy for the future stability of Somalia and the states of the region.

Second, as Clapham suggests, Somalia's problem derives from the imposition of boundaries, creating problems for pastoralists. Whereas most African states defended a colonial frontier, Somalis rejected the colonial frontier because it limited their free movements across such a frontier. Moreover, Ethiopia's maintenance of Ogaden as its 'legitimate' territory helped the Somali nationalists to construct a unified greater Somali state by demanding Somali 'unification'. From this point of view, what the Somalis really sought was not the creation of boundaries but their removal, because wherever the camel goes, that is Somalia. This objective was easily achieved with the merger of British Somaliland and ex-Italian Somalia, but it created endless difficulties in relations with Ethiopia and Kenya, to which boundaries were an essential element in the definition of the sate (Clapham 1996: 240).

Strangely enough the original resentment towards frontier politics was transformed into fight for frontier. Constructed territorial attachments are mobilized into politics in ways that reinforced conflict. Territoriality has two significances: delimitation of boundaries and behaviour within those states. Because of the symbolic stakes that are often invested in international conflicts, territories that are devoid of resources may still become the site of violent disputes because political dynamics are often rooted in the symbolism of territory than its measurable value, like economic or demographic significance (Kahler 2006: 3–7). Territorial specification of group membership – for example 'we' live 'here' – is a particularly powerful and attractive way to coordinate group members to provide a collective defence (Goemans 2006: 26–9).

Nationalism is the process of creating collective national identity or nation-building. One of the historically very relevant fights for the freedom of Somalia and its national identity was the 20-year war of resistance against British and Italian colonial rule in northern Somalia, led by Said Mohamed Abdullah Hassan in 1899. This struggle was anti-foreign and anti-Christian, like the jihad against Abyssinia in the sixteenth century (Menkhaus 2005: 27).

The 'idea of the state', or nationalism, played a decisive role here. The Somali Youth League, founded in 1943, was determined to unify all Somalis in Djibouti, Kenya and Ethiopia in a Greater Somalia, and to create a collective Somali identity. Right after independence in 1960, both Somaliland and Somalia were mobilized and merged by political elites. The identity at the time was based on 'Somaliness'. After the Siad Barre regime took power in 1969, the state was enforced by the power elites.

Mainly based on such artificial nationalism and without other long-term policies, the SYL could gain the support of the Somalis before and after independence. However, most of the parties rallied to the SYL in order to benefit from the state, leading to massive corruption and ultimately to military coup in October 1969 (Prunier 1995: 3). This nationalism was at the same time very fragile. Though the unified Somali Republic after independence was a parliamentary democracy, it suffered from contradicting interests because of up to 60 political parties created along various clan and sub-clan lines.

Based on cultural and religious homogeneity, the post-independence history of Somalia was overshadowed by a sort of obsession for the Somali people in the region (Prunier 1995: 2). National identity-building after independence consisted of two challenges that the Somali government attempted to solve: On the one hand, the political elite in Mogadishu was not able to control the hinterland far away from the capital. Some analysts argue that one of the decisive causes of state weakness or collapse is that the central governments are not able to control the political situations in the hinterland depending on bad communication systems, roads, etc. (Ottaway et al. 2004). The distance between Mogadishu and the North, and the disparity between the nomadic, sedentary minorities and 'modernized' and educated classes made sociocultural integration more difficult. On the other hand, the Mogadishu government and the SYL Party concentrated on the irredentist policy, i.e. to 'redeem' the Somalis in territories under Kenyan, Ethiopian and Djiboutian administration. The Greater Somalia project of unifying all Somalis in the region was not only political, but also anti-Christian. This 'Greater Somalia' policy of unifying all Somali territories in the Horn of Africa was augmented by a perception of a threat from Ethiopian Christian Empire, which neglected the Somalis in Ethiopia (Brons 2001: 168–9).

Whereas the Darod-dominated SYL was a radical irredentist party, its counterpart SNL (Somali National League) was moderate regarding the 'Greater Somalia' policy and concentrated rather on friendly relationships with the neighbouring states and on sociopolitical and economic development. In its political history SYL became more and more corrupt and dominated by clan politics. When Siad Barre came to power on 21 October 1969 (after President Abd ar-Rahid Ali Shirmarke, who won the presidential elections of 1967 and replaced the first president of the Republic (Adan Abdulle Osman), was shot dead in a coup on 15 October 1969), his government of Supreme Revolution Council vowed to abolish clan politics and replace clan networks and non-state institutions with scientific socialism (Brons 2001: 171–3) while pursuing the Greater Somalia project. However, Siad Barre failed in his project. This failure resulted in protest against him and in the formation of different alliances to topple him.

Actors and Alliances

In 1977, because of its strengthened military potential, Somalia began to support the Western Somalia Liberation Front (WSLF – Ogaden region within Ethiopian territory) and the Somali and Abbo Liberation Front (SALF) in the Somali and Oromo inhabited areas of southern Ethiopia (Brons 2001: 182). In spite of the support from the WSLF and SALF, Somalia lost the war against Ethiopia that broke out in July 1977 and lasted until 8 March 1978 (Matthies 2005: 153). This failed irredentism led to the implosion of Somali politics. After their failed coup attempt in 1978, some army officers fled to Ethiopia, where they founded the Somali Salvation Front (SOSAF). In October 1981 the coalition of SOSAF, Somali Workers Party and the Democratic Front for the Liberation of Somalia

(DFLS) was founded as SSDF (Somali Salvation Democratic Front) (Lewis 2002: 252); this included Harti (Majeerten), Warsangeli, Dulbahante as well as Hawiye fighters, intellectuals, politicians and officers (Brons 2001: 185). The Somali Democratic Action Front (SODAF) was formed in Rome in 1976 and coalesced with SSF (Somali Salvation Front), and both consolidated their fight against the regime. The Somali National Movement (SNM) was created in 1981 for Somaliland, as already shown. It was these and similar challenges that pushed the Siad Barre regime to follow the strategy of ethnicization in order to secure his political survival (Matthies 2005: 156–61)

USC was another very significant military opposition that emerged. It was divided into political and military wings. The military wing was led by Farah Aideed, whereas the political wing was led by Ali Mahdi. Only two days after Barre was ousted (26 January 1991) the USC political wing unilaterally declared Ali Mahdi the president of Somalia, after which conflict and violence intensified. The immediate results of these developments were: first, the bloody fight between the political and military wings of the USC, and this intra-Hawiye cleavage continued to have far-ranging political consequences (Lewis 2002: 264) for the years and even decades to follow; second, the SNM declared the secession of Somaliland.

After SNM decided to concentrate on the independence of Somaliland, the two wings of the USC (Hawiye) divided Mogadishu between themselves. The political turmoil of 1990 and 1991 created additional factions. Some clan members of Siad Barre (Marehan) created the Somali National Front, which was trying to recapture the capital, while farming southern communities founded the Somali Democratic Movement (SDM). Meanwhile, the fighters of SPM (Ogadenis) were approaching the capital. The SSDF (Somali Salvation Democratic Front) in the north-east reorganized its forces whereas the Siyadists and other Darod coalition forces counterattacked Mogadishu (Mohamed 2001). The varieties of actors and alliances have been undermining the security of the Somali population. The combination of the ideology of the state, the involvement of numerous actors, and consequently changing alliances has been undermining all chances for security in Somalia.

Security

The main challenge for the security of Somalia was the emergence of various armed factions. Therefore, the web of alliances and antagonisms in the Somali conflict consisted of armed groups such as:

- USC (Hawiye) created in 1987.
- SPM (Ogadeni) created in 1989.
- The Somali Salvation Democratic Front (SSDF) created in 1978 was revived in 1989, representing the three major Majerteen clans: Colonel Yusuf

Abdullahi represented the Omar Mahmood clan, General Mohamed Abshir the Isse Mahmood, and Yusuf Omar al-Azhari the Osman Mahmood.

- An antagonism between two of the main Hawiye clans resulted in early 1990 in the split of the United Somali Congress (USC): Abgal USC supporting Ali Mahdi and a Habr Gidir USC supporting General Aydeed. In January 1991 after Siad Barre left Somalia, the Habr Gidir 'invaded' Mogadishu, not only taking their revenge on supporters of the former regime but also attacking the Abgal and looting their properties.
- The Somali Patriotic Movement (SPM) representing Ogaden clans, created at the initiative of Colonel Omar Jess early in 1990, was soon joined by former Defence Minister Adan Abdullahi Nur 'Gabeeyow, when he realized that the end of Siad Barre's regime was not very far. However, after January 1991, the group split into two: Colonel Jess joined forces with General Aydeed while 'Gabeeyow allied himself with the SNF leader General Mohamed Said Hersi 'Morgan'.
- The Somali National Front (SNF) with General 'Morgan' as its leader consisted of mainly Marehan clan in the latter days of Siad Barre's regime, and General Adan Abdullahi Nur's Ogadeen relatives, a number of Harti and Bimaal from the Lower Jubba and some Kenyan and Ethiopian Ogadenis.
- The Somali Democratic Movement (SDM) was created in 1989 to protect the sedentary peasants of the Digil and Rahanweyn clan families living in southern Somalia from the violent depredations of the warring factions. SDM, which was already weak militarily, split into two during the UNOSOM intervention: a faction led by Mohamed Nur Aleeyow supported General Aydeed, and another led by Abdi Mussa Mayow decided to cooperate with the United Nations.
- The Southern Somali National Movement (SSNM), created in 1991 under the sponsorship of General Aydeed to support Colonel Omar Jess in his fight against the SNF in the area, split during the UNOSOM period into a branch led by its former leader, Abdi Warsame Issaq, who deserted General Aydeed to join the UN, and a branch which remained faithful to the Somali National Alliance (SNA) under minority leader Abd-el-Azziz Sheikh Yusuf.
- The United Somali Front (USF), a political expression of the Issa clans (trying to detach the Issa-populated area from of the rest of the Somaliland but violently crushed down by the SNM forces) in the extreme north-west of Somaliland, and supported by the Republic of Djibouti, which is also politically dominated by the Issa.
- The Somali Democratic Alliance (SDA), which fought for the Siad Barre regime and for the north-western Gaddabursi clans in Somaliland against the SNM. After the independence of Somaliland, it split into a faction of the Alliance, which has become reconciled with the Issaq-SNM-dominated 'government', and another faction, which was based partly in London and partly in Ethiopia's Ogaden region and had remained hostile to the new secessionist authorities in Hargeisa.

- The United Somali Party (USP), a political party created in the late 1950s in British Somaliland by the northern populations belonging to the Darod clan family (Dolbahante and Warsangeli) but insignificant until 1991. After the independence of Somaliland in May 1991 they rasied the issue of the Darod minority in Somaliland. But it was soon divided into 'collaborationists', who accept the role of the Somaliland authorities, and 'opponents', who refuse it and plead for the eastern part of 'Somaliland' to be linked with the SSDF-controlled north-east, populated by the Darod-related Majerteen clans. Moreover, during the UNOSOM period, the 'opponent' faction of USP received support from the United Nations.
- The Somali National Democratic Union (SNDU), created in 1992 to represent the two small Lekasse and Ortoble clans (Darod clan family) of central Somalia area, which was the battlefield between General Aydeed's forces and the SSDF. It was allied with Ali Mahdi and was thus invited to the various meetings of the United Nations in spite of his organization's small size.
- The Somali African Muki Organization (SAMO), created by the southern negroid people who live in the Juba and Wabi Shebelle valleys, despised by the Somalis as 'niggers'. They are sedentary agriculturalists whose villages were burnt down, their grain robbed and their women raped. But the organization lacked weapons and trained military cadres.
- The Somali National Union (SNU), led by Mohamed Rajis Mohamed and Said Omar 'Marino, was an organization (though small and without military means) attempting to fight for rights of the non-Somali people of the Lower Juba in the coastal Barawa Swahili areas, including the Bajuni fishermen of the southern Benadir coastline and some Harti groups.

In March 1991 SNF (Siad Barre's supporters of Marehan clan) launched a military offensive against the USC and SPM forces controlling the capital and farming areas after the ousting of Siad Barre. By November 1992 starvation claimed 1,000 lives per day, and 80 per cent of the food aid was being confiscated by warlords and bands of teenagers (Talentino 2005: 110).

In 1991 Mogadishu became the main battlefield. After Ali Mahdi's forces, which controlled northern Mogadishu, launched attacks against each other on 17 November 1991, flagrant sufferings of civilians took place. Within four months about 14,000 people died and 27,000 were wounded (Human Rights Watch 1993; Lewis 2002: 264). Having taken advantage of the fighting between the two USC wings, the remnants of Barre's supporters of the Marehan sub-clan (from the Darod clan) began an assault on the Hawiye sub-clans of Ali Mahdi (Abgal) and of Aideed (Habr Gedir), whereas Darod were collectively assaulted as the clan of Siad Barre. On the one hand, the SNF (Somali National Front – Siad Barre's supporters) and civilians became targets of the USC military wing's counterattack (Human Rights Watch 1993). In October 1992 the SNF, led by General Morgan

(Said Hersi), Siad Barre's son-in-law, repeated the pattern by committing atrocities against civilians that it suspected of supporting USC.

In north-east Somalia atrocities were committed against prominent Harti civilians considered a potential political threat once the foreign troops had arrived. This assault was conducted by Omar Jess and his SPM (Human Rights Watch 1993). Though the main centre of the conflict and violence was Mogadishu, the whole country was divided and devastated by various factions. The farmland in the inter-riverine lands (Shabele and Juba rivers) was devastated in 1991–92; this brought about the starvation of thousands of Somalis (Brons 2001: 219). This devastation of farming areas began in March 1991 when SNF (Siad Barre's supporters of Marehan sub-clan) launched a military offensive against the USC and SPM forces controlling the capital and farming areas. Besides USC and SPM, there were SDM (Somali Democratic Movement) representing inter-riverine Rewin clans and SSNM (Southern Somali National Movement) of the Dir/Biyamale clan fighting alongside USC/Aideed and SPM. Moreover, since both SDM and SSNM lacked their own military power, they made an alliance called the Somali National Alliance (SNA). Through the help of the SNA the forces of Aideed could defeat Siad Barre's loyalists (SNF) in April 1992 (Brons 2001: 224).

While most of the population welcomed the US initiative, the United Nations International Task Force (UNITAF) and the UN-sponsored peace conference in March 1993 in Addis Ababa, Aideed was reluctant to implement the principles of the conference to establish regional administrations aimed to restore law and order from the grassroots, and to install Transitional National Council (TNC) wherein participant factions and all 18 regions were represented (Mohamed 2001). Since the implementation of these principles would have meant 'civilianization' of politics, Aideed realized that his power was endangered. After having defeated the UN and US, Aideed appointed himself as president of Somalia and was supported by the UN. The military capacity of Aideed was strengthened by the weapons he captured after the fall of Siad Barre; he occupied part of Mogadishu, which was the centre of the attention of the national and international actors; and he got support from Sudan, Libya and various international Islamic organizations (Mohamed 2001).

An uncontrolled flow of arms in spite of the UN arms embargo, Resolution 733, exacerbated this (Matthies 2005:225). In spite of the UN arms embargo and various agreements throughout the various peace processes, arms inflow became a central danger. Rival clans competed for power thanks to arms freely flowing from Ethiopia and Eritrea with backlash on Ethiopian and Eritrean security (Mentan 2004: 238).

In the southern regions fierce fighting continued throughout 1997, and resulted in suffering and food shortage for the civilians. The inter-riverine area was destroyed by the fighting between the new Rewin militia, Rahanweyn Resistance Army (successor of the former SDM militia dissolved in October 1995) and the USC/Aideed forces. The SNF (Marehan militia of Barre loyalists) supported by Ethiopia fought against Al-Itihad al-Islamiya in the Gedo region bordering Ethiopia and Kenya. A new fight for control of Kismayo broke out in April 1998 between Marehan and Majerteen clans. Marehan operated under the command of General Omar Haji Mohammed, whereas Majerteen's Somali Democratic Front created in 1978 to oust Siad Barre was led by Said Hirsi Morgan.

During 1996 and 1997 the warlordism consisted of three decisive events: first, the voted-out president Abdirahman Tour (Somaliland) and his group joined Aideed's administration and stirred up civil strives in Somaliland among Isaaq clans. Second, while the Digil and Mirfile clans of the riverine areas were occupied and persecuted by the militiamen of Aideed, a coalition supported by Ethiopia was created between the factions of the voted-out president of Somaliland (Abdirahman Tour), Darod clan factions, Hawiye leaders including Ali Mahdi and Osman Ali Atto (former financial supporter of Aideed) (Mohamed 2001). Third, a further split occurred within the Hawiye factions of Aideed and Mahdi. In the northern part of Mogadishu, occupied by the forces of Ali Mahdi, Islamic forces were created by Musa Sudi Yalahow. In the southern part of Mogadishu, controlled by Aideed's forces, Osman Ali Atto built up his own forces. After the death of Farah Aideed in August 1996, his son Hussein Aideed took over the leadership of the USC/SNA faction (Brons 2001: 219).

However, the conflict continued, and for example, in 1997 there were 275,000 Somali refugees living in Ethiopia. On 20 March 1998 Hussein Aideed announced his readiness to share presidency with his rival Ali Mahdi, however without dissolving his government. At the same time, militias working for Islamic courts and various warlords continued with their atrocities. Furthermore, the fighting in the lower Shabele region and the south between RRA and Islamic Tribunal, and between Gare and Gido sub-clans continued in the second half of June 2000. The fierce fighting that broke out in 1998 for control of the south-central town of Baidoa between RRA and Aideed forces continued throughout 2000. Militarily the town was controlled by Aideed forces, although the population is Rahanwein whose militias were called RRA. As a first step towards peace, Ali Mahdi, Hussein Aideed and Mohammed Qanyareh Afarah agreed on 3 August 1998 to appoint Ali Mahdi as a new governor of Mogadishu, whereas this move was rejected by Musa Sudi Yalahow and Atto.

The peace process of August 2000 and the transitional government created thereafter were condemned to fail from the very beginning when some warlords repudiated the agreement reached during the peace process as 'invention of the Djibouti government', and they urged a reconciliation conference on 'Somali soil'. Some of these warlords were Ali Atto, Qanyareh Afrah, Hussein Aideed and Omar Jess. The warlord who accepted the TNG of Abd-al Qasim Salad Hassan was Ali Mahdi occupying northern Mogadishu. Towards the end of 2000 and beginning of 2001 the northern area of Mogadishu was overshadowed by fighting between the supporters of Yalahow and the TNG. Moreover, an opposition leaders meeting launched to set up a new 'autonomous regional southern state' contributed to the erosion of a consolidated and inclusive TNG. The meeting was attended by RRA, SPM, SNF, SSNM and SSDF, a coalition which formed SRRC.

A fight between Aideed's forces and pro-government militia on 11 and 12 May 2001 in Mogadishu left at least 50 dead and up to 100 wounded. When Aideed's forces and the pro-TNG government militia attacked each other on 11 and 12 May 2001 in Mogadishu, it left at least 50 dead and up to 100 wounded. A power struggle within RRA left 40 people dead at the beginning of July 2002. Throughout 2001 any political reconciliation process was undermined by the cleft between TNG and SSRC. Even at the beginning of 2002 the tension continued. In February 2002, in order to underscore his legitimacy and wide support of the population, Hussein Aideed (USC/SNA) claimed to have 85 per cent of the support of the Somali people and a militia of 15,000 active men and 1.5 million inactive reserves. As chairman of the SRRC aiming at toppling the TNG, Aideed hoped to create democracy and to enable equality among Somalis. A consolidation of unity was further undermined when south-west Somalia, controlled by RRA, was declared 'autonomous' in March 2002. However, it did not last long; a power struggle within RRA emerged at the beginning of July 2002. As a consequence, the local elders recommended the dissolution of the self-declared autonomous state of south-west Somalia. In June 2002, the rivalry in self-declared autonomous Puntland, between two rival leaders Abdullahi Yusuf and Jama Ali Jama, led to the mobilization of their forces and to renewed fighting between the two sides.

The transitional national government of Somalia was not recognized by all factions. As a consequence, there had been various attacks by warlords against it. Conflicts between warlords Musa Sudi Yalahow and Mohamed Dhereh of SRRC and TNG led to the killing of at least 62 people in June 2002. For the counterattack the TNG received weapons from Djibouti, Libya, Eritrea and Yemen. After the Rahanweyne Resistance Army declared south-western Somalia as autonomous in late March 2002, the fight between the faction of Muhammad Nur Shatigadud on the one hand and the faction of Shaykh Ada and Muhammad Ibrahim Habsade on the other claimed the lives of over 40 people (*Africa Research Bulletin* 2002: 14943).

Fusion and dissolution of alliances between factions is one of the characteristics of warlordism, with grave security implications. This leads not only to deaths and starvation in the country, but also to an endless reorganization of alliances and creation of new administrative regions. Towards the end of February 2003 fighting erupted between the militiamen of Muse Sudi Yalahow and those led by his former right-hand man, Umar Mahmud Muhammad Finish. Two months later Yalahow made an alliance with Usman Hassan Ali Atto to set up a new administration in the Banadir region (Mogadishu and environs).

After having received a detailed report on violations of the Declaration on Cessation of Hostilities signed by all Somali parties on 27 October 2002 in Eldoret, Kenya, on 2 February 2003 the Ministers of the Frontline States (Djibouti, Ethiopia and Kenya), alarmed by the resumption and escalation of violence in Somalia, condemned in Addis Ababa the violations of the Declaration on Cessation of Hostilities that caused great suffering and loss of life. Besides reiterating their stand that appropriate sanctions would be taken against all individuals and groups violating the Declaration, the ministers decided to draw the attention of the United Nations, the African Union, the League of Arab States, the European Union and all other partners of IGAD. Confronted by this worsening security situation in Somalia, the ministers decided to establish a mechanism and committee which consisted of representatives of the Technical Committee and international partners to monitor violation of the Cessation of Hostilities Declaration. The United Nations Security Council also reacted to the violations of the Declaration on Cessation of Hostilities and the dramatic security and humanitarian situation. It unanimously adopted Resolution 1474 (2003) to re-establish a panel investigating the Somalia arms embargo.

While the peace negotiations were approaching their final stages in 2004, various warlords accused and signed statements against Djibouti and Kenya of bias in favour of the Transitional National Government. Signatories to the statement were the Jawhar-based faction leader, Muhammad Habib; Shaykh Adan Madobe of the Rahanwein Resistance Army; Gen Muhammad Sa'id Hirsi Morgan; Abdullahi Shaykh Isma'il; Mahmud Sayyid Adan of the Somali National Front; and Abdiqadir Abdi Hasan. The six belong to the SRRC (Somali Reconciliation and Restoration Council), a group that opposes the TNG. The objective of the group was to protect the Rahanweyn heartlands and control their territory south of the Shebele (Lewis 2002: 302). This warlordism was gradually replaced by Islamism, which concerns not only Somalia but also the region of the Horn of Africa as well as the international community. What now seems to worry regional and global actors is that religious fundamentalism could be a global issue.

The reconciliation conference that began on 15 October 2002 restarted on 2 December 2002 in Mbagathi, Kenya. The objective of the conference was to discuss and draft proposals for the resolution of conflict in Somalia; to identify mechanisms for addressing human rights violations; to pursue clan, regional and leadership reconciliation; and to submit its proposals and recommendations to the Somalia National Reconciliation Conference by 30 January 2003.

Dispute over Laas Caanood (area claimed by both Somaliland since 1993 and Puntland since 1998) led to a violent escalation of the self-declared republic of Somaliland and self-declared autonomous region of Puntland towards the end of 2003. The main trigger of the conflict was the issue of Sool and Sanaag, which fall geographically within the borders of the pre-independence British Somaliland, whereas the people living there are ethnically related to those in Puntland.

During the first half of May 2004 inter-clan fighting in the town of Bulo Hawa, in the north of Gedo Region, south-west Somalia, near the border with Kenya, displaced between 3,000 and 3,500 people (IRIN 2004). In the first half of August 2004 a fight erupted between Dabare and Luway sub-clans in the Bay region of south-central Somalia, resulting in the death of 20 people. In June about 60 people were killed in clashes between rival clans in the south-western town of Bulo Hawa, about 2,500 families were displaced, while 2,000 people fled to Mandera in Kenya.

In the first half of September 2004 the fighting between them and the Juba Valley Alliance militia resulted in internal displacement of some families, whereas about 500 people fled to Kenya. Since Ghedi came from the capital, Mogadishu, where the president lacked support, and was a member of one of Somalia's largest clans (Hawiye), his election as Prime Minister was meant to balance and to bring reconciliation at least between the Hawiye and Darod clans.

'Inclusion' has been the guiding principle since Somalia's collapse. The Cairo accord of 1997 was undermined by tensions among the clans and differences of warlords. It was suggested that a 13-member presidential council should consist of three members each of the major social groups and one of the remaining group; that the Darod, Hawiye and Dir clans would be represented by 113 delegates at a national reconciliation conference, whereas Digil and Murasade would have a total of 113 delegates; and a further 13 places were reserved for minority groups. Further, it was suggested that the president will be chosen from the Hawiye clan to which Ali Mahdi and Aideed belonged; the president from the Darod; and the president of the parliament from among the Rahanwein, in order to end clan-based insecurity.

According to Marchal, however, warlords were providing the Somali civilians with viable levels of governance, public security and social services, rather than simply acting as 'war entrepreneurs' (Marchal 2007b: 1098–9; Menkhaus 2006/7: 102). Warlords are not necessarily violent predatory entrepreneurs that attempt to exploit the 'offers' or opportunities, but also respond to the 'demands' of the society in order to fill the political vacuum (Marchal 2007b: 1096, 1099). However, according to Ken Menkhaus, there are at least two groups of people not necessarily interested in Somali state revival: the first group consists of clan factions or leaders who feel they haven not been represented or rewarded in the government, or they perceive that their economic and political interests are threatened by revival of a functional state. The second group consists of warlords for whom the return of rule of law could result in the marginalization or even arrest for war crimes, illicit business or armed occupation of valuable real estate during the war (Menkhaus 2006/7: 95–6).

The TFG, which was established in October/November 2004 after two years of negotiations, has not been free from clan rivalries. Already in March 2005 the emergence of two rival clan-based camps undermined the political re-stabilization process of the TFG. Under the leadership of President Yusuf and Prime Minister Gedi the 'TFG group' emerged, whereas under the leadership of Speaker of the Parliament Sharif Hassan Sheikh Aden, the 'Mogadishu group' emerged, composed of parliamentarians and cabinet members principally from the Hawiye clan, though he himself is a member of the Digle-Mirifle clan. The militia forces or the 'national army' of the TFG lacked any significant Hawiye leadership, because they were drawn predominantly from Puntland (Majeerteen and other Harti Darod), Ogaden (also Darod) and smaller numbers of other clan groups. Within the Hawiye clan many believed that even the Islamic courts were not representing the interests of all Hawiye. The courts were just a vehicle for Habar Gidir Ayr, which breeds some resentment among other groups, such as within the Abgaal and the Habar Gidir Sa'ad sub-clans (ICG 2006d: 19). This kind of clanism has been successfully exploited by the Somali warlords since the collapse of Somalia, and it contributed to dire security situations.

On 22 June 2006 an agreement was signed in Khartoum between the transitional federal institutions and the Islamic courts to recognize the legality of the transitional federal institutions, the existence of the Islamic courts, to continue talks without preconditions, to try war criminals, not to engage in any provocative or hostile acts or antagonistic propaganda, and to continue discussion on security and political issues. But soon after that the ICU attacked the last remaining ARPCT leader in Mogadishu, Abdi Awale Qeybdid, arguing its military action was directed against an individual, not ARPCT as such (Report of the Secretary-General 2006: 3). It is estimated that some 10,000 civilians have been killed between 2007 and 2008. Such a dire security situation and violence are mainly due to the decades-long militarization of Somalia.

Militarization

Somalia has become a highly militarized and divided society with various warlords and authorities controlling various parts of the country. Even before the emergence of this warlordism, the wars between Ethiopia and Somalia significantly contributed to the arming and mining of Somalia and its borders with Ethiopia. During the Ogaden wars of 1964 and 1977 the Ethiopian border was mined. Similarly strategic facilities, camps and towns were mined in the 1970s and 1980s during the Somali Salvation Democratic Front (SSDF) insurgency in Puntland, and during the 1988–91 war of secession in Somaliland by the Somaliland National Movement (SNM). The minefields increased in the inter-clan fighting after the collapse of Somalia in 1991 (Land Mine Monitor 1999).

Human Rights Watch (HRW) reported that the fighters stole and extorted from rival clans or groups, and their leaders did nothing to discourage them (HRW 1993). 'The looting and banditry prevalent in Somalia is not solely due to individual looters and bandits, but is a direct result of the manner in which the armed factions choose to recruit their forces and wage war. In fact, many observers argue that without the implied promise of fruitful looting, few factions would be able to summon any sizable military support' (HRW 1993). Rival militias had often been fighting for control of toll collection as a major source of revenue for militia leaders.

Siad Barre was partly responsible for the abundance of arms.

> During the reign of Siad Barre, the government of Somalia stockpiled arms with the intent of using them in the armed struggle to create a 'Greater Somalia', which was to include parts of Kenya, Djibouti and Ethiopia. In the course of its armed struggle, the SNM (Somali National Movement) distributed a lot of weapons to sympathizers and auxiliary forces supporting its cause with the help

> of Ethiopia. Militias, which were loyal to different clans, sprung up. A lot of arms also got into civilian hands as the government began to disintegrate in 1980s. (Omar 2002: 18)

The self-declared state of Somaliland as well as the rest of Somalia is affected by arms abundance. Until the Somaliland administration began to centralize its security forces, the United Somali Front (drawing support from the Issa clan armed with small arms and commanded by Abdurahman Dualeh Ali), the Somali Democratic Alliance (supported by Gadbursi clan), the Somali National Movement (supported by 5,000–6,000), drawn from the Tur, Dhegaweyne and Kahin clans were equipped with 5 or 6 medium tanks, 15–20 Fiat armoured personnel carriers, artillery, mortars, anti-aircraft weapons and naval patrol vessels based at the port of Berbera and of United Somali Party (supported by Midigan and Tumaal clans commanded by Ahmed Guure Adan) (Ploughshares 1999).

As we have seen above, all parts of Somalia are occupied and destabilized by various factions of warlords. The Somali Salvation Democratic Front (3,000 militiamen drawn from the Darod clan and commanded by Abdullah Yusuf Ahmed), the United Somali Congress (drawn from Hawiye clan and commanded by Hussein Mohammed Aideed and Osman Atto), the Ali Mahdi Faction (10,000 drawn from Abgal clan commanded by Ali Mahdi), the Somali National Front (2,000–3,000 militiamen drawn from the Darod sub-clan of the Marehan commanded by General Omar Hagi Mohammed Hersi and equipped with mortars, anti-aircraft weapons, recoilless rifles), the Somali Democratic Movement (drawn from the Rahenwein/Digil) and the Somali Patriotic Movement (2,000–3,000 drawn from the Darod clan commanded by Ahmed Omar Jess) have been operating (ISS 2005). The military activity of these warlords, their strategy of ethnicization, the recent developments of Islamism, and even the Greater Somalia policy pursued during Siad Barre as a project of national identity cannot be comprehensively grasped without the regional context of Somali politics.

After the Siad Barre regime was ousted in January 1991, anarchy and war were exacerbated by various militia groups that controlled the top political positions because of their military power (Woods 1997: 1–2). Minorities bear most of the consequences of state failure and militarization of civilians. Internally displaced persons from minority groups such as Rahanwein, Bantu, Ajuran, Jarso, Madhiban and Ashraf lacked political representation; they are discriminated, targeted, displaced and dispossessed by militias. Since competition for political power goes hand in hand with the grabbing of resources, minorities on fertile lands disproportionately suffer from killings, destruction, discrimination, land alienation, obstruction of humanitarian relief, and forced displacement.

When the TNG was challenged by inter-faction wars, it admitted that it could not disarm various militias on its own. Moreover, the proliferation of militias had reached such high proportions that it has asked foreign troops to help disarm these militias. Mogadishu, with an estimated 60,000 militiamen, was divided among and devastated by warlords and arms supply. In order to halt further sufferings of the population, the UN imposed an arms embargo on Somalia.

The availability of an unknown number of light arms is one of the main problems of security in Somalia. The UN sets out evidence of the link between the disintegration of the Somali state, of the easy access of the militiamen to the illegally imported weapons and of terrorism in the region. 'Almost everyone got hold of guns … Armouries are empty. Police have no weapons. There is no army as such. The elders of the clans do not seem to be able to control many of their armed youth, and there are conflicting inter-clan interests, which prevent their elders from acting jointly to improve security' (Woodward 2003:73).

The Somali society has been overflooded by militiamen fighting for clan warlords and working for private businessmen caring for their security on their own. Some 600,000 Somali civilians were internally displaced from and around Mogadishu and 6,000 civilians were killed in fighting in the capital and across southern and central Somalia in 2007 by weapons coming from, for example, the US, China and North Korea. Moreover, about 335,000 displaced Somali refugees fled Somalia in 2007 (IANSA 2008). Ethiopia; Eritrea, Egypt, Djibouti, Iran, Hezbollah, Libya, Saudi Arabia, Syria, Uganda and Yemen were also involved in delivering arms to different parties in Somalia (UN 2006).

Children and minorities are most affected. Somali children are suffering from malnourishment and higher risk of disease. They have been recruited and used by the Transitional Federal Government, Al-Shabab, ICU and other warlords. In 2006 mass recruitment of children in schools took place by militias. Children are increasingly becoming addicted to alcohol and drugs; being sexually abused; have jobs hazardous to their health; and seek volatile protection by joining urban gangs (UN 2008).

The discussion to deploy IGAD troops to secure Somalia for the arrival of the exiled TFG contributed to the political split and militarization of Somalia. When President Yusuf in early 2005 continued to push for the deployment of IGAD troops, 10 Mogadishu-based Hawiye ministers left the cabinet in March 2005. Following this division, the Mogadishu Group announced it would secure Mogadishu on its own. The so-called Mogadishu Security and Stabilization Plan (MSSP) included an impressive proposal for demobilizing 1,400 militiamen and 60 armed vehicles. Further they planned to eliminate militia roadblocks in the city and establish a regional/municipal administration. The MSSP brought together the Mogadishu-based TFG's 'armed ministers', the Islamist leadership (including Hassan Dahir Aweys), TFG rejectionist militia leaders (such as Sheikh Indha'adde), business leaders and civil society groups (ICG 2006d: 4–5).

One important factor for the security deterioration and heavy fighting between the TFG and Islamists was Yusuf's decision to move about 1,000 Puntland militiamen and 60 armed vehicles near Jowhar since September 2005. The deployment of a large number of Majerteen clan militia so close to Mogadishu was viewed by the Mogadishu Group as a military as well as ethnic provocation. The Mogadishu Group responded by mobilizing thousands of militia and arms in Bal'ad, Mogadishu and Bale Dogle (ICG 2006d: 6). For al-Shabab the targets were not only the TFG but also the African Union peacekeepers as well as the Ethiopian troops. For example, al-Shabab attacked and killed an African Union peacekeeper in the first week of January 2009. The TFG, supported by some pro-government militias such as the one led by Barre Hiraale (a member of parliament and former warlord of Kismayo and armed by the withdrawing Ethiopian troops), has been fighting al-Shabab, and attempting to assert itself militarily. Such Islamist developments are increasingly undermining the attempts to reinstate democratic institutions in Somalia.

Democratization

Since the collapse of the Somali state in 1991 there have been various attempts to address the societal, economic as well as political ills. Internal, regional and global actors have been attempting to address violence and economic destruction. Democratization was being pursued as a basis for all of these. Especially since the Islamists have gained power, democratization has become more urgent than ever.

The transitional government after the Arta agreement of 2000 was challenged from the very beginning and was not accepted by various factions. For example the RRA (Rehanwein Resistance Army) and the self-declared autonomous Puntland leaders agreed to form an anti-Abdiqassim administration coalition, and as a consequence they formed a new organization, SRRC, in Awasa, Ethiopia, in March 2001 (Mohamed 2001). Hence, the transitional government has not been in the position to make any steps towards stability and economic reconstruction since its appointment.

In 2004 the TFG was established. It attempted to reconstruct Somalia, which its predecessor TNG was unable to rebuild. President Yussuf Ahmed was president until he resigned and was followed by Acting Transitional Federal President Sheikh Aden Madobe between 29 December 2008 and 1 February 2009. A transitional governing entity with a five-year mandate, known as the Transitional Federal Institutions (TFIs), was established in October 2004. Prime Minister Nur Adde Hassan Hussein succeeded Ahmed Ghedi on 24 November 2007 as prime minister. The unicameral National Assembly originally had 275 seats: 244 members appointed by the four major clans (61 for each clan), and 31 seats allocated to smaller clans and subclans.

Whereas Somali legislators were trying to bring in some 200 moderate Islamists into the parliament in the process to elect the successor of Abdullahi Yusuf Ahmed, who resigned in December 2008, the al-Shabab refused to take part in the peace talks in Djibouti. The hardline Islamists refused to take part in peace talks until the Ethiopian troops left Somali territory. Ultimately the new transitional federal institutions president has been elected, and the parliament enlarged – it now consists of 430 parliamentarians. On 1 February 2009, 293 of them voted for the former leader of the 'moderate' Islamists Alliance for the Re-Liberation of Somalia (ARS), Sheik Sharif Sheik Ahmed, as the new president of

Somalia. This took place in Djibouti because of security concerns in Somalia. He is hoped to rebuild the country together with Prime Minister Shermarke. However, Al-Shabab still does not recognize the transitional government and pledged to continue its fight. The mandate of the transitional government is extended until August 2011. In order to appease Al-Shabab the president has expressed his readiness to introduce Sharia. Nevertheless, the animosity between Al-Shabab and the transitional government continues.

The radical Islamists violently target the democratization process. Freedom of speech and democratization are against their Sharia law. Journalists have been increasingly targeted by various Islamists in Somalia. Between February and August 2007 many journalists and heads of radio stations were killed: Ali Mohammed Omar, a young presenter on privately owned *Radio Warsan*, one of the most listened to in the central-western Baidoa region, was killed on 16 February 2007; Mohammed Abdullahi Khalif, of *Radio Voice of Peace*, in Galkayo, Puntland, was killed on 5 May 2007 while reporting on arms trafficking in the town; Abshir Ali Gabre and Ahmed Hassan Mahad, *de Radio Jowhar*, was killed on 15 May 2007; Mahad Ahmed Elmi, head of the Mogadishu radio station *Capital Voice*, was killed on 11 August 2007; Ali Iman Sharmarke, co-founder of *HornAfrik* radio, was killed on 11 August 2007; Abdulkadir Mahad Moallim Kaskey, a journalist on *Radio Banadir*, was killed on 24 August 2007; and the head of another popular Mogadishu radio station, Bashir Nur Gedi of *Radio Shabelle*, was killed on 19 October 2007 (RWB 2008: 26–7). In early February 2009 the director of Somalia's independent *HornAfrik* radio station, Said Tahlil Ahmed, was shot dead in the capital. Soon after the murder, all radio stations started to play Koranic verses. Ethiopia, Kenya as well as the West are alarmed by these developments. These regional and global dimensions will be shown in Chapter 7. Without inclusion of the regional and global dimensions, Somali peace remains impossible.

Chapter 6
The Intrastate Conflict System in Sudan

Actors and Factors of the Sudanese Intrastate Conflict System

Various actors and factors on societal, political and economic levels have contributed to the Sudanese intrastate conflict system. These factors and actors on the three levels influence each other. Moreover, most actors act not only on one level, but instead they can be multi-level actors. Similarly, almost no factor is limited to one level only. These actors and factors on different levels facilitate the involvement of regional and global actors. In this constellation local factors become regional and global factors; and the local actors act in their own interests as well as in the interests of the regional and global actors. Such constellation is significant for the dynamic of conflict and violence. This chapter discusses the determining actors and factors of the internal conflict system on societal, economic and political levels in Sudan in order to show how they constitute an essential component of the regional and global system in the Sudan.

Society

Ethnic Diversity and Ethnic Marginalization

As the Congolese and Somali cases have shown, while studying conflicts in Africa it is important to discuss how the society is fragmented. One of the characteristics of the fragmentation is the ethnic factor. There are two major forms of this fragmentation: tension between the state identity and ethnic identity, and tension between two or more ethnic groups in a state. Whereas the state constructs its identity in view of the political and historical developments within the context of the international or global system, the ethnic identity is constructed in view of the interaction between ethnic groups and the state itself. Interethnic tensions occur on cultural, economic or political grounds.

Though ethnicity is not in itself an essential conflict factor *a priori*, it can have a huge conflict potential. The Sudanese population of about 40 million consists of more than 50 ethnic groups subdivided into about 570 distinct peoples. The southern peoples constitute 34 per cent of the Sudanese population. The Arabized peoples of northern Sudan (Ja'aliyin Arab, Juhayna Arab, Gezira Arab, Hawawir Arab, Mixed Arab-Nubian and Christian Arab) constitute 40 per cent of the population. The rest are non-Arabized peoples of northern and western Sudan with various subdivisions: Beja, Dar Fur, Nubian and West African such as Fulani,

Hausa, etc. (Lesch 1998: 17). In Sudan these cleavages resulted in tension between ethnic nationalism and territorial nationalism (Lesch 1998: 21). On the state level, the territorial nationalists, mainly Sudanese Arabs, stress uncompromising territorial integrity. They argue that though the south was ruled by the Brits in the 1930s and 1940s separate from the north, there is no historical background for its independence.

In their struggle, originally for independence but later for autonomy, the southerners stressed ethnic nationalism. Ethnic nationalism conceives language, history, culture and religion as a fixed primordial precondition of nationhood. Further, ethnic identity is built on constructed and imagined collective identity. An integral element of this identity is the consciousness of collective suffering. There are various types of suffering in the history of ethnic identity in a multi-ethnic state. The initial formation of the state will be conceived by those suffering minorities as illegitimate and coerced. There are different possibilities to overcome these collective sufferings: secession, autonomy, restructuring of the political system by creating inclusive identity, addressing the social, political and economic imbalances, enabling all peoples to gain a stake in the system, etc. This could be summarized as the ethnic pluralist approach (Lesch 1998: 23). The contention over pluralism (autonomy) or centralism has coined the struggle for state or ethnic identity in Sudan. Hence, the difficult task of state-building is how to reconcile the demands for autonomy and national centralism.

The Sudanese government maintained the violence system in south Sudan through direct military intervention and through re-division of the region into three subregions with the aim of undermining the intra-south solidarity. Khartoum armed the various factions within the south to ignite an intra-south split along ethnic lines. Southern Sudan consists of Nilotic linguistic groups (Dinka, Nuer, Shilluk, Anouak, Acholi, etc.), Nilo-Hamitic linguistic groups (Bari, Kuku, Pojulu, Nyepo, Lokoyo, etc.) and Sudanic linguistic groups (Azande, Muro, Ndogo, Sere, Biri, Madi, etc.).

President Numairi (1969–85) knew about this intra-southern discord and availed himself of it and played persons and groups against each other. For example, the Equatorians had an anti-Nilotic (such as Anuak, Dinka, Nuer, Shiluk) attitude. Especially after the Addis Ababa Agreement (AAA) the number of Dinka increased in government services in the southern region, whereas the Equatorians had prominent roles in the regional executive branches. The inter-tribal rivalry was exacerbated when it became obvious that the Dinka tended to favour fellow Dinka in the provision of services (Lesch 1998: 50).

The rivalry between Abel Alier (a Dinka) – who was a senior Southern Front leader, a cabinet minister, leader of the government delegation to the AAA in 1972, vice president of the nation and Numairi's first choice for president of the High Executive Council (HEC) – on the one hand, and an Equatorian, General Lagu – who was head of the southern military command – on the other hand, was successfully exploited by President Numairi. The move to redress the power balance and counteract the Dinka domination brought Lagu to power after the

1978 elections. After taking power Lagu was opposed from many sides, so Numairi replaced him with Alier in 1980. However, Alier remained insensitive to Equatorian complaints and appointed Dinka to half of the ministerial posts in the HEC (Johnson 2003: 52). Meanwhile, the coalition between Lagu and Numairi intensified. Both advocated re-regionalization of the south by splitting it into three regions in order to undermine the Dinka domination. Whereas Lagu hoped more advantage for his Equatorians after this re-division of the south, the intention of Numairi was to weaken the common position of the southerners. Further, Numairi dismissed Alier and his government in October 1981. At the same time many from the Dinka, Anuak, Nuer and Shilluk (all Nilotic, which Alier himself belongs to) supported the policy of regionalization.

This internal division of the southerners was in the interest of not only Numairi, but also of the northerners who opposed the concessions of the AAA. Even though many Dinka, Anuak, Nuer and Shilluk supported the idea of regionalization, most of the peoples of the Upper Nile and Bahr al-Ghazal were not in favour of the re-division. Moreover, the southern regional assembly had already rejected the idea of re-division (splitting the south into three regions: Equatoria, Bahr al-Ghazal and Upper Nile) in March 1981.

As happened towards the end of the 1970s, splits within the Sudan People's Liberation Movement (SPLM) began in 1983. Throughout 1984 independently acting units of Gai Tut and Akwot (Nuer and Dinka officers respectively) continued to cooperate by receiving military help from the government of Khartoum. Ultimately, the split and the violence related with it resulted in the death of both by the end of 1984. As a consequence, both units were combined and became allies with the Nuer majority led by William Abdallah Cuol. The Sudan People's Liberation Army (SPLA) consisted of Dinka as well as Nuer. When the confrontation between SPLA and Cuol's forces (Anyanya-2) occurred, it developed into a Nuer civil war (Johnson 2003: 66).

Although the 40,000-strong SPLA controlled half of the rural south already, between 1985 and 1986 its presence in Equatoria was very marginal since the SPLA was often considered as a Dinka-dominated movement. For decades this intra-south split was successfully exploited and exacerbated by the Khartoum government. The government of Sudan knew that hatred against the Dinka had increased since the AAA. As mentioned above, the number of Dinka in administrative positions increased and, after all, Garang himself was a Dinka. To ignite more divisions within the south, the national government began to supply arms for raids against the Dinka and Nuer, who were supporting the SPLA. To obtain this military aid, Cuol moderated his platform of separation to federation (Johnson 2003: 68).

A further factor that hampered the coordination of southern interests emerged during the period prior to the 1986 elections. A number of parties briefly flourished with tribal restrictions. The People's Progressive Party (PPP) and the Sudan African People's Congress (SAPCO) concentrated in Equatoria and were suspicious of the Dinka. Moreover, the Sudan African Congress (SAC) and Sudan People's Federal Party (SPFP) were both confined to the Upper Nile; however, the SPFP was restricted to the Nuer. Moreover, the Sudan African National Union (SANU) was revived and the Southern Sudan Political Association was created (Johnson 2003: 73).

In addition to the split itself, the Nasir faction of the SPLM/A received military aid from the government. Through the direct and indirect help of the Nasir faction, the government army regained territory in Jonglei and eastern Equatoria by July 1992. While the government concentrated its troops by July 1994 between Juba and Numele, the SPLA defence was hampered by defections and a lack of supplies (Johnson 2003: 100). From the end of 1991 and beginning of 1992 onwards, the government encouraged SPLA-Nasir (SPLA-United, led by Riek Machar) to attack SPLA-Torit (SPLA-Mainstream, led by Garang) in the Bor and Kongor districts. The SPLA-Nasir (mainly Nuer) devastated Dinka villages, destroyed properties and caused the displacement of the population, which ended up as a hot war between these two factions (Lesch 1998: 159).

Fear of the Dinka dictatorship emerged after the January 2005 peace agreement between the SPLM/A and the Khartoum government. The same problem emerged in Southern Kordofan. Khartoum continued to promote a pan-Arabist and Islamist ideology and divisions between the Nuba and Arab tribes in Southern Kordofan. The Misseriya and other Arab tribes have been instrumentalized to maintain a buffer zone between northern Sudan and the oil fields adjacent to the Nuba homelands. The Nuba are the largest group in the region, which consists of a diverse population that speaks more than 50 languages and follows Islamic, Christian and traditional faiths. They are of indigenous African origins and mainly sedentary. Baggara pastoralists (ethnic Arab, cattle herders) are the second largest group of tribes. They are divided into the Misseriya (western Southern Kordofan), the Hawazma around the central Nuba hills and the Awlad Himaid (eastern Southern Kordofan). Other minority communities include the Berno, Bargo and Hausa, Arabic-origin camel herders such as Shanabla, Ma'aliya, Kebabish, Kenena, Beni Jerar, and the Jellaba from the north, who have historically controlled trade through large parts of Sudan and owned large agricultural lands. Besides the fertile soil and grazing land in the eastern and southern part of Southern Kordofan, the Abyei territory (rich in oil, and inhabited mainly by Ngok Dinka together with other minority tribes, including Arabs) has amalgamated the cultural, economic and political dimensions of the conflict (ICG 2008a: 1).

At different times Khartoum started settling Arab tribes in West Kordofan to undermine the Nuba majority. The Nuba protested against the inclusion of the Arab-settled parts into the Southern Kordofan state. Many Nuba view the addition of new Arab-dominated localities into their state as a strategy to produce demographic change and weaken Nuba chances to obtain constitutional and political gains from the eventual popular consultation (ICG 2008a: 3). Moreover, the National Congress Party (NCP) and the SPLM disagree on the name of the state, with the former pushing for South Kordofan and the latter for Nuba Mountains. For the SPLM the majority are non-Arab and its supporters. Hence, the name 'Nuba Mountains' might help them win support from non-Arabs and non-Muslims, in particular the

Nuba, in the general elections. For the NCP, the general census of 1994 confirmed that the majority are Arabs, hence its supporters. Therefore, 'South Kordofan' would reaffirm the party's local pre-eminence (ICG 2008a: 3).

Sudan has been attempting to create a kind of common national identity by bridging the differences based on languages, cultures, histories and religions within the state territory. These common denominators determine the process of *imagination* of the nationhood as its *preconditions* of the state. Common culture, history, religion and language are usually believed to be primordial preconditions for nationhood (Lesch 1998: 5). Through such imagination Sudan has been trying to construct and institutionalize a nation and mobilize the people by stabilizing their sense of belonging and identity. Islamization has been one such mobilizing factor. Warburg concludes that as long as the ethnic–religious divide, led by a northern Muslim elite, continues to play a central political role, there is little hope for peace in Sudan (Warburg 2003: 143). For the Islamists in Khartoum the Sudanese identity has been inseparable from Islam.

Religion

Religion has always played a key role in the whole peace process. It was not only the southern demand for autonomy or the discovery of oil or the issue of grazing land and water that have been stumbling blocks for peace. Indeed, the abolition of regional autonomy in 1983 was a major issue. But the Islamization that started in September 1983 continued to exacerbate violence. At the time of independence and afterwards, the political stakeholders represented the notion that only a Muslim state can legitimately rule over a Muslim majority (Idris 2005: 49–51). The Turabi, in favour of Islamization and Sharia, argued that since Muslims in Sudan are the majority, their will should prevail. According to Turabi, Islam is religion *and* state; a non-Islamic political system is the work of despotic rulers with Western culture and orientation; an Islamic constitution would be a rule of *sacred* law (Warburg 2003: 147f).

Chapter III, Art. 14 and 15i of the AAA of 1972 clearly states the rights of the southern regional assembly to vote in order to request the president of the republic to withdraw any bill before the national assembly that adversely affected 'the welfare, rights or interests of the citizens of the southern region'. Numairi's undermining of the AAA was based on his political calculation: Islamization seemed to yield more political benefits than alliance with the south. Through Islamization he could attract Islamic political forces from various Islamic parties. In 1980 he decreed the unification of civil and Sharia courts, culminating in the presidential order of 8 September 1983, which made Sharia the sole guiding force behind the law of Sudan (Idris 2005: 53; Lesch 1998: 55). Politically he pleaded and decided to fight for abolishment of the dynastic attitudes of the Mahdists and their Ansar (Sudanese Sunni Muslim religious sect and followers of the Mahdists). Moreover, the fight and reform focused on the al-Mirghani's Khatmiya (the Brotherhood, which opposed the Mahdi family in the last century) and on the

Islamic sectarians (National Islamic Front – NIF). Since its creation in 1949, the NIF remained uncompromising regarding the introduction of Sharia.

In 1945 the Mahdists founded the Umma Party, which was the political organization of the Islamic Ansar movement strictly following the teachings of Mahdi, who ruled Sudan in the 1880s (De Waal 2004: 4). The Umma's political agenda basically does not object to Sharia as the law of the country. Although al-Mirghan's Khatmiya was ready to negotiate and compromise regarding Islamic law as it did in 1988 with SPLA/M, Mirghani felt uneasy when the SPLA demanded the abrogation of Sharia.

After the SPLA advanced into the north and on the backdrop of Garang's insistence on a secular constitution, the chair of the Transitional Military Council (TMC) (after the ouster of Numairi), al-Dhahab and al-Mahdi officials began to mobilize some Arab states, such as Libya, Iran and the Gulf States, against the SPLA, accusing it of poising to overrun the north and attempting to impose African identity on the Muslim Arab majority (Lesch 1998: 93). Similarly, since the Darfur war started, any criticism against the Khartoum regime by Africans or the West has been rejected as an attempt to demonize Arab cultural and Islamic identity (De Waal 2005: 200f).

Al-Sadiq al-Mahdi (1966–67; 1986–89) as prime minister reasserted that the Arab–Islamic ethic is the feature for Sudan. Accordingly, Sudan can preserve its pride and prestige only under Islam and its overpowering expression, which is Arab. The 1968 draft constitution underscored Islam as the official religion and Arabic as the official language. Besides ruling out the possibility of the regions to select their own governors or control their own security forces, the constitution made Sharia the basic source of civil and criminal law (Lesch 1998: 42). The Koka Dam Declaration of 24 March 1986, which urged the repeal of the 'September 1983 Laws' and adoption of the constitution based on the 1956 Constitution, was a major success for the SPLM/A after 1983. Further, the 1988 agreement between the DUP and the SPLA/M demanded the freeze of the Sharia laws of September 1983.

However, when the NIF took power in 1989, the situation changed dramatically. Islamic texts and the establishment of an Islamic state became the programme for a comprehensive Islamic call of the NIF regime (Warburg 2003: 201; Johnson 2003: 85; De Waal 2005: 191). This programme of a complete wrapping of politics by religious *messianism* is summarized in a speech by Bashir, who declared:

> We will gain nothing from relinquishing the sharia because he who seeks people's satisfaction by causing God's indignation loses everything … Our existence is originally linked to the implementation of his sharia. Therefore, it is a matter of principle for us … It is better for us to die in the cause of that principle and we are ready for that. (Lesch 1998: 129)

Even if Darfurians are Muslims, they are probably too black for the Islamist movement (El-Din 2007: 105). Even Darfur's Arabs were not recognized as equals to the Islamist–Arabists of Khartoum. They were denigrated as uncivilized, second-rate Arabs (De Waal 2005: 197).

In southern Sudan the Islamist ideology already had a long tradition of violence. The second phase of Islamization and Sharia and Arabization has had a considerable impact on the Sudanese inter-ethnic and inter-religious relations and conflicts, especially since 1983. For example, in response to the TMC's letter to the SPLA/M in July 1985 regarding the intention of convening a national conference, the SPLA/M stuck to abolition of Sharia as the precondition for the conference (Tetzlaff 1993: 14). In the face of the expansion of Islamization and Sharia law as well as Arabization, the SPLM/A was not ready to compromise. For the Arabizing ideology, Arabs were the standard-bearers for religion, culture and civilization. Claims to Arab genealogy became significant for legitimation: Arabic brought literate tradition, and Islam is associated with Arab language and culture (De Waal 2005: 185). The government policy and support for the Arabs cemented the popular assertions common among the nomadic that blacks and farmers are of low cultural status (ICG 2004: 10).

Theoretically, the implementation of the comprehensive agreement of January 2005 is the cornerstone of peace in Sudan. According to the Protocol (26 May 2004) between the government of Sudan and the SPLM/A on the Resolution of the Abyei conflict, Abyei will be accorded special administrative status, in which (*inter alia*) residents of Abyei will be citizens of both Western Kordofan and Bahr al-Ghazal, with representation in the legislatures of both states; Abyei will be administered by a local executive council, elected by the residents of Abyei. Pending the election of the executive council, its initial members will be appointed by the presidency; and net oil revenues from Abyei will be divided six ways during the interim period: the national government (50 per cent); the government of Southern Sudan (42 per cent); Bahr al-Ghazal region (2 per cent); Western Kordofan (2 per cent); locally with the Ngok Dinka (2 per cent); and locally with the Misseriya people (2 per cent). However, the heavy fighting that broke out on 20 May 2008 shows that the comprehensive peace agreement is far from being implemented, and an administration has yet to be set up in Abyei, which is claimed by both the north and south. Whereas an army spokesman said the former rebel SPLA had attacked with tanks, a senior southern politician in the town blamed the violence on the army, which he said had massed troops in the town; therefore, the objective of the SPLA is first of all to defend themselves and the area, and to defend the people. In the second half of February 2009 an armed conflict between the southern army and a militia led by Gabriel Tang, who was backed by Khartoum during Sudan's 21-year north–south civil war, took place.

Defending national or cultural identity and defending natural resources have become increasingly inseparable. The current economy of Khartoum and that of the Southern government depend on these natural resources.

Economy

Natural Resources: Grazing Land, Water and Oil

Grazing land has been one of the most important conflict factors of the Sudanese conflict (Johnson 2003: 44). The land issue was decisive in the emergence of conflicts in different parts of Darfur since various Arabs migrated there (De Waal 2005: 188). Since most of the grazing land for the cattle of the Arab nomads was within the southern territory, many northerners, including the Umma Party, opposed the very idea of autonomy for the south.

At the root of the conflict between Darfur and Khartoum, and Khartoum and South Sudan has been the competition over fertile land and water, exacerbated by desertification in northern Sudan and the drought that has affected Darfur and the whole of the Horn of Africa since the 1970s. This drought pushed nomadic groups from the northern semi-desert areas southwards. This ecological decline and a lack of development in the entire region have combined to impoverish Darfur people (Perry 2008: 137).

The principles of the protocol between the government of the Sudan (GOS) and the Sudan People's Liberation Movement/Army (SPLM/A) in Naivasha, Kenya, on 26 May 2004 on the resolution of the Abyei conflict acknowledge that Abyei is a bridge between the north and the south, linking the people of Sudan. Therefore, the Misseriya and other nomadic peoples would retain their traditional rights to graze cattle and move across the territory of Abyei, whereupon residents of Abyei would be citizens of both Western Kordofan and Bahr al-Ghazal, with representation in the legislatures of both states.

The grazing land dispute has been a serious cause of the war in Darfur. In South Darfur in a clash between the Fallata and Salamat ethnic groups on one hand and the Habaniya on the other, 150 people died in early December 2008. Similarly, about 100 more died in clashes between two groups from the Gimir group. The conflicts were mainly caused by disputes over access to grazing land and water. Khartoum has been trying to take advantange of such conflicts and arm groups from some of the groups involved in the fighting, buying their allegiance and using them as proxies against the rebels.

Similarly, the issue of water is another major conflict factor in Darfur (Perry 2008: 140–42). The role of Nile waters as a conflict factor is significant not only within Sudan but also regionally. Relations between Ethiopia on the one hand and Sudan and Egypt on the other depend heavily on the Nile water politics (Johnson 2003: 47). The 1959 Agreement between Sudan and Egypt on full utilization of Nile waters angered Ethiopia. The Khartoum regime continued to exploit the resources of the south by proposing to build the Jonglei canal, whose primary objective was to ensure the flow of 4.7 billion cubic meters of water of the Nile River annually, to be equally distributed between Egypt and Sudan (Sa'oudi 2001). The government tried to convince the southerners by naming some minor benefits of the canal for the local area to improve transport, cash-crop production schemes, drinking water and drainage. However, the main intention of the government was to fulfil the needs of the northerners and to develop mechanized agricultural farming and industry, such as the Kenana Sugar Refinery (Johnson 2003: 49). The discovery of oil contributed to the exacerbation of the conflict.

Oil

Oil has increasingly been the most important natural resource attracting global actors into Sudan. The discovery of oil in 1980 substantially contributed to the undermining of the 1972 AAA agreement (Johnson 2003: 45). Having realized that oil has been playing a key role in the Sudanese conflict, the SPLA/M forced Chevron to suspend drilling for oil in February 1984.

The Nuba Mountains faced two problems: the discovery of oil; and determining whether the population belongs to the north or to the south. Whereas Khartoum wanted to see the Nuba Mountains, the Blue Nile and the Abyei region in Southern Kordofan as part of northern Sudan (as agreed in February 1972), for the SPLM/A the population in these regions are black people and constituent parts of the south. Besides this, the oil discovery played a big role when Numairi rescinded the semi-autonomous status of the south with government in Juba, which was agreed in the 1972 peace agreement (Prunier 2002).

As part of the Popular Defence Force, NIF students have been mobilized and armed beside the Arab militia for the 'purity of religion', the 'integrity of the homeland' and to intimidate the 'foes of Allah'. The people of the Upper Nile area continue to be attacked and displaced by government-backed militias because of oil exploration in the area. In the south various militias, which consist of the government's Popular Defence Force, press-ganged youths, Islamist zealots, government-paid Arabic speakers, etc., to attack and chase civilians.

Throughout 1997 and even afterwards the grievances in Bahr al-Ghazal continued. In this region, although the SPLA mainstream had made significant military gains and slight improvement of security, it was not able to protect civilians from the widespread militia raids of the PDF (Popular Defence Forces). Moreover, the SPLA mainstream was not able to cope with the scale of the relief effort that was needed during the famine of 1998 (Johnson 2003: 108).

When the government was unable to regain the Nuba controlled by the SPLA, it attempted to render the area beyond the reach of relief by cutting off all access after bombing, burning villages and raping civilians in the Nuba Mountains in May 2000. The issue of the Nuba Mountains (besides the Southern Blue Nile, and lucrative oil fields of Heglig, Unity and Bentiu) continued to be a cause of conflict and a stumbling block to the peace process. These oil fields are especially important because, for El Turabi, the Sudanese government had only two allies: 'Allah and oil'.

The principles of the Protocol of Naivasha, Kenya, on 26 May 2004 between the government of Sudan (GOS) and the Sudan People's Liberation Movement (SPLM) on the resolution of conflict in Southern Kordofan/Nuba Mountains and

Blue Nile as regards wealth-sharing state that the oil-producing state is entitled to 2 per cent of the oil produced in that state, as specified in the Wealth Sharing Agreement. In allocating the funds to the war-affected and least developed areas, the National Reconstruction and Development Fund (NRDF) shall use the effects of war and level of development as the main criteria. The parties agree to allocate 75 per cent of the total fund to the war-affected areas, particularly to Southern Kordofan/Nuba Mountains and Blue Nile States, while the remaining balance shall be earmarked to the least developed areas.

Similarly, the principles of the protocol between the government of Sudan (GOS) and the Sudan People's Liberation Movement/Army (SPLM/A) in Naivasha, Kenya 26 May 2004 on the resolution of Abyei conflict state that net oil revenues from Abyei would be divided among the national government (50 per cent), the government of southern Sudan (42 per cent), Bahr al-Ghazal region (2 per cent), Western Kordofan (2 per cent), locally with the Ngok Dinka and Misseriya people (2 per cent) each. Administratively, the interim period for Abyei ends with a referendum either to retain its special administrative status in the north or be part of Bahr al-Ghazal.

On 7 January 2004 an agreement was reached in Naivasha, Kenya, on wealth-sharing during the pre-interim and interim period. The agreement elaborated guiding principles as the basis for wealth-sharing. It was agreed to share the wealth of Sudan equitably so as to enable each level of government to discharge its legal and constitutional responsibilities and duties, whereupon it would be expected from the national government to fulfil its obligation to provide transfers to the government of southern Sudan. Accordingly, it was agreed that at least 2 per cent of oil revenue would be allocated to the oil-producing states/regions in proportion to output produced in such states/regions. Moreover, the principles state that after the payment to the Oil Revenue Stabilization Account and to the oil-producing states/regions, 50 per cent of net oil revenue derived from oil-producing wells in southern Sudan shall be allocated to the government of southern Sudan (GosS) as of the beginning of the pre-interim period and the remaining 50 per cent to the national government and states in northern Sudan. Clashes in March 2008 between an Arab militia and the SPLM, and in May 2008 between northern and southern forces in the disputed oil-rich area of Abyei on the north–south divide, show that Sudan is not yet able to solve oil-related conflicts in spite of the fact that this was a key sticking point in the 2005 Comprehensive Peace Agreement. In the 2008 conflict some were killed and about 90 per cent of the Abyei population was displaced.

Non-poverty-reducing expenditure (defence, security, domestic and external debt repayments and debt services; delegations and conferences; subscription in regional and international organizations; custom duties; and government travel and hospitality) dominate at 85 per cent of total federal expenditure. Non-pro-poor spending, including defence spending, had declined in 2005 but rose to 72 per cent of non-pro-poor wages and salaries in 2006. According to the World Bank, non-poverty-reducing expenditure made 85 per cent of total federal expenditure in 2007. For example, in 2000 the government of Sudan collected $1.3 billion in revenue, $6.9 billion in 2006, and an estimated $9.1 billion in 2007. On the federal level pro-poor spending increased over the same period, but in absolute terms only from $0.137 billion to $0.364 billion. In 2005 and 2006 there was an increasing pro-poor effort, but there were higher transfers to northern states. This transfer accounted for more than half of total pro-poor spending in 2006 (Ezekwesili 2007: 38–44).

Structural Marginalization

In the western Sudan grievance, fighting over natural resources and structural marginalization resulted in the current destruction of Darfur. Except for the Zaghawa, who are mainly camel herders, the indigenous black African groups in the Darfur region depend on subsistence farming and animal husbandry, while groups of Arab extraction live on camel herding in northern Darfur and cattle herding in southern Darfur (De Waal 2005: 184). The rapid escalation of violence is caused by a systematic undermining of development efforts in the region and the shortcomings of government administration, including its deliberate weakening of the native administration (ICG 2004: 5). In the mid-1980s the government of Sadiq al-Mahdi armed the Rezeigat tribesmen of south Darfur, which contributed highly to the conflict between 1987 and 1989 over access to grazing lands and water sources between nomads of Arab origin and the sedentary Fur (ICG 2004: 6). In the same way Khartoum armed the Arab tribes of Darfur to turn back the 1992–93 attack by the Fur, aided by the SPLA. There were violent conflicts between the Zaghawa and groups of Arab background in 1994 and 1997 (ICG 2004: 7; De Waal 2005: 197; Prunier 2007b: 54–7).

The 1994 division of the Fur region into Northern, Southern and Western Darfur – splitting the centrally located fertile plains of Jebel Marra among the three states by Khartoum – led to the chronic economic and political marginalization of black African groups (De Waal 2005: 195). With a strategy to weaken the black Africans and to strengthen the influence of the Arabs in the region, Mohamed al-Fadul, then governor of Western Darfur, decided on 13 March 1995 to divide the traditional homeland of the Massalit into 13 Amarat (principalities), of which 5 were allocated to Arab groups. This is considered to be the trigger for the 1996–98 conflict in the region (ICG 2004: 7).

Similar policies by the Sudanese government have been pursued in the Southern Kordofan state, resulting in conflicts and distrust between the Nuba and Arabs. Some of the reasons for distrust between the Nuba and government-supported Arab ethnic groups include: developmental projects for roads, water reservoirs, clinics and hospitals were built mostly in the Arab-dominated areas, whereas the Nuba-dominated areas of the east and south-east Nuba Mountains were neglected; only 30–40 per cent of the 115 contracts signed under the initiative of the SPLM Governor Galab have been implemented three years after the CPA was signed; a contract worth $17.7 million was awarded to a company (Ithar) known to be run by the Popular Defence Forces (PDF) to drill 100 surface water reservoirs. In 1996–97 fertile lands used for mechanized farming were redistributed to Arabs (ICG 2008a: 6–10). This means that the political weakness of Sudan is a result of social and cultural fragmentation and economic and structural marginalization (Prunier 2007b: 58–60. Hence, a stable politics depends on social and economic fairness.

Politics

All the military and political measures undertaken by the Khartoum regime did not deter the southern fighters. However, after observing the international stance regarding a possibility of secession, most of the fighters moved from the adamant position of secession for the south towards a more compromising position of federation; this means compliance with the Organization of African Unity (OAU) charter of 1963 that opposed any alteration of territorial boundaries demarcated by the colonizers. Hence, although the Anyanya convention of August 1967 set up the southern provisional government in areas controlled by the Anyanya with the intention of secession, almost one and half years later (March 1969) it attenuated its insistence on secession by reserving the option of establishing a federation if acceptable terms could be negotiated (Lesch 1998: 43).

Throughout this period violence escalated. At Juba on 8–9 July 1965 about 1,400 people were killed, and between 1963 and 1966 a total number of more than 500,000 southerners were killed (Idris 2005: 52). The second half of the 1960s was overshadowed by an intensified war in the south, especially after the government security forces received a 'free hand' in June 1965 to 'restore law and order'. In July 1965 unarmed demonstrators were killed in Juba and half of its residential areas were razed. A few days later the army killed 76 people (including 49 southern government officials) during a wedding. After killing 187 civilians in the following month, the army put the editors (led by Clement Mboro and Abel Alier) of the Southern Front newspaper on trial for publicizing those killings. In the summer of 1967 the army killed 15 tribal chiefs in Bor where they were detained (Lesch 1998: 43). The war continued in the south and left hundreds of thousands of people dead by the late 1960s and forced several hundred thousand into forests or refugee camps in neighbouring countries.

The originally secular, socialist and pan-Arab government of Numairi that came to power through military coup in 1969 pursued a policy of inclusion. Industrialization, capital-intensive agriculture and government-directed development programmes characterized the focus of its economic policy. Numairi's original concept of political integration of the south was the most accommodating so far. He was committed to finding a lasting solution by enhancing the rights of the southern Sudanese to regional autonomy within a united Sudan. After crushing traditional Islamic parties and suppressing the left, both of which opposed concessions to the south, Numairi paved the way for a realistic chance of peace between north and south. This position of Numairi was further augmented by the insight of Colonel Joseph Lagu, who had commanded the Anyanya since July 1970, that though the guerrillas could prevent the government from controlling the south they would not achieve a decisive military victory against the government (Lesch 1998: 46). In January 1971 Joseph Lagu proclaimed the creation of the SSLM (Southern Sudan Liberation Movement). He could unite Anyanya leaders behind him, and could secure the support of the exiled southern politicians. Whereas the SSLM dominated the governing infrastructure, the Anyanya held the real power.

The peace agreement negotiated in February 1972 in Addis Ababa was an unprecedented success for south Sudan. It resulted in the Regional Self-Government Act for the south. Although many opposed the conflict settlement terms of the AAA, it was a remarkable solution at the time. According to the agreement, autonomy for the southern region composed of the three provinces of Equatoria (Al-Istiwaiyah), Bahr al-Ghazal and Upper Nile (Aali an Nil). Regarding the administration of the region, an elected Southern Regional Assembly recommends a regional president who will be appointed by the national president. The High Executive Council (HEC – cabinet) would be named by the regional president. The HEC would be responsible for all aspects of government in the south except defence, foreign affairs, currency and finance, economic and social planning and interregional issues (Idris 2005: 52). Another component of the agreement was the incorporation of the qualified Anyanya veterans into a 12,000-man southern command of the national army represented by equal members of northern and southern officers.

After the AAA of 1972, 27 March was celebrated as National Unity Day. Although the AAA and the Regional Self-Government Act were incorporated into the constitution of 1973, most of the constituent elements of the agreement remained unfulfilled. Numairi continued to intervene in every election for the president of the HEC. During his presidency of the HEC between 1972 and 1978, Alier was more or less a puppet of Numairi. Between 1972 and 1983 the southern region did not exercise any significant autonomy in the economic sector and education, and the regional ministries were mere prolonged departments of the central government ministries (Johnson 2003: 42).

The violent incidents that broke out in Juba in 1974, Akobo in 1975 and Wau in 1976 after the southern soldiers refused the transfer to the north were an anticipation of the decisive turning point of the incidents of January 1983. The implementation of the AAA was circumvented and ignored in the course of the Numairi government. The integration of the Anyanya army into the national army led to another decisive standstill of the accords. According to the accords, half of the troops stationed in the south would be drawn from the Anyanya; the police and prison guards would be recruited entirely from the south. However, the government replaced southern troops at Bentiu with northern soldiers to ensure Khartoum's control over the oil fields (Lesch 1998: 49). The integration was incomplete and many ex-Anyanya were left without livelihood (Johnson 2003: 64). In January 1983 the Dinka soldiers in the Bor and Pibor garrisons in Jonglei (Upper Nile) refused orders to move north. Moreover, attacks on the south Sudan in Bor and Pibor led to the defection of about 2,500 southern soldiers by July 1983, opening the way for the second Sudanese civil war.

After the clashes between the northern and southern soldiers on 16 May 1983, several hundred southerners left the Bor and Pibor garrisons and regrouped in Ethiopia. The Anyanya II became stronger after some of the deserters joined it. However, most of them joined the SPLA led by John Garang de Mebior, who was director of research in the army before he deserted (Lesch 1998: 49). After these revolts of May 1983 President Numairi abolished the Southern Region (Johnson 2003: 42). The Khartoum government has been using different strategies to destroy the peace and autonomy of the south. First, as shown above, it was moving the southern soldiers to the north and northern soldiers to the south. Second, it diluted the solidarity of the southerners by re-dividing the south Sudan into three regions: Equatoria, Bahr al-Ghazal and Upper Nile. Third, it armed different southern groups fighting one another.

After the 1983 southerners' military revolt against Khartoum, a very important aspect of the war was the new political strategy of the south. For various reasons the political and military goal of Garang as leader of the southern rebels (Sudan People's Liberation Movement/Army [SPLM/A]) was unity rather than separation. One reason was that the OAU charter clearly stated that colonial boundaries had to be respected. Another reason was that Ethiopia, the SPLA's main foreign supporter at the time, would have opposed any secessionist move fearing similar secessionist desires within its own territory. Having recognized that the regional and international atmosphere was not favourable for cessation, the SPLA and its leader Garang stuck to unity within a new Sudan. But still the pressing issue was how to reconcile the 'new Sudan' proposed by the southerners, and this could be possible in the face of the strong Islamizing ambitions of the Khartoum regime.

In 1988 talks took place between the SPLA/M and the DUP, and both agreed on 16 November 1988, and agreed among others abrogation of the interstate military pacts, ending of the state of emergency, constitutional conference, etc. But these demands were rejected by both the NIF and Umma. However, in the course of time, popular pressure for peace increased considerably, and a broad-based government was formed. Consequently, the national assembly endorsed the agreement of the DUP-SPLA/M on 3 April 1989. The agreement came to a decisive stalemate when the government was overthrown by NIF officers on 30 June 1989. This was the worst thing that could have happened to the Sudanese peace process at the time.

Since 1985 the government of Sudan has created an Arab civil militia against the population of southern Sudan. When the SPLA killed unarmed Arab residents of Gardud in 1985 the government responded brutally by arming the Baggara Arab militias (Muslims of Arab descent and considered racially acceptable to the regime), which perpetrated a catastrophic retaliation (Lesch 1998: 91). The *Baggara* (22–30 per cent of the Darfur population) militias were created to counter the SPLA (Prunier 2007a) when the SPLA made its first incursion into the Nuba Mountains in July 1985. The *Baggara* had stolen cattle and burnt Nuba villages, and consequently more than 60,000 people were displaced from the oil-rich area. More than 10,000 children from this area were kidnapped and kept in military training camps and indoctrinated in Islamic fundamentalist principles. Moreover, the *Murahaliin* militia was created as a pro-government force against the SPLA. Around the mid-1980s the *Misiriya* militia was reorganized into a paramilitary force. Towards the end of the 1980s the NIF government in effect legitimized the *Murahaliin* militia. The politics of the civil militia and the militarization of the civil society was exacerbated after the creation of the Popular Defence Force (PDF) in 1990, which became the 'school for national and spiritual education' in order to 'remould the mind' and enhance 'religious consciousness' (Gebrewold 2005: 200–202).

Nevertheless, during its five-month offensive in 1992 in the south the government army suffered heavy losses of about 20,000 casualties (Lesch 1998: 160). Besides training almost the same number of new recruits in order to compensate these losses, the government of Sudan began to arm Arab militias (Popular Defence Forces), which resulted in successes of the Sudanese government and continued throughout 1993 and 1995. With their attacks in October 1989, the 3,500 Miriyya Arab militias in south Kordofan had already contributed to the military successes of the government of Sudan. By 1993 about 250,000 Nuba were displaced; their land was sold to Arab entrepreneurs, handed to government and military officers and Arab militias (Lesch 1998: 162–3).

When after decades of negotiations and sufferings of the southern Sudanese a prospect for peace started to arise, the Darfur war started. Idris suggests that the Darfur tragedy is not to be seen in an 'Arabs vs. Blacks' model of explanation (Idris 2005: 79). Similarly Alex de Waal argues that 'Darfur's historic identity has been "African" and "Arab" with no sense of contradiction between the two' (De Waal 2007: 2). The crisis in the Darfur is a manifestation of the national crisis wherein Sudan has found itself for decades moved by a single vision of nation through the state's policy of Arabization and Islamization (Idris 2005: 79–92). Gérald Prunier suggests that the true reasons of the genocide in Darfur are racial and cultural (Prunier 2007a). But this cultural and racial violence is a manifestation of an underlying problem rather than the first cause. This violence implies that the state as a whole is in a fundamental crisis, partially caused by diminishing natural resources (Perry 2008: 140–43), ethnic and cultural conflict, negotiations and peace agreements in southern Sudan without addressing the grievances of other regions, and the poor relationship of the national government with impoverished and marginalized groups throughout the country (Kasfir 2005: 197; Tubiana 2007: 70–71).

The Janajawiid militias are armed and mobilized by the Sudanese government to strike back against the Fur, Zaghawa and Massalit peoples, who are collectively accused of supporting the rebellion in the Darfur region. 'Janajawiid soldiers' salary comes from what they can grab and the promise of land cleared of those now living on it: in other words, the booty captured in raids on villages. The government has thus handed the Janajawiid a licence to terrorize the population of Darfur and given them the means to do so' (Prendergast 1999). The Janajawiid militias, supported by the government, burn villages and kill people.

The Darfur conflict system consists of political discord between the people of Darfur and the central authority in Khartoum; ethnic divergences combined with conflict over means of subsistence; the Khartoum regime's covert exploitation of local differences; and its claims of these differences to be *the* real cause of Darfur's ailments (El-Din 2007: 93). De Waal suggests that there are three persistent problems in Sudan: the hyper-dominance of the national capital or the centre; inability of any one elite faction to establish unchallenged political dominance over the state; and ethnically targeted violence involving regular army units and tribally mobilized paramilitaries (De Waal 2007: 4–8). And Kasfir makes a very significant conclusion that solutions should focus not only on ending violence in Darfur, but also on ensuring that violence is not repeated elsewhere. De Waal argues that the main cause of violence in Sudan is not necessarily the criminality of the political leaders but instead the structure of the Sudanese polity, the Sudanese political condition that selectively permits those with certain skills and proclivities to assume power; difficulty in obtaining a consensus of any bold, sustained, or proactive policy; dealing with the immediate manifestations of the problem without regard to the long-term consequences (De Waal 2007: 32–3).

Because of this crisis, the conflict system in Darfur is made up of multiple actors and factors that go beyond Darfur and include the whole country. Sometimes government-aligned forces, government militia, rebels and various Darfur communities fight each other in different parts of the country, and sometimes they support each other. For example, when the conflict in Darfur began in 2003, SPLA officers trained some SLA fighters (Kasfir 2005: 198). Moreover, the Beja Congress from eastern Sudan allied itself with the SLA. The differences between the SLA and the Justice and Equality Movement (JEM) leadership encouraged the former to join the National Democratic Alliance (NDA). Darfur's diversity has always been successfully exploited by Khartoum, and in the mid-1980s it manipulated ethnicity in the interests of central politicians and their provincial allies. Understandably the government of Sudan has been trying to prevent the fighters from unifying their movements. Even on the interregional level, an alliance between the Darfur rebels and an independent oil-rich south Sudan would be the major threat to the Sudanese state (Prunier 2007a).

The armed militia, Janjawiid, is systematically orchestrated by the Sudanese government. Some sources suggests that Janjawiid consists of various categories of pro-government armed groups and groups *formally* armed by the government: the 'Peace Forces' (Quwat al Salaam), the Nomad Protection Forces, the Um Bakha Irregular Forces, the Um Kwak attacker forces, the Popular Defence Force (Difaa al Sha'abi), the Popular Police Force (Shorta al Sha'abi), Border Intelligence Guard (created in early 2003 in a public ceremony and armed by and annexed to fight the rebels), Nomadi Police (Shorta Zana), the National Army, and official Police Force, the latter two not directly related to Janjawiid but working closely with it (Haggar 2007: 113; Kasfir 2005: 195).

The Abbala of Northern Rizeigat militia encouraged by the GoS to gain land and regional power with money and weapons available from the government have been vital to the GoS's military strategy and became the backbone of the Janjawiid. They later integrated into official organs such as Border Guards (formerly known as Border Guard Intelligence Unit) in north Darfur, Popular Defence Forces in west Darfur, and in south Darfur Nomadic Transhumance Routes Police (ICG 2007a: 9).

There have also been tribal militia groups *informally* or indirectly armed by the government including Musa Hilal's infamous El-Khafif, El-Sariya, El-Muriya (the *light*, the *fast*, the *fearful*) based in the Kebkabiya area of north Darfur (ICG, Africa Report No. 89, 8 March 2005: 8). Musa Hilal has been fighting not only for the government but also for the Arab cause between the Nile and Lake Chad (Daly 2007: 266). The government supported creating another group called *Khorbaj* in south Darfur, a predominantly Arab group designed to organize Arabs into a political movement. In the same way in the Western Kordofan a new group, al-Shahamah, emerged in October 2004 led by Musa Ali Muhamadein, a former follower of Popular Congress Chairman Hassan el-Turabi and a former leader of the Popular Defence Forces in Western Kordofan, vowing to fight the Nuba Mountains/Southern Kordofan Agreement reached between the GoS and the SPLM, on the grounds that this would dissolve Southern Kordofan. Another anti-government group, called the Sudanese Movement for the Eradication of Marginalization, emerged in December 2004 attacking government oil installations, such as at Sharef at the border of South Darfur and Western Kordofan (ICG 2005a: 13).

The DPA (Darfur Peace Agreement) of 5 May 2006 (signed between Mini Minawi's SLM and the Government of Sudan) required complete, verifiable disarmament and demobilization of the Janjawiid militia by mid-October 2006 and restrictions on the movements of the Popular Defence Forces and their downsizing. Right after signing the DPA, Minawi's SLA in North Darfur committed so many atrocities that many of its members defected to the non-signatory rebel movements (ICG 2007a: 11). Further, such a step towards peace was complicated not only by the divisions between government-supported Arabs and non-Arabs. The conflict dynamic evolved due to the rise in intra-Arab conflict. Arab groups who had been actively supporting the government and supported by the government began to fight against each other. The conflict between the small Targam tribe and Abbala Rizeigat (Arab camel herding tribes); between the Salamat militia and Habaniya against the Abbala in general and the Salamat in particular in Dar Beni and Halba and Habaniya; Beni Halba's rejection of the government strategy to recruit Popular Defence Forces from its members; a decreasing influence of 'Arab Gathering' formed in the late 1980s to represent the political interests of Arabs in the central government and the NCP (ICG 2007a: 4). Such divisions among Arabs suggest that the NCP cannot represent all Arab groups in Darfur and cannot be depended upon to unify the Arabs and non-Arabs (ICG 2007b: 13).

On the one hand the intra-Arab divisions have been emerging, and on other hand the Arab and non-Arab coalitions are surfacing in spite of the attempts of the NCP to hinder this. Whereas in 2007 relations between the Fur/Massalit and the Arabs were established after the latter realized that over the long term they have to cooperate with the sedentary groups such as the Fur, the relations between the Arabs and Zaghawa are strained by a successful government campaign to isolate the Zaghawa (mostly living in North Darfur), exacerbating the already existing resentment of the population against the Zaghawa who the Fur consider as betrayer of their cause since Minni Minawi, a Zaghawa, signed the Darfur Peace Agreement (ICG 2007a: 4).

Some factors that continue to obstruct any chance for peace in Darfur are: arms pouring into the IDP (internally displaced people) camps in the Darfur, politicization of the IDP communities, re-emergence of alliances between Arabs and non-Arabs, fragmentations within Arabs, some rebel leaders such as Abdel Wahid commanding great popularity among Fur IDPs, etc. This means that both political and economic root causes of the Darfur conflict have to be addressed: economic and political marginalization of Darfur from the centre; political manipulation of the region by the NCP, land ownership, grazing rights and water, traditional structures of governance which monitored these systems, desertification, access to small arms (ICG 2007a: 7; Perry 2008: 140; De Waal 2005: 192–200; Prunier 2007b: 54–8).

From a continuing conflict in Darfur the GoS can only profit, at least in the short term: the elections and the southern referendum will be postponed; there is no unified opposition in Darfur; in the current state the losses for the GoS if national elections are held by 2009 are certainly low; Khartoum-linked non-Darfur forces control Darfur states' security committees; the governors are co-opted and corrupted by the GoS; Khartoum hindered deployment of a more powerful international force (rejecting UN Resolution 1706); supporting Chadian rebels to undermine hostile Chadian policies; helping settle friendly tribes (mostly Abbala) in border areas in order to isolate the non-Arab groups from their kin; Chad as a strategy to contain the tribes affiliated to the rebels such as the Fur, Massaleit and the Zaghawa, and settling Arabs in the areas inhabited by the Fur and Massaleit to create a demographic change crucial for manipulation of elections (ICG 2007a: 8–10).

Militarization of Sudan poses another challenge for peace. The southern conflict claimed more than 2 million lives and displaced more than 4.5 million between 1984 and 2000 (Jack 2002: 52). Both the government and the rebels in the south militarize the civil population under the disguise of self-defence. The real motivation behind 'maintaining security and self-defence' has been to organize an allied, armed tribal militia. 'This has a long history dating back to 1960s when the government launched civilian national guards (Haras al-watan) and the Anyanya created civilian armed groups that they referred to as scouts' (Wassara 2002: 12).

The civil war initiated several self-defence groups: the Arab, Dinka, Ferit, Mandari, Murle Nuer and the Toposa militias, among others. Most militia formations are politically motivated (Wassara 2002: 12). The strategy of the government of Sudan as well as of the main rebel group (among others) in the process of the militarization of the civilian population is multifaceted. It is estimated that there are at least 25 government-backed militias in Sudan. Civilians are armed either to fight the respective 'enemy' with the argument of self-defence, or they are armed to attack and pre-empt the alleged supporters of the Sudanese government or of the SPLA. These militias have adopted tactics aimed at denying a civilian base of support for the adversary. Wherever there is a suspicion of civilian support, the civilian population will be looted of its property, houses burnt, women kidnapped, etc.

In the view of the southern population the Sudanese government was not simply arming the Arab pastoralists, but also orchestrating ethnic cleansing. This led to the formation of armed alliances and the purchasing of arms in the south for their self-defence. Coordinated destruction of villages, attacks on travellers, persecution, the arrest of educated people, hanging, amputation and stoning to death have been part of the Arabization and Islamization. The government tried to cover up the assaults against the people as ethnic conflicts exacerbated by weapons flowing in from Chad and Libya. For example, in April 2002 the Arab militias burned over 600 homes and killed 17 civilians. The government forces knew but did not want to intervene or act on behalf of the people.

The strategy of duplication of conflict actors and factors has been successfully implemented by the Khartoum regime for decades. The conflict maintains itself as a system by providing itself with necessary actors and factors. In the Darfur conflict that broke out in January 2003, the Sudanese government uses the same strategy of arming and igniting divisions. It is difficult to know the exact number of those killed, raped, enslaved, forcibly married or abducted as child soldiers in Darfur (Darfur Consortium 2008). It is estimated that between 2003 and 2007 about 400,000 have died in the conflict. During the same period 2 million people have fled Darfur in northwest Sudan, 250,000 of them between August 2006 and March 2007 (Prunier 2007a); in 2007 alone over 260,000 were displaced and, as of early 2008, 2.5 million in the West, North and South Darfur States (UNHCR 2008).

Conflict and Contradicting Interests of Rebels

The divisions and animosities among the Darfur rebels could not only exacerbate the various forms of violence against the civilians, but it has also been exploited by the Khartoum regime. A self-defence militia led by ethnic Furs, the Sudan Liberation Army (SLA), emerged from the unrest that followed a devastating famine in 1987 as a reaction to the Arab alliance that was established and armed by Khartoum to oppose the non-Arabic-speaking farming communities like the Fur, Zaghawa and Masalit peoples. The Darfur Liberation Front emerged in February 2003 and later renamed itself the Sudan Liberation Movement (SLM) (Daly 2007: 279–80). The SLM, instead of coordinating its struggle against the Khartoum regime, split along tribal lines in November 2005 after a power struggle between Minni Arkou Minnawi (an ethnic Zaghawa who controlled the military arm) and

Abdel Wahid Mohamed Ahmed el-Nur (an ethnic Fur who controlled the political wing). The split within the insurrection deepened when the SLM-Minni faction signed the 2006 Darfur Peace Agreement (DPA) and Minni accepted a government role as special adviser to the president, triggering fears of a sell-out to Khartoum. As a consequence, his group began losing strength as fighters formerly loyal to Minni deserted to join other anti-Khartoum groups (among them the Greater Sudan Liberation Movement, G19, Abdel Shafi Faction and SLM-Unity). Abdel Wahid, leader of the SLM-Abdel Wahid faction, refused to sign the 2006 DPA and refused to attend the Libyan peace talks unless the United Nations–African Union peacekeeping force was deployed first to guarantee security in Darfur, and some of his men joined with Ahmed Abdel-Shafi, the splinter group known as SLM-Classic, or SLM-Shafi faction, which broke away in July 2006. The SLM-Unity faction led by Abdallah Yehya was predominantly made up of ethnic Zaghawa in North Darfur and drawn from the Group of 19 (G19) commanders who served under Khamis Abdullah Abakr and cooperated with Abdel Wahid el-Nur until the 2006 peace talks when they split. This faction was blamed for much of the violence in the second half of 2007 and for killing 10 AU soldiers. Further SLM splinters include factions calling themselves Free Will, the Greater Sudan Liberation Movement/Army, and the National Movement for the Elimination of Marginalization.

The Justice and Equality Movement (JEM) founded by Darfuri Muslims loyal to Islamist leader Hassan al-Turabi and led by Khalil Ibrahim Muhammad has been fighting against the disproportionate numbers of Arabs in powerful positions. However, it fragmented into several disparate groups, including a faction led by Idris Azraq which broke away in January 2007 when he fell out with Mr Khalil over the marginalization of non-Kobe Zaghawa (Khalil's clan) within the JEM; the National Movement for Reform and Development (NMRD) broke away from the JEM in 2004 and was led by former JEM Chief of Staff Jibril Abdel Karim Bari, also known as Tek, fighting for separation of religion and state and, like the SLA, opposing the JEM leadership's Islamist tendency. He not only served as a colonel in the Chadian President Idriss Déby's republican guard, but also supports NMRD (made up of Zaghawa sub-clans) because its leadership has been closer to President Déby than that of the JEM (which is also largely made up of Zaghawa) (ICG 2005a: 12). Tek and his NMRD co-founders, Hassan Abdullah Bargo and Mahamat Ismail Chaibo, are ethnic Zaghawa from the Kabka clan. Other JEM splinters include the JEM Peace Wing, Field Revolutionary Command, Popular Forces Troops, the JEM Collective Leadership (JEM-CL created on 4 October 2007), etc. Since the JEM leadership could not agree who would sign for it, it contributed to the failure of the N'djamena talks. The National Redemption Front, an umbrella grouping of Darfuri rebel groups (including JEM and the G19), was founded with Eritrean backing in June 2006 and it opposed the DPA of May 2006, but later the commanders of this group also split into factions, out of which groups such as SLA/Unity emerged.

There have also been other groups: the Revolutionary Democratic Front Forces (RDFF) of the Baggara Rizeigat, but it later splintered; the United Revolutionary Forces Front created in the second half of 2007 by some elements from Abbala Northern Rizeigat; the Group for Development of Grievances; the 'Mother of all SLAs', Sudan National Liberation Movement; the United Front for Liberation and Democracy (helped and created by Eritrea after the National Redemption Front disintegrated), an umbrella organization comprising SLA/Khamees, SLA/Unity, RDFF, NMRD, and the Sudan Federal Democratic Alliance (SFDA). In late 2007 some of these factions joined together (two SLA factions, one JEM faction, the NMRD and the URFF) and created the United Resistance Front. Abdel Wahid and Khalil Ibrahim remained away, and SLA/Unity continued to pursue a separate unification process inside Darfur.

The Sudan Liberation Movement/Army (SLM) of Khamis Abakr, SLM-Field Leadership led by Adam Bakheit, the Movement for National Reformation and Development (MNRD) led by Jibril Abdel-Karim, the United Revolutionary Front led by Ibrahim al-Zubaidi, and the Justice and Equality Movement (collective leadership) led by Bahar Idriss Abu Garda signed a deal and created the United Resistance Front in Juba on 8 April 2008. But towards the end of July 2008 Adam Ali Shoggar, Khamis Abdella Abakr and Ibrahim al-Zubaidi decided to withdraw from the United Resistance Front (URF), accusing their partners of working to sabotage their leadership and failing to materialize the merger of the five groups during a period of six months.

Contradictory Ideologies within the Southern Rebels

When the revolt against Khartoum emerged, especially after the Addis Ababa Agreement of 1972, there also emerged different organizations with different orientations, including the National Action Movement (NAM) led by Akuot Atem and Samuel Gai Tut; Movement for the Total Liberation of Southern Sudan; Anyanya Absorbed Forces Underground Movement (opposed to the Addis Ababa Agreement, including Captain John Garang de Mabior); Anya Nya Two Movement led by Gordon Kaong Chol; Abyei Liberation Front led by Deng Alor Kuol and Chol Deng Alaak; and the Student Revolutionary Group headed by Pagan Amum Okiech among others. The initial intention of the southern rebels was to fight for the independence of the south. However, it was Ethiopia that played a big role by pushing them to move away from the southern liberation to Sudan's liberation. But Akuot Atem, suspicious that the Ethiopians wanted to impose John Garang on them as an absolute leader of the new movement to be launched, ordered his supporters to collect their belongings and move back to the Sudan, where they launched the People's Revolution for the Liberation of the Southern Sudan. Forces loyal to Akuot and Gai and other forces that joined them in Bukteng Camp together with Gordon Kaong soldiers inside the Sudan border launched an alternative movement they would also call SPLM/SPLA. The rest of the Anyanya II and those who disagreed with Akuot Atem and Samuel Gai Tut joined the John Garang de

Mabior group. John Garang de Mabior was elected by the group that remained loyal to him in Itang in August 1983, as Chairman of the SPLM and Commander in Chief of the SPLA. At the end, while trying to convince each other, Anyanya II and SPLM/A continued to operate as two separate organizations fighting amongst each other. This stubbornness brought untold sufferings and setbacks to the people of Southern Sudan that they had wanted to liberate (Arop Madut Arop 2008).

Garang made the SPLA a military organization with strong centralization and cohesion. As a consequence, many officers (such as Lam Akol, Riek Machar, Kerubino Kuanyin Bol, Arok Thon Arok) began to complain against this centralized structure. Besides its policy of cohesion and centralization, the SPLA began to approach the Anyanya II in Jonglei and Upper Nile between 1986 and 1988 to negotiate with Arab militias along the Bahr al-Ghazal border with Kordofan and Darfur in 1990 and create peace between Nuer, Anuak and Murle in the same year (Johnson 2003: 93). However, the initiative to remove Garang continued. When the Mengistu government of Ethiopia began to scramble, the military resources for Garang became very tight. In this precarious time for Garang the leader of the Anyanya II (Cuol), Riek Machar and Lam Akol (commanders in Nasir) on 28 August 1991 declared their aim to overthrow John Garang (Johnson 2003: 97). The outcome of the power struggle was the split of the SPLA, defection of several non-Dinka leaders and creation of SPLA-United (largely a Nuer-led group), which was later renamed SSIM (Southern Sudan Independent Movement/Army). It was led by Riek Machar and mainly restricted to the Upper Nile area. The split damaged the SPLA enormously in comparison with the advances it had made since the end of 1989. The alleged authoritarian rule of Garang was not the only reason for the intra-SPLA revolt. The insistence of Garang on a new united Sudan seemed to the contenders illusionary. The hard-line attitudes of Bashir and Turabi especially cemented the fear of those criticizing Garang's belief in a new Sudan. Before the split, the SPLA achieved a lot of gains: it retook the town of Kurmuk at the end of 1989; it launched a decisive attack around west Juba in February 1990; it controlled western Equatoria in November 1990; the Fartit militia in western Bahr al-Ghazal and Murahlin in Southern Kordofan and southern Darfur decided to negotiate with the SPLA (Johnson 2003: 85). However, the success was countermanded by the split that resulted in significant government advances between 1992 and 1994. It was after the SPLA split that Garang met the Nasir group in Torit in early September 1991. After this meeting Garang did not rule out, for the first time, the possibility of secession if the Sudanese government maintained its centralized Arab–Islamic system (Lesch 1998: 158).

In addition to the support Khartoum provided for the dissidents in order to weaken the southern movements, the government made false promises of independence of the south to the SPLA-United. Its only strategy was just to isolate the SPLA-United from SPLA-Torit. It was a paradox that an uncompromisingly unitary government support a group that was fighting for independence. From the background of the strategy and stance of the government of Sudan the Frankfurt accord of 1992, which did not rule out any option of the national aspirations of the southerners, was a mere tactic to obstruct any rapprochement between the SPLA mainstream and SPLA-United.

Until about the middle of 1995 the success of the Sudanese army was considerable. This success depended largely on the split of the SPLA-United that began towards the end of 1994. The SPLA mainstream took advantage of the differences between Akol, Kerubino and Nyuon on the one hand, and Machar on the other. SPLA mainstream convinced Machar to sign a ceasefire in February 1995, and with Nyuon on 27 April 1995 (Lesch 1998: 163). On 25 March 1995 they could regain Nasir from SSIM of Machar, an ally of the government of Sudan until then. In the same way it occupied Pariang in June 1995 and Kaya on 13 August in spite of the peace talks brokered by Jimmy Carter. However, on 25 October 1995 the Sudanese army suffered heavy setbacks, losing 10 garrisons and the town of Parajok to the SPLA, with the side effect of massive civilian suffering.

Besides the peace charter of 10 April 1996 between the SSIM, Kuonyin Bol of Bahr al-Ghazal-SPLA and the government of Sudan, a strategy of coordination of struggle against SPLA mainstream started. The peace charter, which was expected to be the final peace agreement, was further endorsed by former vice president Joseph Lagu (on 16 May 1996), Lam Akol of SPLA-United (on 22 June 1996), Equatoria Defence Forces (on 30 June 1996) and Philip Gabash of the Sudan National Party (on 30 June 1996). In addition to this rapprochement between the rebels and the government, which damaged the SPLA, the defection from and criticism of some Nuba and Bor against the SPLA considerably increased its crisis.

Another factor that characterized the inability of many southern rebels to coordinate their aims and interests was a new proliferation of renamed groups after the 'government's internal peace initiative of 1997'. In addition to the SSIM and SPLA-Bahr al-Ghazal of Kerubino, the Independent Bor Group, Independent Bahr al-Ghazal Group, Equatoria Defence Force, and SPLA-Independent Group (Nuba Mountains) emerged. Paulino Matip renamed his forces the Southern Sudan United Army (SSUA) after breaking away from Machar in 1998. Furthermore, SSDF-II (Juba), SSDF-United and SSDF-Friendly to the SPLA (western Nuer) were created by replicating the SSDF of Machar (Johnson 2003: 126).

It has often been suggested that land and water play a key role in the Darfur conflict. Tubiana argues that the land issue became prevalent only in 2005. Until then leaders of the JEM as well as the SLA focused on national issues rather than 'tribal' issues because many leaders such as Abdel Wahid Mohamed al Nur (a lawyer from the Fur tribe) and Minni Arkoy Minawi (a Zaghawa teacher) were young urbanites who had lived outside Darfur for long stretches of time, and not always in tune with local issues. As a consequence, they were weakly rooted in the local tradition that they pretend to represent (Tubiana 2007: 72). SLM/Minawi signed the DPA and became the adviser of the state president, whereas SLM/Wahid and JEM refused to sign the DPA on the grounds that it did not represent the real interests of the Darfurians, such as inadequate security arrangements and compensation (Daly 2007: 310). If there is a discrepancy between the real concerns of the affected population and the issues the rebels pretend to represent, an important element for conflict system is created because the various rebel leaders cannot agree which issue is the most pressing issue. Such conflicts usually undermine any chance for democratization, without which no sustainable peace will be possible.

Democratization

The 'nation-building' process in Sudan has been accompanied by crisis of democratic citizenship and violence against those considered as 'blacks', 'heathens', 'primitive' and non-Muslims (Idris 2005: 4, 13). Al-Turabi opposed Western as well as communist (people's democracy) models of democracy on the grounds that they were governed by men (not by God) and led to suppression (Warburg 2003: 182). Moreover, according to al-Turabi, since the Sudanese Islamic state enjoyed popular support, anyone who opposed it would be fought in a Jihad (Warburg 2003: 195). The government of Sudan and the SPLM/A met in Machakos, Kenya, from 18 June 2002 through 20 July 2002, and agreed on principles that would enable peace in a united Sudan based on free will of the Sudanese people; accountability, equality, respect, justice, and democratic governance taking into account cultural, ethnic, racial, religious and linguistic diversity; participation of the south Sudanese in the national government; the South's right to control and govern affairs in the region; south Sudan's right to self-determination through a referendum. The agreement established the basis for free and fair general elections during the interim period; a bicameral national legislature with equitable representation of the people of Southern Sudan in both legislative chambers; the National Civil Service and Cabinet Ministries be representative of both north and south; holding a census during the interim period; reviewing the constitution; how to share revenue and national resources.

The peace process was hoped to achieve an equitable and fair formula for sharing power and guaranteed by a responsible, just, transparent, people-led and integrity-based governance which functions only if decentralization and empowerment of all levels of government are implemented. The government of southern Sudan would play its role as the linkage between the national government and the states in southern Sudan. On the horizontal level, the principle of linkage within the southern autonomous states would be based on cooperation, open communication and coordination of functions. Regarding the legislature and executive authority of southern Sudan, the SPLM shall be represented by 70 per cent, whereas the NCP and the other southern political forces shall be 15 per cent each. The judiciary of southern Sudan would consist of Supreme Court, Courts of Appeal and any such other courts or tribunals as deemed necessary to be established in accordance with the southern Sudan Constitution and the law.

In the same way the general principles of the protocol of Naivasha, Kenya, 26 May 2004, between the government of Sudan (GOS) and the Sudan People's Liberation Movement (SPLM) on the resolution of conflict in southern Kordofan/

Nuba Mountains and Blue Nile states consider human rights and fundamental freedoms guaranteed to all individuals in the state as prescribed in the interim national constitution as basis for political, administrative, economic and social solution to the conflict in Southern Kordofan/Nuba Mountains and Blue Nile. The agreement plans to develop and protect diverse cultural heritage and local languages of the population of the state. Moreover, it states that development of human resources and infrastructure shall be the main goal of the state to meet human needs.

Another decisive aspect of the protocol was popular inclusion. According to the protocol, a just, fair and comprehensive peace agreement presupposes a popular consultation which is democratic right and mechanism to ascertain the views of the people. Therefore, the comprehensive agreement shall be subjected to the will of the southern Kordofan/Nuba Mountains and Blue Nile states. It was agreed that the legislatures of the two states would each establish a Parliamentary Assessment and Evaluation Commission to assess and evaluate the implementation of the agreement in each state. The two commissions would submit their reports to the legislatures of the two states by the fourth year of the signing of the comprehensive peace agreement. Moreover, an independent commission will be established by the presidency to assess and evaluate the implementation of the comprehensive Peace Agreement in each of the two states. The commission would submit its reports to the national government and the governments of the two states, which would use the reports to rectify any procedure that needs to be rectified to ensure faithful implementation of the agreement. The security arrangements of the agreement state that in the two states 55 per cent of the executive and legislative posts would be allocated to the National Congress Party and 45 per cent to the SPLM.

As long as there is systematic marginalization, there cannot be a genuine democracy. Khartoum has been systematically marginalizing the Nuba in favour of the Arabs. In April 2008 a census took place whose outcome the Nuba consider inaccurate because of insecurity and with many IDPs not yet back home. Therefore, according to the Nuba, the census results do not accurately reflect their size and their claims to representation. The NCP was deliberately creating conditions that made their return difficult. Nuba IDPs who failed to return before the census were not counted as living in the Nuba Mountains. This would make Arabs appear in the majority in the region, where the Nuba Mountains were subjected to a scorched earth policy and forced migration between 1991 and 1996.

According to the Comprehensive Peace Agreement of 2005, the next elections will be held no later than July 2009. But the overall situation in Sudan still seems very gloomy – neither the elections of 2009 nor the referendum of 2011 can take place. Viable democracy without a free media is not possible. Democratization and an active and critical civil society are the basis for viable peace and security (Annual Report of Reporters Without Borders 2008). The media environment in the southern part of the country is relatively stable, though there are no existing laws that explicitly guarantee freedom of the press (Freedom House 2008).

According to Reporters Without Borders, security agents in Sudan are increasingly suffocating, raiding, confiscating, sabotaging and closing privately owned media, forcing their editors to live in fear and depriving them of significant resources (Annual Report of Reporters Without Borders 2008). Media covering sensitive issues, including the conflict in Darfur, is punished by a code of criminal procedure. One alarming method of controlling the media is that media outlets are required to employ security personnel to review stories prior to publication; in addition, the government-influenced National Press Council is responsible for licensing and has the power to suspend journalists and newspapers. Moreover, entry into the media is difficult unless one is a supporter of the government. According to Freedom House, throughout 2007 journalists faced harassment, attacks, intimidation and direct censorship at the hands of both government and non-governmental forces (Freedom House Report 2008).

However, the Sudan case shows that peace and democracy cannot be possible unless the complex conflict consisting of various internal, regional and global actors is addressed. As shown above the lack of unity among rebels, ethnic groups and regions has been one of the main sources of violence in Sudan. Such inter-rebel division has caused external involvement as well as it has been caused and exacerbated by external involvement. In the following chapter this external dimension will be shown.

Chapter 7
Regional and Global Conflict Systems in the Horn of Africa

Regional conflict system consists of alliances and antagonisms of the states of a region. It is not just a spill-over effect, but also a deliberate mutual destabilization by directly involved actors and factors of the region. In the same way, the global actors like China, France, etc., are not the causes of the conflict in the individual states as well as in the region but their global calculations, ambitions and actions have become part of the overall conflict system. The roles of the US, China and France in the Horn of Africa are decisive in solving as well as exacerbating the conflicts. Fighting terrorism and piracy, as well as natural resources, have brought the regional and global actors together. Therefore, this chapter deals with the impact of the regional and global actors on the Somalia and Sudanese conflicts. The first section of the chapter deals with the regional constellation, whereas the second section discusses the global aspect.

Regional Conflict System in the Horn of Africa

In the regional conflict system, internal conflicts are externalized and external conflicts are internalized going beyond the interstate level to the regional level. In this region economic interdependence among neighbours is very low; whereas, capacity to mobilize the population easily for war is very high. Political power is frequently contested by force (Buzan 2005: 224). As a consequence, states in the Horn of Africa live in tension with each other.

Kelman and Fisher suggest four propositions on international conflict: First, international conflict is a process driven by collective needs and fears rather than entirely a product of rational calculation of objective national interests on the part of political decision-makers. This is based on psychological needs such as identity, security, recognition, autonomy, self-esteem and a sense of justice. Second, international conflict is an intersocietal process, not merely an intergovernmental or interstate phenomenon. Therefore it includes economic, psychological, cultural and social-structural dimensions. Third, international conflict is a multifaceted process of mutual influence wherein each party seeks to protect and promote its own interests by shaping the behaviour of the other. Fourth, international conflict is an interactive process with an escalatory, self-perpetuating dynamic (Kelman and Fisher 2003: 316–20).

Sudan and the Regional Conflict System

The key concept of conflict system analysis is to demonstrate how a minor change in the political constellation of one state or between two states influences the political developments in another state or in the region. From this point of view, an interesting phenomenon in the regional conflict system had already happened in the nineteenth century. For example, one decisive factor in the Egyptian–British south expansion into Sudan was that the Sudanese Mahdi army that defeated the Egyptians in 1885 could expand its power even more, incited by the Ethiopian victory over the Italians on 1 March 1896 in the battle of Adwa. Consequentially, the Egyptian army trained by Britain began to advance southward to take full control over the Mahdist state of Sudan. In this way the Ethiopian military success led to the recolonization of Sudan (Oliver and Fage 1966: 192). The Ethiopian success against the Italians had negative impacts on Sudan even though the conflict was not between these two states. Further, the Sudanese conflict since its independence in 1956 has had a devastating impact on the region. The Horn region has been trying to address the regionalized conflicts through its InterGovernmental Authority on Development (IGAD) mechanism. But its success and influence has been limited despite the Comprehensive Peace Agreement (CPA) of 2005. Why? IGAD has not been able to address internal and regional conflicts simultaneously.

IGAD and Sudan

The regional conflict system presupposes intrastate conflict system as shown in the chapters of case studies. It is not by chance that the positive peace process in the north–south conflict of Sudan and the outbreak of violence in Darfur took place in 2003, as shown in the chapter on Sudan. The non-Arab elite in Darfur realized that their cause was sidelined by the peace process for north–south negotiations of Sudan. From the intrastate point of view, Turabi as well as Garang supported them.

The IGAD member states (Djibouti, Eritrea, Ethiopia, Kenya, Somalia, Sudan and Uganda) are directly or indirectly involved in the Sudanese conflict, and they constitute the regional conflict system. The Sudanese Islamist uncompromising position led not only to its isolation, but also to confrontation with neighbouring Eritrea and Ethiopia. This has affected IGAD negatively. In the 1990s IGAD was not in the position to make any breakthrough despite various attempts at negotiation (Terlinden 2004). Unity of purpose within IGAD was stated as the basis for the success of the peace process. However, some members of the IGAD were supporting the SPLM/A, whereas the rest supported the GoS. Throughout 2000 and 2001 the IGAD achievements were insignificant for the Sudanese peace process. The progress was very slow and the conflict between the GoS and the SPLM/A were continuing – and at least partly the IGAD member states contributed to it.

The Ugandan Dimension of the Sudanese Conflict

By 1969 the military capacity of the southern Sudanese rebels began to expand extensively. The rebels could recruit from the refugees in the forests as well as in the neighbouring countries. They created foreign contacts to obtain weapons and supplies. Israel, which was attempting to weaken Egypt and Sudan, trained Anyanya recruits and shipped weapons via Ethiopia and Uganda (Daly 2007: 265).

President Obote was overthrown in 1970 and Idi Amin, a friend of the Israelis (who were supporting the SSLM) came to power in Uganda. As a consequence, the SSLM and its leader Lagu started to enjoy a convenient regional and international atmosphere. Moreover, during the Egyptian–Israeli war Sudan sided with the Egyptians. Hence, Joseph Lagu, leader of the SSLM, could benefit from the anti-Sudanese governments of Israel, Uganda and Ethiopia and could expand his Anyanya armed forces. After the Anyanya command was unified, its military supply was secured; when its military strength and activity became significant, the Sudanese government was forced to accept the SSLM's demands and to negotiate with it as an equal partner. Enhanced by these regional conflict constellations the conditions for negotiations for SSLM appeared to be far better than they had ever been before (Johnson 2003: 37). At the same time the Ugandan rebels were operating from the southern Sudanese territories with the support of Sudan.

The existence of the Sudanese and Ugandan rebels in the southern Sudan and northern Uganda had had significant impacts on Sudanese–Ugandan relations. In 1995 both broke off their diplomatic relations, accusing each other of supporting rebel groups. In February 1998 Sudan accused Uganda of preparing to invade Sudan in support of the SPLA. In order to fight against the West Nile Bank Front and Lord's Resistance Army (LRA) operating on the border areas between Sudan and Uganda, and ADF (Allied Democratic Forces) on the border areas between DR Congo and Uganda, Uganda spent 30 per cent of its budget for military goals in 1996 and 1997.

ADF, as a joint Sudanese–Zairian initiative made up of four groups of Muslim fighters, has had considerable regional implications because it involved various armed groups: members of the Tabliq sect from Pakistan, who took refuge in Zaire in the spring of 1996 after a failed attempt to set up a resistance movement near Lake Albert; the unemployed from various tribes (Baganda, Banyoro, Batoro); Bakonjo fighters fighting every Ugandan government for forcible inclusion in the Kingdom (now District) of Toro by the British in 1904; and Rwandan Hutu supporters of the former government responsible for the 1994 genocide (Prunier 1997).

In mid-April 1998 the SPLA claimed that Ugandan rebels (LRA, WNBF, ADF), Sudanese government troops and former soldiers of Mobutu were operating against Uganda in Garamba, northeast of the DR Congo, with the help of the NIF of Sudan to overthrow Museveni and destabilize the allies of the SPLA in the Great Lakes region. Further the Uganda National Rescue Front II was operating in the Uganda–Zaire–Sudan corner. It is a group that broke from the West Nile Bank Front in 1996 including members of the original UNRF (Uganda National Rescue

Front), which was composed of former supporters of Idi Amin formed to oppose Milton Obote during his second term (1980–85) as president of Uganda. It did not make peace with Museveni until it signed a formal ceasefire with the government on 24 December 2002. It operated mostly in Aringa County, Arua District, out of bases in southern Sudan, and received support from the Sudanese government (the National Islamic Front), in retaliation for Ugandan government support for the Sudan People's Liberation Army.

The LRA has been a destabilizing factor in the common corners of DR Congo, Southern Sudan and Uganda. The LRA leader has lived in a jungle hideout in north-eastern DR Congo for the last few years. Moreover, the LRA supported by the NIF regime was engaging the SPLA during the latter's struggle against the NIF until the CPA of 2005. The South Sudanese government has been acting as negotiator between Museveni's regime and the LRA. In June 2006 the Juba peace process started, and it produced five signed protocols in 21 months, designed to conclude 22 years of conflict and guarantee the disarmament and reintegration of one of the worst human-rights-abusing insurgencies ever. However, key issues such as northern Ugandan grievances over marginalization and victimization by the Ugandan regime's National Resistance Movement (NRM) government have not been addressed. Kony knew how to exploit the grievances though he does not represent those in northern Uganda. The peace process, which seemed to be progressing rapidly, collapsed when LRA leader Kony failed to appear to sign the Final Peace Agreement (FPA) on 29 November 2008. After that the armies of Uganda, South Sudan and DR Congo carried out a joint offensive against the rebels in mid-December 2008. In retaliation LRA rebels hacked to death more than 400 people, mostly women and children, and cut them in pieces in a Catholic church in the Democratic Republic of Congo on 26 December 2008.

The Ethiopian Dimension of the Sudanese Conflict System

The Ethiopian–Sudanese conflict system was substantially related to the Eritrean rebellion against Ethiopia. The UN General Assembly passed Resolution 390 A(V) on 2 December 1950 to federate Eritrea with Ethiopia as an autonomous unity under the sovereignty of the Ethiopian Crown. And the Eritrean Constitution was adopted on 10 July 1952 and the Federal Act was ratified by the Emperor on 15 September 1952. As a result, Eritrea became an officially autonomous region of Ethiopia (Iyob 1995: 82–7). However, in October 1952 the first protests emerged against the imperial policy. Through growing undermining of its autonomy, Eritrea was declared part of the Ethiopian Empire on 14 November 1962. Various rebel movements began to emerge: Eritrean Liberation Movement in 1985, Eritrean Liberation Front in 1960, Eritrean People's Liberation Front in 1973, Eritrean Democratic Movement in 1976. After various alliances as well as fragmentations of the Eritrean oppositions in various times, the EPLF emerged as the most enduring representative of Eritrean nationalism (Iyob 1995: 123; Negash and Tronvoll 2000). For the success of the Eritrean rebellion against Ethiopia, Sudan's contribution was substantial. The rebels used Sudanese territory as a launching pad against Ethiopia (Johnson 2003: 87).

The Sudanese relationship with the Mengistu regime was undermined by the Sudanese support for the Eritrean rebels fighting for independence. On the other hand the SPLA/M launched their revolution against Sudan from Ethiopian territory in 1983. Ethiopian authorities made a village called Adura in Gambela region (inhabited by the people of Southern Sudanese origins, Gajaak and Gajok Nuer and the Anyuak) available for John Garang de Mabior and his group, and others such as Deng Alor Kuol and Chol Deng Alaak, two leaders of Abyei Liberation Front. Similarly the leaders of the Student Revolutionary Group and the leaders of the battalions 105 and 104 that revolted in Bor, Ayod, Pachalla, Pibor and Waat led by Kerubino Kuanyin Bol and William Nyuon Bany arrived in Ethiopia. The outbreak of war in south Sudan attracted hundreds of thousands of Southern Sudanese into Western Ethiopia in order to join the revolution. Ethiopian political, financial and logistic support was crucial for effective launch of the movement (Arop Madut Arop 2008).

Though Ethiopia was supporting the SPLA, it opposed the initial intention of the movement for secession. Ethiopia's Mengistu feared similar desires within its own territory, such as the Eritreans or the Tigrayans. The initial position paper of the movement stated that it was a socialist-oriented movement fighting for the total liberation of the Southern Sudan. The Mengistu regime fundamentally opposed this position. As a consequence the movement was asked to write another paper. The movement wrote its manifesto, which stated that the Sudanese Movement containing all the scattered rebel forces will start the liberation struggle of the whole country and fight for the creation of a socialist-oriented united secular Sudan (Arop Madut Arop 2008).

For Ethiopia Sudan was a decisive enemy until Eritrea became independent and the Ethiopian People's Revolutionary Democratic Front (EPRDF) coalition took power in Ethiopia in 1991. Numairi's support for the Eritreans, and subsequently for anti-Derg forces, continued despite Ethiopian warnings in 1976 that it would support various Sudanese dissidents (Johnson 2003: 59). Similarly, in the 1990s Sudan supported the Oromo Liberation Front (OLF) and Benshangul Liberation Front (Shinn 2005: 101). Moreover, Hassan Al-Turabi had been attempting to export Islamic fundamentalism to Ethiopia.

In 1992 the NIF's agents began trying to recruit all those who were unhappy with the Ethiopian government's policy of ethnic federalism, mainly the Oromo. The main objective of the NIF's action was to advance gradually its Islamization process in the region through non-Ethiopian Muslim minorities, such as the Somalis; through setting up purely Islamic Oromo organizations, like the Islamic Front for the Liberation of Oromiya (IFLO); through Al-Ittihad al-Islami (Islamic Union), which was founded around 1991 and declared a jihad against 'infidel' Ethiopia, carrying out murders and attacks in Ethiopia as an indirect agent of Sudanese Islamizing foreign policy instrument (Prunier 1997). As a reaction to this, the Ethiopian government also began actively cooperating with the Sudanese opposition in exile. They were already established and supported by Eritrea. Sudan has been supporting Islamists in Eritrea. Ethiopia is still keeping a watchful eye on the possibility of renewed efforts by Sudan to export Islamic fundamentalism; both are currently in a good mood.

Paradoxically, the war between Eritrea and Ethiopia between 1998 and 2000 brought peaceful cooperation between Sudan and Ethiopia when Khartoum used to support Ethiopian insurgencies, such as the OLF (Achieng 1998). For Ethiopia and Sudan since 1998, Eritrea became a common enemy. In case Ethiopia and Eritrea should solve their conflicts peacefully, the next problem would be how it would affect future Sudanese and Ethiopian relations, as long as the internal conflicts of Ethiopia are not solved. Conflicts in Ethiopia will be exacerbated if Sudan resumes supporting the armed opposition in Ethiopia. The conflicts in Sudan will be worse than they are currently if Ethiopia and Eritrea supported the armed oppositions in Sudan.

The Eritrean Dimension of the Sudanese Conflict System

Since the conflict in Sudan was regionalized, it is hard to imagine a purely Sudanese agreement (Prunier 2007b). After its independence in 1993 Eritrea, a radically secularist state, began to oppose Sudan, a former supporter of Eritrea in its struggle for independence. Eritrea started to support Sudanese armed and non-armed oppositions, such as the National Democratic Alliance of the Sudan, whereas Sudan supported Islamist armed Eritrean oppositions. However, Eritrea during the independence war and for a while afterwards had good relations with Sudan, but because of the NIF's Islamization policy it remained cautious. In January 1994 Eritrea intercepted a first commando from Sudan on Eritrean territory. Asmara threatened reprisal. The Sudanese opposition began arriving in Asmara; and the Eritrean government broke off diplomatic relations with Khartoum and handed over Sudan's embassy in Asmara to its opponents (Prunier 1997).

Eritrea has successfully been involved in the global and regional coalition against Sudan's NIF. Eritrea has been backing the opposition National Democratic Alliance (NDA) wanting a secular constitution. Early in January 1998 Sudan accused Eritrea of amassing troops on the common border in preparation for an invasion. In February 1998 the Asmara-based Sudan Alliance Forces attacked Sudan from the east, and Sudan claimed that the attackers included Eritrean soldiers. When the war between Ethiopia and Eritrea broke out, Sudan rejoiced and El Turabi said, 'God had ensured that Ethiopians and Eritreans turned on each other the weapons the USA has sent to destroy Sudan' (*Africa Confidential* 1998, 39(11): 1). Nevertheless Ethiopia turned to Sudan, whereas the enmity between Eritrea and Sudan intensified. In October 2002 Sudanese rebel forces such as Arab fighters, members of the Sudanese Alliance Forces, SPLA as well as Eritrean army regulars were attacking Sudan from Eritrean territory (Prunier 2002), and Eritrea intensified its alliance with them. Eritrea has been providing arms, logistical support, military training and political support to both the JEM and the Sudan

Liberation Army (SLA), including training of JEM and SLA in Eritrea on the Eritrea–Sudan border (United Nations 2006: 3).

Given its tense relationship with Ethiopia and its conflict with Djibouti, Eritrea attempts to be recognized as an important regional player by helping create rebel groups in Sudan, such as the National Redemption Front and the United Front for Liberation and Development (UFLD) in July 2007. Eritrea was unhappy that Darfur mediation was conducted by the UN/AU, not by Eritrea, and that Libya, not Eritrea, was chosen as the venue for the peace talks (ICG 2007b: 19). This Eritrean dimension is further exacerbated by the Chadian dimension of the Sudanese conflict.

The Chad Dimension of the Sudanese Conflict System

The crisis in Chad and Sudan has got a strong regional dimension because Chad and Sudan have been destabilizing each other since 2004 by conducting proxy wars. Until then they were allies (ICG 2004: 4). This Chadian–Sudanese constellation has further affected the Central African Republic. This suggests that the obstacle to the peace process in Darfur as well as Chad is not only their internal crisis but also their regional aspects, which are essentially interconnected with the internal crisis. The regional system of violence in Chad, Central African Republic and Sudan has connected different actors and issues, creating a regional conflict nexus (Marchal 2007a: 196). The Darfur violence has become cause and consequence of the conflict in Chad. It has been directly threatening the regimes in both Sudan and Chad and inspiring insurgencies in other parts of the country.

Before the modern state concept was introduced in the region, tribes in the area used to move to places where grass was abundant. However, this does not mean that there were no conflicts before the modern state system. But the extent of violence has been exacerbated by the concept of state territory and the abundance of weapons coming from abroad. The armed exiles of the opposing state have been using the other's territories as a training ground for ethnic militias. In 1982, the former president of Chad, Hussein Habre, launched his armed rebellion from Darfur. Similarly, President Idriss Déby, who replaced him in 1990 after a similar campaign, used the 'Sudanese' territory during his military training. He escaped into the Sudanese territory through Libya after the failed coup of 1 April 1989 against Habré. However, in December 1989 he successfully overthrew Habré's regime with the help of Sudan and Libya (ICG 2004: 5; Marchal 2007a: 178). Haggar suggests that all the major conflicts in Darfur for decades have been associated with the presence of armed Chadian groups. In addition, a significant number of armed Chadians are in the Janjawiid militias (Haggar 2007: 113).

The Darfur conflict has a considerable impact on the Chadian–CAR–Sudanese relations. This means that the conflict itself is at least to some extent the result of the bad relations between CAR/Chad on the one hand and Sudan on the other hand. The government of Bozize in the CAR has been accusing Sudan of supporting a rebel group in the CAR, *Union des forces démocratiques pour le rassemblement* (UFDR).

In June 1998 more than 6,000 Sudanese refugees had already fled into neighbouring Chad; at least 235 were killed, 43 injured and 74 villages ablaze – caused by a conflict between the Fur community and Arabs. In spite of the February 1996 agreement between CAR (Central African Republic), Chad and Sudan, CAR and Chad have been supporting Sudanese rebels, while Sudan has been supporting the rebels of CAR and Chad. Armed oppositions against the Chadian regime had been using CAR as a refuge. Chad supported General Francois Bozizé's coup in March 2003. About 100 Zaghawa soldiers from the Chadian presidential guard have guaranteed the survival of the Bozizé regime and his personal security. Moreover, the Chadian armed forces are deployed along the border between Chad and CAR in order to neutralize the activities of the rebel groups from both countries. In 2005 the Chad rebels intensified their attack against N'Djamena, resulting in redeployment of the Chadian armed forces from the Chadian–CAR border areas to eastern Chad. This move of Chad indirectly forced CAR to normalize relations with Khartoum conscious of Sudan's support for CAR rebels. Khartoum tried to force CAR to accept some preconditions unacceptable to Chad, including Bozizé's visit to Khartoum to sign a separate peace treaty; to reopen the border with Sudan closed in solidarity with Chad; and to deploy a CAR–Sudan force on their common border. But Bozizé was vetoed by Déby, who threatened to withdraw his military support. Through its presence and intervention in CAR, Chad wants to demonstrate that it is a regional counterweight against Sudan; it can protect the oil production zone against rebels setting up bases in southern Chad and northern CAR. Moreover, Chad wants to prevent any Khartoum–Bangui axis.

The Central African Republic regime has indirectly profited from the current Darfur crisis through which it gained international attention and aid. The CAR got more aid in 2006 alone than in the period between 2003 and 2005. During the first half of 2007 it got more funds than in the whole of 2006. On 26 October 2007 donors promised aid totalling nearly $600 million (ICG 2007c). In spite of such business with war and humanitarian tragedy a comprehensive peace accord was signed on 21 June between the government and the rebel groups of the *Armée pour la restauration de la république et la démocratie* (APRD) and the *Union des forces démocratiques pour le rassemblement* (UFDR). Initially it was hoped that the armed groups that had not signed the agreement, including the *Front démocratique du peuple centrafricain* (FDPC) and the *Union des forces républicaines* (UFR), were ready to sign it in Tripoli between 21 and 23 July 2008; however, this could not happen owing to an amnesty law for crimes and offences committed by APRD, FDPC and UFDR (United Nations 2008b: 4).

When Déby assumed power in 1990 he not only profited from the deterioration in relations between N'Djamena and Khartoum, but also Khartoum allowed him to establish a base in its territory and import arms and funds that his troops received from Libya through Sudan. Between 2003 and 2004, Déby planned to send troops to Darfur and encourage dissidents within JEM to support Zaghawa in Darfur with equipment and to give them the right to set up base camps in eastern Chad. But only gradually the situation started to change. Until mid-2004 relations between N'Djamena and Khartoum were not yet that bad. However, in May 2004 Khartoum reacted with support for an attempted coup against Déby. In December 2005 Chad declared a state of war with Sudan (*Africa Confidential*, 47, 7, 31 March 2006). Khartoum tried to unite the Chadian rebels. It supported the Chadian rebellion with arms, equipment and training to overthrowing Déby.

In December 2007 an alliance between UFDD (United Force for Democracy and Development) led by Mahamat Nouri (from the Gorane clan as Chad's former President Hissene Habre), RFC (Rally of Forces for Change – led by Timane Erdimi, President Déby's former chief of staff and president's own uncle), and UFDD-Fondamentale, a splinter group of UFDD, headed by Abdelwahid Aboud Mackaye, was created, attempting to overthrow Déby. Likewise, Concord National Tchadien (CNT) has been fighting the Chad government with the same objective.

Thanks to the help of France and Sudanese rebels (JEM – mainly from the Zaghawa ethnic group) who crossed into Chad to prevent Chadian rebels reaching N'Djamena, the Déby regime survived the coup attempt of late 2007 and early 2008 (Tull 2008: 2). The JEM and the (Sudanese) MNRD (National Movement for Recovery and Development), supported by Chad, strengthened the operational capacities of the Chadian armed forces and fought at its side against the Chad rebels, preventing the overthrow of Déby in February 2008 by Chadian rebels supported by Sudan. Moreover, the survival of the JEM is dependent on the survival of Déby.

The rebels have been fighting Zaghawa dominance, a corrupt regime and the president who changed the country's constitution in June 2005 to run in May 2006 for an unprecedented third term in office. In April 2006 the Chad government experienced a coup attempt by rebels SCUD (*Socle pour le Changement, l'Unité et la Démocratie* – Platform for Change, Unity and Democracy), formed in October 2005 and FUC (*Front Uni pour le Changement* – United Front for Democratic Change), formed in December 2005 consisting of eight different rebel groups.[1]

Though Sudan has denied backing the Chad rebels, it is well known that weapons, vehicles, uniforms and medical supplies have been coming from there as a revenge to counter the threat that it faces from Darfur rebels that are from the same ethnic group as President Déby of Chad. Hidden as well as open support of rebel groups makes the peace process between these two states and in the whole region very unpredictable. For example, on 13 March 2008 Chad and Sudan signed a peace agreement in Senegal only hours after Chad accused Sudan of backing a rebel attack, and in April 2008 Chad repeated the accusations that the Sudanese army was fighting alongside Chadian rebels. Whereas Chadian rebels supported

1 RDL (*Rassemblement pour la Démocratie et la Liberté*, formed in August 2005 by former members of the military of Chad who deserted and united under their founder and current leader, Mohammed Nour); CNT; SCUD; FIDEL; FNTR; Group of 8 December; CNR; FRRRT.

by Sudan almost toppled the government of Chad in early 2008, on 10 May 2008 Sudanese rebels supported by Chad went as far as Khartoum and attacked government installations there. This shows that Chad and Sudan are fighting a proxy war using each other's rebels to achieve their military objectives. As a result, Chad is caught in the Darfur crossfire (Prunier 2008). Therefore, Sudanese Foreign Minster Deng Alor was right by suggesting that Sudan cannot solve the problem in Darfur without Chad, since the JEM as well as the NMRD, which broke away from the JEM, have been supported by Chad (Marchal 2007a: 192).

Libyan Dimension of the Sudanese Conflict System

Similarly, Libyan influence in Chad has had at least indirect repercussions on the current conflict between Chad and Sudan. When the Chadian Arabs were fighting for independence from Chad, both Libya and Sudan supported Chadian Arabs militarily and politically. When repressive Chadian governments targeted Chadian Arabs, they got refuge in Libya and Sudan (Haggar 2007: 116). The Front for the National Liberation of Chad (FROLINAT) and the Chad National Front (CNF) were established in June 1960 and supported by Libya aimed at the overthrow of the Tombalbaye regime (1960–75), the reduction of French influence in Chad, and closer association with the Arab states of North Africa, such as Libya (Daly 2007: 206; Prunier 2007a: 43). The rebellion was suppressed in the early 1970s with French support. However, the rebellion continued throughout the 1970s despite the overthrow and death of Tombalbaye in 1975, whom Libya as well as the FROLINAT described as a Christian dictator (Haggar 2007: 119). By creating the Islamic Legion in the early 1970s, by attempting to create an Arab homeland across the Sahara (Marchal 2007a: 178–9), by expanding the definition of 'Arab' to any Bedouins that aspire to become Arabs, and by willing to enlist any Muslims who were ready to fight, Gaddafi in his Jamaheeri system and revolutionary necessity had been playing a key role in Chad as well as in Sudan (Haggar 2007: 121–2; Prunier 2007a: 44).

Libya supported Goukouni Oueddei's regime (1978–82), and in support of the president Libya invaded Chad to pursue Hissene Habré. This ended in a proclamation of union of the two countries in January 1981. On the other hand, the French supported Habré's FAN (Armed Forces of North), which prevailed after heavy fighting in 1983–84. After Habré assumed power in 1982, Libya continued to interfere in Chad by supporting exiles opposed to Habré (Daly 2007: 218). When Libya launched incursions deeper into Chad in 1986, the territory could be recovered only with the support of France and the United States in early 1987. When Déby's forces, the Movement for Chadian National Salvation, assumed power after the French-supported Habré was overthrown on 1 December 1990, France accommodated itself with the new regime of Déby, who had been supported by Libya (Marchal 2007a: 179). After Numairi (who had fallen out with Gaddafi) was overthrown in 1985, Al-Mahdi and Gaddafi established good relationships, and Libya stopped supporting the SPLM/A in south Sudan, whereas

Sudan stopped supporting Chad's Habré, who was hated by Libya. However, the Fur self-defence forces emerged and were supported by Habré (Daly 2007: 242, 267; Prunier 2007a: 45).

Libya wants to play a considerable role in all parts of Africa. The Libyan role is considerable in the Sudanese regional conflict system. Right after President Numairi of Sudan was overthrown in the first half of July 1985. the defence minister Major General Oman Abdallah Mohamed of Sudan signed a military pact with Libya, worrying the United States, which had given more aid to Sudan than to any other sub-Saharan country (Prunier 2007a: 54). Between 1970 and 1977, Gaddafi militarily and financially supported Sadiq Mahdi's military opposition against Numairi (Marchal 2007a: 180; Prunier 2007a: 45–8). During the war between the Fur and Arabs in Darfur between 1986 and 1989, Libya and the new government of Sudan backed the cattle- and camel-herding Arab tribes in Darfur (Tubiana 2007: 70). In October 1988 Libya (which was in conflict with Egypt because Egypt had been dominating politics of the region [Daly 2007: 206]) and Sudan signed a unity proposal. Sudan's calculation was that a hostile Libya could be a threat to Sudan; moreover, it hoped to secure a continuous flow of oil, to gain military aid and the support of Libyan fighter planes flown by Libyan pilots in its fight against the SPLA (Perlez 1988).

Libya has been playing a role as intruder as well as 'peacemaker'. For Chad, Libya brokered an agreement which was signed on 25 October 2007 in Sirte between the *Union des forces pour le développement et la démocratie* (UFDD), the *Rassemblement des forces pour le changement* (RFC), the *Concorde nationale tchadienne* (CNT), and the *Union des forces pour le développement et la démocratie fondamentale* (UFDD-F). The agreement included ceasefire and amnesty, provisions for disarmament, reintegration and appointment of rebel leaders to government posts. Probably one key motivation of Libya as mediator was to enjoy international recognition and to ward off a possible deployment of UN forces along its borders (ICG 2007a: 18–19). Similarly, Libya brokered an agreement between the government of Chad and RFC dissidents under which the latter signed the Sirte agreement of 25 October 2007, resulting in the integration of 50 former RFC fighters into the national security forces. But on the other hand, on 14 August 2008, the MDJT (*Mouvement pour la justice et la démocratie au Tchad*), an armed group from the northern part of the country, joined the Alliance Nationale led by the UFDD leader, Mahamat Nouri; and on 16 August 2008, Timan Erdimi, the leader of the RFC, threatened to join other armed opposition groups to reinforce its military capacity (UN 2008b: 2). Gaddafi opposes the expansion of Wahhabism in the region and does not want the involvement of Saudi Arabia in resolving the conflict between Chad and Sudan. As a result, the Riyadh Agreement of May 2007 between Sudan and Chad made Libya nervous because of possible Wahhabist influence in the region. Sudan has been trying to support rebellion against Déby in order to install an Islamist regime in N'Djamena.

Gaddafi's fear is that Wahhabism could be spread through Sudan. On 10 May 2008, strongly armed with military equipment received from Chad and Libya,

JEM columns pushed as far as Khartoum. Gaddafi seems to have asserted his role in this conflict. The mid-March 2008 agreement between Chad and Sudan provided among others for creating a contact group chaired by Congo Brazzaville and Libya. When the establishment of an African buffer force on the border was discussed, Libya announced it would provide $2 million for the force. Libya opposed the involvement of the UN and EU by organizing and deploying this force. Gaddafi wants to extend his own sphere of influence. Gaddafi ignored the UN arms embargo (Resolution 1556 (2004) of 30 July 2004) on all non-governmental groups (the Sudan Liberation Movement/Army (SLM/A) and the Justice and Equality Movement (JEM) and provided them with arms, ammunition and/or equipment as well as financial, political and other material support (United Nations 2006: 3).

Gaddafi has sponsored three agreements so far for Chad: the 'Tripoli Agreement', signed by the Chad government and the rebel leader Mahamat Nour on 8 February 2006; the 'agreement between the Republic of Chad and the FUC', signed in Tripoli on 24 December 2006; and the 'Sirte Agreement', signed by the Chad government and the four main rebel groups in eastern Chad on 25 October 2007, which was signed by rebels such as the UFDD's leader Nouri, not for seeking genuine peace, but instead because they were under pressure by Gaddafi. Nouri maintains links with Saudi Arabia. Gaddafi does not want the UFDD's Mahamat Nouri, supported by Khartoum and Saudi Arabia, to take power in N'Djamena. Nouri's assuming power in Chad would mean the diminishing influence of Gaddafi in the region. But Gaddafi supports not only the rebels against Khartoum, but in some cases even those against Déby. The former FUC leader, Mahamat Nour Abdelkarim, found refuge in the Libyan embassy in Chad in November 2007 after being dismissed as defence minister by Déby (ICG Africa Report No.144, 24 September 2008: 28–30).

Out of its national security and economic interests Egypt wants a non-disintegrated Sudan, a non-Islamist and non-extremist Sudan which guarantees Egyptian access to the Nile water. As a consequence, Egypt has been against sanctions against Sudan. It blames the ill in Darfur on the non-signatories of the DPA of May 2006. Egypt supported Sudan when Sudan opposed the deployment of non-African forces in UNAMID. As a consequence, the Sudanese rebels rejected Egypt as a biased mediator (ICG 2007a: 20). These Egyptian interests also have an impact on the Somali conflict system in general and its Ethiopian dimension in particular.

In July 2006, former TFG of Somalia Prime Minister Ghedi lashed out at Egypt, Libya and Iran, accusing them of supporting Islamic 'terrorists' in Somalia (ICG 2006e: 20). Egypt has historically competed with Ethiopia for influence over the Somali peninsula (ICG 2006e: 20). Even Somali and Sudanese conflicts are indirectly interconnected. The Somali conflict case is another good example for this regional dimension.

The Regional Conflict System of Somalia

Since the collapse of Somalia its conflict is systemically interconnected to the political situation in its neighbouring states in general and Ethiopia in particular. IGAD has been taking various initiatives to rebuild Somalia. But at the same time the IGAD member states have been playing double roles in Somali politics. On the one hand, they pretend to seek regional solutions to regional problems; on the other hand they pursue their own national interests, which are usually against the interests of affected states.

Immediately after the collapse of Somalia various actors attempted to initiate a peace process for Somalia. An international conference convened and sponsored by the Djibouti government appointed Ali Mahdi as interim president in 1991. The regional initiative that began in Djibouti continued in 1993 in Ethiopia sponsored by the UN. Another conference was called in Cairo in 1997 for Somalia. However, the conference ended without results because of the split of the so-called Sodere-group (in Ethiopia) into two. Through the help of IGAD, the TNG was established but with limited power.

IGAD and Somalia

When Ethiopia and Eritrea were still friends, they opposed the Cairo accord of 1997, while the accord was backed by Kenya and Uganda, and fully embraced by Sudan and Libya. Eritrea and Ethiopia opposed it on the grounds that the accord was infiltrated by Islamic fundamentalism, which could threaten the stability of the Horn of Africa.

The 11 January 2002 IGAD summit on Somalia called upon Kenya, Ethiopia and Djibouti to coordinate under the IGAD chairmanship their efforts for national reconciliation in Somalia, and it decided to hold a reconciliation conference on Somalia in Nairobi and to continue the Somalia peace process. Since combating terrorism was a central issue, the summit urged the Transitional National Government of Somalia and other parties to commit themselves to combat terrorism, and called upon the international community to join IGAD for establishing peace in Somalia. A technical committee comprising the frontline states (Kenya, Ethiopia, Djibouti), chaired by Kenya and including the IGAD Secretariat was established to work out modalities to facilitate the proposed national Somali Reconciliation Conference, which opened on 15 October 2002 in Eldoret, Kenya.

Again through the initiative of IGAD, the Transitional Federal Government (TFG) was established in 2004. It again proved weak. IGAD appealed to the United Nations Security Council to take immediate action to deploy a peacekeeping mission to Somalia to assist in securing the new government and also help train security forces. Furthermore, the African Union took the initiative in conjunction with IGAD to convene a meeting of military and political experts to assess the needs of Somalia with regard to deployment of a peace restoration and protection force. The IGAD executive secretary's proposal for a new mechanism for IGAD participation in the post-conflict phase in Somalia, known as the IGAD Monitoring and Follow-Up Team (IGADMOFT), was adopted.

Two significant regional obstacles to the Somali peace process are 'the proliferation of similar and non-complementary initiatives, and the uncompromising hard line of warlords who, in spite of the opportunity extended to them to save Somalia by constituting a government, have never agreed upon lasting settlement of conflict or even the implementation of the decisions taken by consensus' (IGAD 2002). Moreover, IGAD states' mutual destabilization has been crucial. This has become worse since the outbreak of the war between Ethiopia and Eritrea. Both Ethiopia and Eritrea are heavily involved in the Somalia conflict, resulting not only in the destabilization of Somalia, but also in long-term mutual damage of both Eritrea and Ethiopia. Therefore, in the following subsections we shall see different regional dimensions of the Somali conflict system.

The Ethiopian Dimension of the Somali Conflict System

The existence of Somali ethnic groups in Kenya, Ethiopia and Djibouti has been one of the key factors of the conflicts in Ethiopia and Somalia. Mainland Somalis see the Somalis in neighbouring countries not as neighbours but instead as their kinsmen with false citizenship separated by 'indiscriminate boundary arrangements' (Lewis 2002: 179). In 1950 Somali tribesmen in the Northern Frontier District of Kenya announced to join their kinsmen in Somalia. In 1956 the British government intended to allow and support the unification of Somaliland and Somalia. Ethiopia was alarmed by that. Kenya underscored that the district is an inalienable part of Kenya. However, for Somalis, pan-Somalism was a 'legitimate regional application of the general and wider principle of Pan-African unity' (Lewis 2002: 195). Escalation between Ethiopia and Somalia resulted in an open war in 1977–78, preceded by the 1964 war in Ethiopia's Somali-inhabited Ogaden region.

It was from the very beginning the ambition of Siad Barre to create a Great Somalia through irredentism. Ethiopia, because of a huge area populated by the Somalis, was the main target of Siad Barre's irredentism politics. The Siad Barre regime had been arming the Ogaden National Liberation Front and Oromo Liberation Front in Ethiopia, whereas the Ethiopian government had been arming the Somali National Movement (SNM). In 1977, because of its strengthened military potential, Somalia began to support the WSLF (Western Somalia Liberation Front) and the SALF (Somali and Abbo Liberation Front – in Somali- and Oromo-inhabited areas of southern Ethiopia) (Markakis 1978, cited by Brons 2001: 182). In spite of its integration of the WSLF and SALF into the regular army, Somalia lost the war, which broke out in July 1977 and lasted until 8 March 1978. This military defeat resulted in the political crisis in Somalia. A failed coup attempt by army officers after the lost war caused some officers to flee to Ethiopia, where they founded the Somali Salvation Front (SOSAF). In October 1981 the coalition of SOSAF, the Somali Workers Party and the Democratic Front for the Liberation of Somalia (DFLS) was founded as SSDF (Somali Salvation Democratic Front), which included Harti (Majeerten), Warsangeli, Dulbahante as well as Hawiye fighters, intellectuals, politicians and officers (Brons 2001: 185).

Carment et al. argue that the most intense period of conflict between Ethiopia and Somalia occurred when Ethiopia experienced significant internal upheavals of the Tigrayan, Eritrean and Oromo liberation struggles (Carment et al. 2006: 15, 76). Somalia wanted to exploit this opportunity. On the other side in Somalia, the failure of democratization, public welfare and social justice forced Siad Barre to externalize the internal problems by attacking Ethiopia. Moreover, the war between Somalia and Ethiopia not only sent Oromos and Somalis from Ethiopia into Somalia, it also contributed to the conflict between the refugees belonging to the Somali Ogaden clan and Isaaq clan (Lewis 2002: 248). The defeat in the war, the flow of refugees, the conflict between the Ogadeni refugees and the Isaaq clan, the unexpectedly low financial support for Somalia from the US ($8.7 million instead of the requested $47 million) after the defeat, Western governments' emphasis on civilian rather than military aid for Somalia in order to discourage Somali irredentism and to improve Somali–Kenyan relations, all led to the continuous weakening of the state and one-man state of Somalia (Lewis 2002: 248). Moreover, the Ogadeni rebellion was split between two Ogaden movements: the Siad-controlled Western Somali Liberation Front, and the anti-Siad, Ogaden National Liberation Front (ONLF), which considered WSLF as a Siad-puppet (Lewis 2002: 254, 257).

The Somali Democratic Action Front (SODAF), which was formed in Rome in 1976, was an organization opposing Barre's regime. After SODAF and SSF formed a coalition called the Somali Salvation Democratic Front (SSDF), they consolidated their fight against the regime through support from Ethiopia and Libya. In 1982 SSDF was commanded by Colonel Abdullahi Yusuf Ahmed, who was commander of the SNA (Somali National Army) during the Ethiopian–Somali war. After deserting from the Somali army, he began to fight together with Ethiopia against the Somali army. Although the SSDF encompassed various factions, politicians and intellectuals who were against the regime, it was dominated by the Majerteen clan. A dispute with Ethiopia over the Balumbale and Goldogob districts heavily undermined SSDF's leadership until its relationship with Ethiopia normalized and it became active again towards the end of the 1990s.

Ethiopia also supported the SNM (Somali National Movement, supported mainly by the Isaaq clan in Somaliland). After the 1988 negotiations and agreement between Ethiopia and Somalia, the Somali government hoped to isolate the SNM when each party expressed its obligation to terminate support for the other's Somali dissidents (Lewis 2002: 262). However, in spite of the rapprochement between Ethiopia and Somalia (which means loss of support from the Ethiopian side), the SNM could consolidate its military power.

The activities of the Somalia-based Al Itihad Al Islamiya (AIAI), with the aim of irredentism (incorporating Ethiopia's Somali population) in an Islamic Somali state, was considered by Ethiopia as a major regional threat (Shinn 2005: 101). The growing threat of Islamic fundamentalism is worrying Ethiopia and has forced it to intervene militarily in Somalia and search for and identify those responsible for Islamic terrorist attacks in Ethiopia or in the region, such as Sheikh Hassan Dahir Aweys and other AIAI leaders, whom Ethiopia blamed for terrorist attacks in the mid-1990s in Ethiopia, including the bombing of two hotels and the attempted assassination of a cabinet minister. Sheikh Hassan Dahir Aweys asserted that he remains committed to an Islamic state in all Somali territories, including the Ogaden of Ethiopia (ICG 2006e: 17). More alarmingly, a leading ICU (Islamic Courts Union) figure, Sheikh Sharif Sheikh Ahmed (current president of Somalia), urged all the Somali people to wage holy war against the Ethiopians.

Already in the 1990s there had been Islamist threats against Ethiopia. Ethiopian forces defeated Islamist fighters in the Somali town of Luuq in 1996. After Islamic courts took control of the Somali capital in June 2006, Ethiopia's involvement in Somalia increased dramatically. In July 2006 Ethiopian trucks and armoured cars crossed into Somalia, and Ethiopian military trainers in the country were helping the interim government. On 21 July 2006 the ICU leadership ordered a 'holy war' against Ethiopians in Somalia. On 25 October 2006 Ethiopian Prime Minster Meles Zenawi said Ethiopia was 'technically at war' with the ICU, and on 30 November 2006 Ethiopia's parliament passed a resolution authorizing the government to take all legal and necessary steps against what it terms as any invasion by the ICU. On 25 December of the same year Ethiopian aircraft bombed Mogadishu airport. Between 26 December 2006 and 1 January 2007, driven out by forces of the Somali transitional government and Ethiopian troops and jets, ICU lost control of the towns of Burhakaba, Jowhar, Mogadishu, Kismayo, and other areas of southern and central Somalia and were forced to retreat.

A coalition supported by Ethiopia was created between the factions of the voted-out president of Somaliland (Abdirahman Tour), Darod clan factions, Hawiye leaders including Ali Mahdi and Osman Ali Atto (former financial supporter of Aideed). This neighbourhood policy of Ethiopia intensified after the fighting between Eritrea and Ethiopia erupted in 1998. Ethiopia's worry has been that Somalia will become a safe haven for militant opponents of the Ethiopian government, such as the Ogaden National Liberation Front of the ethnic Somalis in east Ethiopia, and the Oromo Liberation Front.

The UN Security Council Committee reported on 27 January 2003 on the situation in Somalia and the regional arms-related destabilization. Ethiopia, Eritrea, Yemen, Djibouti, Sudan and Egypt are all said to have violated the embargo over the last decade by giving arms, equipment, militia training and financial support to Somalia factions. In 1999 two Mogadishu faction leaders (Ali Mahdi Mohamed and Hussein Mohamed Aideed) called Ethiopia 'the enemy', trying to disintegrate Somalia, interfering in its internal affairs and obstructing the formation of a central government by giving arms and ammunitions to Somali factions (Somali Salvation Democratic Front, United Somalia Congress Peace Movement and Rahanwein Resistance Army).

The situation in Somalia is a combination of proxy war and inter-state war since Eritrea and Ethiopia waged war against each other. In 1999 a ship from Eritrea carrying arms for Aideed and purchased by the OLF docked at Marka port in southern Somalia, which caused Ethiopia to plan to bomb if more arms were offloaded there. Despite the UN arms embargo on Somalia in 1992, there are arms supplies for various factions from the usual suspects: the Anti-Transitional National Government forces, such as Muse Sudi, supported by Ethiopia; the TNG and its allies backed by Eritrea, Egypt, Yemen and Libya. Further, the OLF and ONLF could purchase their weapons on the Mogadishu market.

When Aideed in 1994 declared himself president of Somalia and formed a government, a coalition of various factions countered him by forming the so-called Sodere (Ethiopia) group in Ethiopia in 1996, including all Darod clan factions and some Hawiye leaders, such as Ali Mahdi and Osman Ato. The idea of the support for Sodere-group by the Ethiopian government was to organize and consolidate the factions fighting Aideed's claims. The Sodere-group planned to hold a conference at Bosaso (northeast Somalia) in November 1997, with the aim to weaken Hussein Aideed, son of Farah Aideed, who was killed on 31 July 1996.

The fight between the RRA (Rahanwein Resistance Army – supported by Ethiopia) and USC/Aideed (supported by Eritrea) intensified to control the town of Baidoa in mid-1999. When the Transitional National Government was created in 2000, various opposition groups (RRA, SPM, SNF, SSNM and SSDF) met to create a new 'autonomous regional southern state' and established the SRRC, supported by Ethiopia.

Ethiopia has been continuously accused of providing Somalia with arms. Between November 2005 and April 2006, Ethiopia provided at least three separate consignments of arms, including mortars, machine guns, assault rifles, anti-tank weapons, and ammunition to the TFG (Shinn 2006). Ethiopia's military presence inside Somalia and arms supply significantly inflamed Somali nationalism and antagonism towards Ethiopia (ARB 2006: 16835). Ethiopia's intervention bolstered not only the elements of the UIC, Al-Shabab, but also many non-radical Somalis. The terrorist organization Al-Shabab, which grew from the Union of Islamic Courts, took control of Kismayo, Merca and large parts of southern Somalia. The presence of Ethiopian troops inside Somalia kept united various factions of the ARS (Alliance for the Re-liberation of Somalia), the government, or indeed Al-Shabab. Gradually, Al-Shabab has expanded its military and territorial strength. The already weak TFG has started to cease to exist since Ethiopia started to pull out towards the end of 2008. Islamist militiamen have taken over a number of abandoned police stations in the Somali capital as Ethiopian troops withdrew from the city. The militiamen include those militia who support the ARS (Alliance for the Re-liberation of Somalia – whose leader succeeded President Ahmed Yussuf); Al-Shabab; Ahlu Sunna Wal-jamaah, etc.

However, Ethiopia's presence in Somalia cannot be explained merely by its interest in fighting the Islamic terrorists or the enemies of the TFG. Ethiopia is also indirectly fighting Eritrea, which is using the failed state of Somalia as a launching pad against Ethiopia and supporting the Islamists.

The Eritrean Dimension of the Somali Conflict

The impact of Eritrean politics on Somali politics is considerable. In Eritrea there are about 50 per cent Muslims and 50 per cent Christians. Eritrea has been an important player as an ally of the ICU, though the Eritrean government does not support any religious fundamentalism in Eritrea. Eritrea's interest is rather to harm Ethiopia. Eritrea expects and calculates that its involvement in Somalia would divert attention from the border conflict or put additional pressure on Ethiopia. Whereas Ethiopia has been supporting the TFG of Somalia, Eritrea has been providing the ONLF, the ICU and the governor of Lower Shabelle with substantial quantities of arms to counter the support provided by Ethiopia to the TFG (Shinn 2006; ICG 2006e: 20).

Crucial for this competition between Eritrea and Ethiopia in Somalia is the border conflict between them. Beginning in August 1997, the prime minister of Ethiopia and the president of Eritrea exchanged letters regarding the issue of the border because there were some border clashes between Ethiopia and Eritrea at the time. Moreover, the two countries were no longer on good terms due to disagreements about trade-related issues (Negash and Tronvoll 2000: 26). Before the war began, two-thirds of Eritrea's external trade was with Ethiopia; access to Eritrean ports was significant for Ethiopia's trade (ICG 2003: 14). But it is wrong to reduce the conflict between Eritrea and Ethiopia to only a conflict between governments; it is also a people-to-people conflict (Negash and Tronvoll 2000: 83; ICG 2003: 3).

There is a deep-rooted hatred between the peoples of both countries (Trivelli 1998: 284–5), not just between Eritreans and Tigrayans (Peninou 1998: 509), nor merely between Meles and Issaias (Beehner 2005). There are superiority complexes, exceeding stubbornness and irrevocable commitment on both sides (Russel 1999: 8). Eritreans feel more civilized than the Ethiopians (Abay-Asmerom 1999; Tadesse 1999: V). Milkias quotes that Eritreans believe that 'Eritrea has always been a grade above other African nations' (Milkias 1999: 39). One Ethiopian claims in his article that Eritreans have a deep-rooted superiority complex, and quotes an Eritrean saying:

> Oh Eritrea, I wish you were somewhere away from the evil Agames [Tigrayans] and donkeys [Ethiopians] you are an island of the chewa [noble] in an ocean of scams; you are an island of pride in an ocean of pathetic. You are an island of hard working people, in an ocean of beggars. I wish you were somewhere in the middle of an ocean, far, far away from the corrupt afro man, far, very far away from the savage Ethiopian … (Tsegaye 1999)

For outsiders, Ethiopians and Eritreans are brothers. The difference between brothers is less than that between all other degrees of relations (Girard 1986a: 61). Girard asserts that 'twins are in a sense reinforced brothers whose final objective difference that of age, has been removed; it is virtually impossible to distinguish between them' (Girard 1986a: 61). Further, Girard says that, 'ultimately, the insufficient difference in the family relationships serves to symbolise the dissolution of family distinctions; in other words, it *desymbolises*' (Girard 1986a: 62). Tigrayans incorporate the incomplete similarity or the negation of distinction between the Ethiopians and Eritreans that creates identity crises to both Ethiopia and Eritrea. As Girard says, '... because fraternal relationship implies only a minimum of differences, it obviously constitutes a vulnerable point in a system structured on differences' (Girard 1986a: 63). This system and structure in the international relations is the state. According to international law, both Ethiopia and Eritrea are sovereign states, but the difficult historical experiences have rendered this distinctness of these two entities problematic, and the existence of Tigray between these two adversaries blurred the distinctness. Girard says that the representation of non-difference ultimately becomes the very exemplar of difference, which he calls a 'classic monstrosity' that plays a vital role in violence. This monstrosity implies the combination of two distinct entities: difference and non-difference as the same, because a monster is a mythological being which incorporates two different beings such as man and animal. Difference as a sign of distinction will be replaced by violent reciprocity (Girard 1986a: 64).

Hence, conflicts cannot always be explained by material causes. There are three different perceptions n a conflict: the one of the outsider (mediator) and two contradictory conceptions of the adversaries:

> Each side in a conflict usually blames the other's aggressive behaviour for the fight. An observer, however, might conclude that the adversaries are mistaken: each is actually seeking to defend what it has and its defensive efforts are incorrectly perceived by the other as threatening. Or the observer may believe that the basis of the conflict is a natural consequence of the imbalance of power among the adversaries and the lack of agreed-upon procedures for managing issues in dispute between them. (Kriesberg 1998: 30)

Through structural and psychological commitments to mobilize group loyalty and diabolization of the enemy, the conflicts will be maintained as a source of profit, power, status or *raison d'être*. Through the formation of mirror images and the resistance of images to contradictory information, the processes of conflict formation will be perpetuated. Both parties tend to develop parallel images of self and other, except with the value reversed. The enemy is aggressive, hostile and evil, whereas self is peaceful and ready for compromises (Kelman and Fisher 2003: 322–6). Kelman and Fisher suggest that 'the concept of selective exposure, selective perception, and selective recall all point to the fact that our attitudes help determine the kind of information that is available to us' (Kelman and Fisher 2003: 327).

The main spark that ignites a bloody conflict is fear by both sides: fear of the perceived growing power of the other. The speeches by both sides will be replete with immensely negative stereotypes and caricatures of each other by constructing strong enemy images. Cottam et al. suggest that conflicting parties are driven by fear, perceptions or misperceptions of each other that make war inevitable (Cottam et al. 2004: 258). Further, Cottam et al. claim that offensive and defensive capability are indistinguishable from one another. Warring parties on all sides try to reassure one another that their mobilizations were purely for defensive purposes only; each falls into the security dilemma. Actions taken by one state to increase its own security has the effect of simultaneously decreasing the security of neighbours. 'Because the true motivations of their neighbours could not be determined with certainty, each state was left to make decisions based solely upon their beliefs about their neighbour's motivations and capabilities' (Cottam et al. 2004: 259). Cottam et al. suggest that a major conflict factor is what they call 'the fundamental attribution error' creating a 'diabolical enemy' (Cottam et al. 2004: 52). As a consequence, wars make national identities. Such new identities of self and the enemy justify the $2.9 billion spent by Ethiopia during the war between 1998 and 2000 against Eritrea (Teigist Lemma 2006: 7), whereas millions suffer from chronic or acute hunger, diseases and poverty.

The EPRDF government of Ethiopia recognizes Eritrea's independence, whereas most of the Ethiopian population does not. For many Ethiopians every inch that is lost to Eritrea is painful (ICG 2003: 10). Eritrea knows this and has been creating various strategies to weaken Ethiopia: internally (by supporting Ethiopian armed opposition) and regionally (by supporting the anti-Ethiopian Islamists in Somalia).

Ethiopian opposition parties for their part denounced the handling of the entire border conflict, the Boundary Commission decision, and accused the Ethiopian government of violating Ethiopian sovereignty (ICG 2005b: 6). But for Meles Zenawi as well as the Ethiopian opposition, the conflict is more than merely about the disputed territory, as the following interview with Meles Zenewi in October 2003 shows:

> **Q:** But Eritrea will never accept that Badme is in Ethiopia. It seems an intractable situation.
>
> **A:** Well justice will have to prevail, fairness will have to prevail. When we were told Asmara is not Ethiopian, when we were told Assab is not Ethiopian we said – sure, if the Eritrean people think that Assab is Eritrea and not Ethiopia, that's alright. And we went to Asmara and celebrated with the Eritreans the independence of Eritrea. Badme is not bigger than Asmara. Badme is not more important than Assab by any stretch of the imagination. It is some godforsaken village. So it's not about territory. According to the latest rendition of the

> Boundary Commission, Badme would be 800 metres inside Eritrea. What's 800 metres in a country as big as Ethiopia? What's 800 metres compared to what we willingly and happily gave up as Eritrea? It's nothing. But it's 800 metres which we are told is something it has never been, and something that it will never be. That's the point. That's the crux of the matter. (IRIN 2003)

In spite of the international community's claim that the piece of land fought for is geographically small and economically valueless (Clapham 2003), 'remote pockets of barren land' (Blair 2005), the 'dusty border village of Badme' for which 'two bald men fighting over a comb' now acquired a 'symbolic importance entirely out of proportion to its size and population' (ICG 2005b). In spite of such derogation, Ethiopia and Eritrea extended their fight for Badme to Somalia. But for Ethiopians and Eritreans, as Medhanie argues, Badme is 'fertile', 'rich', as significant as Adwa, Maichew, etc., where Ethiopian nationalism was resurrected and celebrated (Medhanie 2007). Girard has beautifully shown that 'it is no longer the intrinsic value of the object that inspires the struggle; rather, it is the violence itself that bestows value on the objects, which are only pretexts for a conflict' (Girard 1986a: 144).

Walter suggests that:

> territorial attachments and people's willingness to fight for territory appear to have much less to do with the material value of land and much more to do with the symbolic role it plays in constituting people's identities and providing a sense of security and belonging … Territory is important, not so much because land contains important natural resources that can be translated into tangible assets, but because it plays the more important role of defining one's social, spiritual, and communal world. (Walter 2006: 288–9)

Further, she maintains that:

> measuring the value of land strictly in terms of its tangible assets and making these assets the basis of a settlement is unlikely to bring long-term peace. To date, policymakers have tended to focus disproportionately on issues related to border length and placement, and the distribution of natural resources, rather than dealing with the deep psychological bonds that may exist among individuals at the local level. (Walter 2006: 294)

Georg Simmel, in his very important analysis on conflict, has suggested that antagonism is usually involved as an element in intimate relationships. As argued above, Simmel asserts that it is similarity that exacerbates a conflict, not difference. A conflict is more passionate and more radical when it arises out of close relationships: deepest hatred grows out of broken love, which results in denial of one's own past and of change of feeling. It will be recognized that the past was an error or a failure of intuition; the past splits the security and unity of our self-conception; our current intolerable feelings mean that we pay for our failures (Simmel 1964: 45–8).

Many diplomatic attempts and recommendations strive to come up with new solutions to solve the problem between Ethiopia and Eritrea, such as delimitation and demarcation of the border based on colonial treaties of 1900, 1902 and 1908; this means looking for objective solutions based on 'objective' facts. But the international negotiations do not take into account that the problem between Ethiopia and Eritrea is not only based on 'objective' problems, as Meles himself said.

Eritrea tries to destabilize Ethiopia within Ethiopia as well as from outside. Within Ethiopia it attempts to militarily support organizations such the OLF; from without, it utilizes the collapsed state of Somalia. Strangely enough, the Eritrean government, which is intolerant towards religions, is collaborating with the Islamic fundamentalists in Somalia to destabilize Ethiopia. Thanks to Eritrea, Somalia now has more arms than at any time since Somalia collapsed in 1991. An unknown number of surface-to-air missiles, suicide belts and explosives with timers and detonators are flown from Eritrea to Somalia, with the main objective being to destabilize Ethiopia from Somalia where, according to a report, at least 2,000 Eritreans are operating (Reuters 30 July 2007).

Eritrea continues active support by arming and training Ethiopian opposition movements, such as the Oromo Liberation Front (OLF), reinforcing Ethiopian concerns after which a number of violent incidents have made the Ethiopian government nervous. Ethiopia, Sudan and Yemen have been the main supporters of the Alliance of Eritrean National Forces; whereas Eritrea aids the Oromo Liberation Front (ICG 2003: 1). The OLF strives for the independence of the Oromo region. It was accused of the bomb blast of 24 June 2002 on the railway between Addis Ababa and Djibouti. On 11 September 2002 another bomb exploded in the Tigray Hotel in Addis Ababa, which again the OLF was blamed for. For the Ethiopian government, the OLF is a 'terrorist' organization.

The Ogaden National Liberation Front supported by Eritrea seeks ethnic national independence or autonomy (Connell 2005). Since its establishment in 1984, the ONLF has been fighting against the Ethiopian government, which it says marginalizes the Ogaden region. In April 2007 it killed 65 Ethiopians and 9 Chinese oil workers in the area. The discovery of gas and oil has brought new economic hopes, but at the same time fuelled the conflict. At least 4 people were killed and 40 injured in a series of explosions in Addis Ababa on 13 May 2006, for which Eritreans or Somalis were suspected. Ethiopia accused Eritrea for the blast that killed 7 people and wounded 10 on 13 March 2008 in the Ethiopian-controlled town of Humera on the Ethiopia–Eritrea border. In April 2008 three people died in explosions at two petrol stations in Addis Ababa and on 20 May 2008 three people were killed and nine wounded in a bomb explosion on a minibus in Addis Ababa. Ethiopia blamed them on the Eritrean-backed Oromo Liberation Front (OLF). According to Ethiopia, Eritrea is supporting not only the OLF but also Somalis opposing Ethiopia's military presence there. Therefore, Ethiopia wants to create a buffer region to stop Somalis and the OLF from using Somalia as a rear base. Paradoxically, groups that oppose the independence of Eritrea, such as the Ethiopian People's Patriotic Front (EPPF) and the Tigray People's Democratic Movement (TPDM), are supported by the Eritrean government. While Ethiopia and Eritrea support elements of the other's opposition, they however underestimate the will of the other to hold together if there is a new military confrontation (ICG 2003: i).

Peace talks on Somalia of 16 May 2008 broke up without any face-to-face discussions between the government and the main opposition alliance (Asmara-based Alliance for the Re-liberation of Somalia, which includes leaders of the Union of Islamic Courts (ICU)) because it did not want to engage in direct negotiations until the government agrees a timetable for Ethiopian troops to leave Somalia.

The Djibouti Dimension of Somalia's Regional Conflict System

Djibouti was on the Ethiopian side during the war between Eritrea and Ethiopia between 1998 and 2000. At the same time, it has supported both the TFG as well as the Islamists. Djibouti has been enjoying more freedom of action than Ethiopia and Kenya in dealing with the Islamic Courts. In December 2005 Djibouti provided the TFG with 3,000 military uniforms. Djibouti, unlike Ethiopia and Kenya, had been urging the TFG and Islamic Courts to reach an understanding at the talks in Khartoum in 2006. Djibouti did not support a peacekeeping mission in Somalia and urged 'outsiders' not to interfere in Somali internal affairs. In 2006 Djibouti received a senior delegation from the Islamic Courts. Its approach to Somalia seemed to be in line with the positions of Eritrea and Sudan, since Djibouti is an overwhelmingly Muslim country and increasingly dependent on Arab investment. Therefore, its attitude towards the Islamic Courts does not seem to contradict the position of the Arab League substantially (Shinn 2006).

The Conflict System of Somali–Kenyan Relations

Contrary to the agreement of 29 January 2004, six faction leaders signed a statement accusing Kenya and Djibouti of mismanaging the talks and of bias in favour of the Transitional National Government during the Somali national reconciliation conference, which took place in Nairobi, Kenya, in February 2004. Kenya fears that the Somali-dominated north-eastern province could be a breeding ground for rising Islamism. This means that radicalization in Somalia could have spill-over effects on Kenya. The Islamists can exploit years of underdevelopment and marginalization by successive Kenyan governments. Those marginalized could embrace the well-organized Salafist groups funded by Saudi charities as well as any other Islamists that promise to address their grievances. Out of this fear of the rising Islamic Courts, Kenya joined with Ethiopia in strong support of the TFG. Kenya has been favouring an African peacekeeping force in Somalia. The

ICU's alleged intention to incorporate Somali-inhabited territory pushed Kenya to support the TFG (Shinn 2006).

As seen so far, the IGAD peace process for Somalia reflects different interests of the regional member states. However, such interests and involvements of actors go beyond immediate neighbourhoods. The Arab–Islamic identity is an important regional factor.

The Arab–Islamic Identity as a Factor of the Somali Conflict System

The involvement of various actors from the Arab and Islamic world has exacerbated the Somali conflict system. Influential powers such as Egypt and Saudi Arabia are major players. In 1977 during the conflict between Ethiopia and Somalia, Saudi Arabia had already started its support for Somalia and promised £230 million when it learned that Somalia broke off its relations with Russia (Lewis 2002: 234). Egypt and Saudi Arabia consider themselves as regional Islamic powers and, therefore, have the interest and power to be involved in the future of Somali politics. Since the collapse of Somalia, Egypt, Libya, Yemen, Sudan and Saudi Arabia have been direct or indirect stakeholders (Shinn 2006). In spite of the devastating political situation, in 1993 Sudan, Libya and various international Islamic organizations were supporting Aideed (Ahmed 2001). This was mainly because of the rivalry between Ethiopia and these states in order to be recognized as the dominant regional power. In November 1997 Arab countries such as Libya, Egypt and Yemen insisted on convening a conference in Cairo and including Aideed, who was opposed by Ethiopia and by the Sodere-group (an anti-Siad Barre coalition created in Ethiopia). Prior to the Cairo Conference various efforts were made in Nairobi in October 1996, Sodere/Ethiopia in January 1997, Sanaa in May 1997 and Cairo in May 1997.

The place of a conference usually reflects the role each hosting country wants to play. Ethiopia considers Egypt as a rival power in the region. The Nile Water, which partially originates in Ethiopia, makes it a strategic regional danger for Egypt if Ethiopia were to invest in heavy hydro-projects that consume a considerable amount of the Nile Water. Egypt is interested in a weak Ethiopia and anxious to check any extension of Ethiopian power in the region in general and in Somalia in particular (Lewis 2002: 309). Ethiopia is, moreover, a state dominated by Christians, whereas Egypt attempts to emerge as an Islamic regional power. Whereas Ethiopia opposed the Transitional National Government in Somalia for its Islamist elements and support for the Ogaden National Liberation Front in Ethiopia, Egypt assisted it by providing training and uniforms to its police force and military (Grosse-Kettler 2004: 28). However, Arab states attempted to expand their sphere of influence in Somalia through TFG institutions as well. IGAD countries, particularly Ethiopia, suspect the Arab League of being overly sympathetic to the Courts (ICG 2006e: 25).

The League of Arab States organized a meeting attended by transitional federal institutions (led by President Abdullahi Yusuf Ahmed, Prime Minister Ali

Mohamed Gedi and Speaker Sharif Hassan Sheikh Aden) and the Islamic Courts in Khartoum on 22 June 2006. However, soon after signing the 22 June 2006 agreement in Khartoum, each side accused the other of violating its provisions: according to the Transitional Federal Government, the Islamic Courts had violated the agreement by taking military action against the ARPCT (supported by Ethiopia) and its leader in Mogadishu, Abdi Awale Qeybdid; whereas for the Islamic Courts the Transitional Federal Government violated the Khartoum agreement and the sovereignty of the Somali people by inviting Ethiopian troops into the country. The Arab League has been sponsoring the reconciliation talks between the TFG and the Islamic Courts, and Bashir is in charge of the process (Shinn 2006). Under the auspices of the League of Arab States the second round of peace talks between the two parties opened in Khartoum on 2 September 2006 and signed an agreement two days later, *inter alia*, to reconstitute the Somali national army and national police force and work towards reintegration of the forces of the Islamic Courts, the Transitional Federal Government and other armed militias; practise the principle of peaceful coexistence between Somalia and its neighbours; to discuss power-sharing and security issues; form a technical committee consisting of the Arab League General Secretariat, the Arab League Committee on Somalia and others from the Transitional Federal Government and the Islamic Courts (United Nations 2006a: 3). Rising Islamism is the main reason for Ethiopia's invasion of Somalia in 2006. Ethiopia's invasion of Somalia in December 2006 has underlined not only the regional systemic dimension of the conflict in Somalia as well as Ethiopia, but also its global aspect.

Global Conflict System in the Horn of Africa

The intrastate and regional factors seen above and in the previous chapters are not enough to explain the conflict dynamic in Sudan and Somalia. The failure of Somalia, a clan-based pastoral society, is 'greatly exacerbated by the incorporation of the Somali state into regional and global conflicts, the massive supply of external weaponry, and the methods resorted to by the Siad Barre regime in its desperate attempt to cling to power…' (Clapham 1996: 247)

Egypt has had an interest in the Somali coast dating back to the nineteenth century, and it shares a membership with Somalia in the Arab League and Organization of Islamic Conference (OIC). Ethiopia, the source of 86 per cent of the Nile, opposes the Egyptian and Sudanese treaty that divided all of the Nile water between the two of them. Egypt can use Somalia as a pawn in order to put pressure on or weaken Ethiopia: it supported Somalia, for example, during its war against Ethiopia in 1977. The TFG during Yussuf Ahmed included Egypt among those countries that assist terrorists in Somalia (Shinn 2006).

Libya does not share a border with Somalia and does not have an historical connection with Mogadishu, but it does not want to be left out of the competition to mould the Somali future, especially since Somalia is a member of the Arab

League and OIC. The TFG accused Libya of interfering in the internal affairs of Somalia and providing arms and cash to the Islamic Courts. Yemen has been one of the supporters of the TFG as well as the ICU. In January 2006 it sent 15 pickup trucks and military clothing to the TFG police and money to the Islamic Courts, though Yemen's foreign minister denied that his government provided any aid to the Courts. Islamic charities funded by Saudi Arabia, such as al-Haramain, were active in Somalia until they were shut down by the Saudi government, probably because of links to terrorist organizations (Shinn 2006). It is not only funds from private Saudi sources that continue to support fundamentalist activity in Somalia. According to the UN Monitoring Group on Somalia, Saudi Arabia provided the TFG with military uniforms, medical equipment, medicine and anti-gas protective clothing in December 2005. The United Arab Emirates has become the financial centre for Somali businesspeople. Iran's leaders have been showing special attention to the rising Islamic movement in Somalia. Iran even opposed any peacekeeping force to Somalia. The TFG had earlier accused Iran of providing arms to terrorists in Somalia (Shinn 2006).

The ICU sent a 720-strong military force to Lebanon to fight alongside Hezbollah against Israel during the Israel–Lebanon war in July and August 2006. In exchange, Hezbollah facilitated the Iranian–Syrian arms support to the ICU. ICU co-leader Sheik Hassan Dahir Aweys and the ICU's hard-line security chairman Sheik Yusuf Indahaadde believed that the war against terrorism is a war against Islam. The Islamists have instituted in Mogadishu and other ICU-controlled areas Taliban-style rules banning drinking, cinemas, dancing and women swimming, as well as curbing the press (Dealey 2006: 37–8). Such radical Islamism has raised concerns with the global actors.

The global conflict system has at least four manifestations: first, multinational intervention even in the name of humanitarian intervention can provide merely thin cover for self-interested states to attain their national interests (Talentino 2005: 29). Consequently, debate on relations between self-interest and humanitarian intervention by states or international institutions is considerable. For understanding the impact of intervention on the intervened-in state, it is important to take into account not only the direct military intervention but also the *structural intervention*. Neighbouring and global actors intervene at different times in different forms with considerable implications for the peace and security of the intervened-in states. For example, Chinese policy towards Sudan can be seen as intervention because it has a considerable structural impact on the conflict dynamic in Sudan. China disregarded the catastrophic situation in Darfur and used its position on the UN Security Council to dilute repeated resolutions; it successfully prevented almost any mention or threat of sanctions against the Sudanese government. Even during the resolution of March 2005 (1591), when world opinion focused on Darfur, China only abstained from voting instead of supporting the Security Council Resolution that referred the possible war crimes and charges of genocide there to the International Criminal Court (ICC) and set in motion a UN study of possible sanctions (Lake et al. 2006: 44).

A second manifestation of the global conflict system is a proxy war: client states or groups in the third world fight each other. For example, UNITA (pro-US) fought the MPLA (USSR-Communist) in Angola. Until UNITA leader Savimbi was killed in 2002, Angola was the theatre of one of the world's brutal civil wars. When Ethiopia (pro-Communist) and Somalia (pro-US) fought in the 1970s, it was explicitly the war for Ogaden, but implicitly a war between the friends of communism and capitalism.

Third, structural impact of competition for resources and influence between two or more global powers on a third country (competition between Russia, China and the West for natural resources such as oil in Sudan or DR Congo) exacerbate existing conflicts or cause new ones. Even if states like Russia or the US no longer attempt to gain allies at all costs (Talentino 2005: 48), they secure their national interests globally at all costs.

Fourth, the global actors that have the power to intervene might not be interested in intervention because either there is no immediate gain from the intervention or it would undermine their interests. In the Sudanese war of the Darfur region, the US Congress half-heartedly declared it as genocide. The UN Commission Report of January 2005 concluded that there was no genocide (UN 2005: 160). Russia and China oppose any uninvited intervention (Talentino 2005: 280–1). Both Russia and China have been benefiting from the conflict: the former by selling weapons, the latter providing Sudan with hard currency earned by buying the oil. However, China's role is not limited to oil exploitation. Chinese arms were captured by the JEM in early 2008 when JEM forces ambushed a Sudan Armed Forces convoy travelling northwards on the main road from El Geneina to Kulbus Western Darfur (United Nations 2008c: 21).

According to Talentino, intervention 'does not happen when it should, and sometimes it happens when it should not. And even if it is true that humanitarian interests motivate action, states will try their hardest to ignore or avoid those interests because the problems are big, the costs are too high, and the logistics too complex' (Talentino 2005: 284). Further, Talentino concludes that 'humanitarian concerns and institutional action are laudable, but only if they can deliver the goods they are intended to achieve' (Talentino 2005: 301). Whether moralistic and humanitarian intervention would achieve lasting solutions is, however, not that clear. Michael Walzer is right when suggesting that intervention out of moral reasons does not exist in political life because states always consider the interests of their own people (Walzer 2002: 26). Morality at least in this context is a weak approach because it presents the others as mere profiteers of the intervention and the intervener as mere actor out of morality and selflessness.

Morality as a basis for humanitarian intervention to rectify wrongs and protect the innocent (Nardin 2003) diverts attention away from addressing the root causes that necessitate intervention. Morality becomes hypocrisy if structures and systems that create such conditions are not addressed and changed. For example, the pre-independence policy of the UK has had some impact on the post-independence conflict between Ethiopia and Somalia. When Italy was defeated during the

Second World War, in the first months of 1941 Italian Somalia was occupied by the UK, which advocated a form of unification of the Somali lands (British Somaliland, Italian Somalia and Ogaden of Ethiopia) under British jurisdiction. The UK argued that it would be better for the nomads in the area to move freely where there is more water and grazing land. In Somalia the British move was on the one hand rejected because the UK was going to maintain and consolidate power in Somalia; on the other hand it was favoured because it would support the aim of the Somali Youth League (SYL), founded in 1943, to fight tribalism and create a Greater Somalia. This move would have meant independence and unity in the 1940s. Though the ambitions of the SYL were at least indirectly supported by the UK, the big powers were against the Greater Somalia project, as it would have caused similar desires in many colonized African regions. In spite of the protest from the SYL and other Somali elites, the UN General Assembly on 21 November 1949 assigned the former Italian Somaliland to Italy as a trusteeship for a 10-year period.

The 1998–2000 war between Eritrea and Ethiopia emerged from the interaction of various internal, regional and global factors: Italian colonialism; Ethiopian victory over the Italians in 1896, which enhanced Ethiopia's prestige; establishment of Italian colonial rule in Eritrea between 1890 and 1941, which created Eritrea; Ethiopian hegemonic ambitions and the post-1945 international state system, which favoured Eritrea's autonomy rather than independence. For Eritrea, Ethiopia between the 1940s and the 1980s manipulated and exploited the Eritrean rivalries to gain imperial control (Iyob 1995: 3). Arguing from the pre-colonial situation, Ethiopia insisted that Eritrea is part of Ethiopia, whereas Eritrea saw itself as a colonial creation of a separate political entity (Iyob 1995: 14).

Italy attempted revenge against Ethiopia for its defeat by contributing to factors that indirectly destabilized Ethiopia, such as by promoting regional and national affinities in Eritrea and Somalia. Somalia and Eritrea were exploited by Italy as a springboard to destabilize Ethiopia. In order to launch another attempt to attack Ethiopia and to take revenge for its continuous embarrassments in Ethiopia, Italy ignited a conflict at the Somali border. It accused Ethiopia of violating the Somali borders, though Italy was already on the Ethiopian side. But this was just a pretext by the Italian government for invasion of Ethiopia on 3 October 1935 and occupation on 5 May 1936.

In spite of the international Gas Protocol of 1925 (which banned the use of poisonous gas in wars), Italy began to use gas against Ethiopians in December 1935. For the first time in a colonial war, gas and air bombardment were used. London and Paris abandoned the Ethiopian crisis and were busy with the German danger after Hitler's occupation of the Rhineland on 7 March 1936 (Sbacchi 1985: 26). After Ethiopia was proclaimed conquered by Italy in May 1936, the violent occupation began. Ethiopian oppositions were treated as bandits; the bombings began in June 1936 in western Ethiopia. Ethiopians were despised and made second-class citizens in their own country as colonial subjects (Sbacchi 1985: 67–75). Until early 1937 Italian persecution and destruction continued and paradoxically it became the source of Ethiopian moral strength (Sbacchi 1985: 185). But this Italian aggression had an impact on Ethiopian national identity. This brutal violence left strong memories in the Ethiopians, and it has been abused to legitimate intrastate violence at different times in the name of national unity and identity. Since the 1970s, during the war against Somalia, conflict with Sudan, which was supporting the Eritrean and Tigrayan rebels, and the campaign against these and the Oromo rebels, the war songs against the Italians in the 1930s and 1940s were sung.

The global conflict system in this case is twofold: first, Italy occupied Ethiopia in order to create East African Italian Empire because France and the UK established their own de facto empires in Africa by occupying those huge colonial territories. This means the colonial ambitions of Italy were initiated primarily by the rivalries of the European powers in order to demonstrate national grandeur through colonization. The second aspect of the global conflict system was that because of the German aggression in Europe, France and the UK were not interested in helping Ethiopia against Italian aggression.

During the Cold War, especially from the early 1970s, the two super powers played a key role in the political developments in the Horn of Africa. Somalia, Ethiopia and Sudan were the scenes of the Cold War proxy wars. The initially East-minded President Numairi of Sudan and his government, which came to power through a military coup on 25 May 1969, intensified military activity in Sudan by importing various fighter jets, anti-aircraft guns and transport aircraft from the Soviet Union according to the arms agreement concluded in 1968. In the first two years of the 1970s, the Soviet Union delivered an impressive array of equipment to Sudan, such as T-54, T-55, T-56 and T-59 tanks, and BTR-50 and BTR-152 light armoured vehicles (Global Security 2008).

Russia as global actor played an important role in the dynamic of the conflict in Ethiopia as well. Moreover, both Ethiopia and Russia were monarchies. In the nineteenth century Ethiopian emperors (Emperor Tewodros II, 1855–68, and Emperor Yohannes IV, 1872–89) sought cooperation or aid from Russia against the Turks/Egyptians. When the British, Italians and French were scrambling for the Horn of Africa, Emperor Menelik of Ethiopia (1889–1913) sought Russian aid, such as military assistance mainly against the Italians, who distorted the Wechalé Treaty document between Ethiopia and Italy of 2 May 1889 by claiming that Ethiopia, according to the Treaty, will be obligatorily represented by Italy for all communications with other powers. The opening of the Ethiopian Embassy in Russia in 1895 implied that relations between these two empires developed rapidly. Russian military support for Ethiopia was decisive for the Ethiopian victory over the Italians at the battle of Adwa on 1 March 1896. Some factors helped to expand Ethiopian–Russian ties: in addition to military support, Ethiopian students were educated at Russian universities; the Russian Red Cross Mission and Hospital was established; Russia supported Ethiopia while it was trying to establish its borders in the south, east and west. In spite of these ties, Russia ultimately collaborated with Italy when it invaded Ethiopia in 1935–36. But in 1943 Russia again sought ties with Ethiopia, mainly in competition with Britain, which was a 'liberator' of Ethiopia against fascist Italy. Though Russia was displeased by Ethiopia's move to give the US a communication base outside Asmara and to enter into the Korean War on the side of the UN between 1950 and 1953, Russians praised Emperor Haile Selassie for his support against the colonizers in other parts of Africa, and Russia granted at least $100 million to Ethiopia when the Ethiopian Emperor visited Russia in 1959 (Pankhurst 2007: 219–28).

The disagreement between the West (mainly the US, Italy and West Germany) and the East (Russia) on how much army the newly independent Somalia (1960) needed to maintain law and order led to the first impact of the East–West conflict on the Horn of Africa. Whereas the Russians suggested a 10,000-man army, the West suggested a 6,000-man army. Subsequent Russian military and monetary support for Somalia and American military and monetary support for Ethiopia cemented the competition between Ethiopia and Somalia as well as between the US and Russia in the Horn of Africa. The US military assistance to Ethiopia was $308 million within 25 years (60 per cent of the total military to Africa), whereas the Russian aid to Somalia was estimated at $1,000 million. Between 1972 and 1976, Ethiopia and Somalia imported arms for a total of $110 million and $321 million respectively (Pankhurst 2007: 229).

Concerned with the growing military expenditure and power of Somalia, its ties and Friendship Treaty with Russia on 11 July 1974, the Ethiopian military junta, which deposed the emperor on 12 September 1974, started an intense diplomatic initiative with Russia, the German Democratic Republic and Cuba, and proclaimed its revolutionary ideology. But Russia was cynically supporting both Ethiopia and Somalia in order to keep both countries as its clients; for this reason Somalia unilaterally abrogated its Friendship Treaty with Russia on 13 November 1977 and transferred its strategic base at Berbera from the Russians to the Americans. Angered by this move, Russia increased its massive military support to Ethiopia and moved its military advisers from Somalia to Ethiopia. In addition to this, with the support of 20,000 Cuban and South Yemeni troops, Ethiopia defeated Somalia and signed a Treaty of Friendship and Cooperation with Russia on 20 November 1978 (Pankhurst 2007: 232).

In order to assert its stance the Numairi government of Sudan relied on its military power. After an attempted pro-communist coup in 1971 against Numairi, he abandoned his socialist–communist ideology and intensified its alliance with the US. The reward was, among others, an abundant supply of weapons from the US. The US did this deliberately to counterbalance the Soviet-backed Ethiopian army and its weapons after the overthrow of Emperor Haile Sellassie. The SPLA's dependence on Communist Ethiopia caused Washington to support Khartoum militarily (Johnson 2003: 59–60). Having profited from the arms supply and economic aid, the Numairi government waged an uncompromising war against the southerners, instigating the second civil war of Sudan in 1983.

The Sudan/Egypt–Israel conflict had considerable impact on Ethiopia. Ethiopia in the 1960s and 1970s was walking a tightrope in its relations with Israel. As a founding member of the OAU, it had to respect the sovereignty of Egypt when the latter fought against Israel in 1954, 1967 and 1973. On the other hand, Ethiopia has cultural and historic ties with Israel because of the Ethiopian Jews as well as because the Ethiopian emperors derive their authority from their Solomonic Dynasty. An alliance with Ethiopia was key to Israel's economic and diplomatic gains in Africa, since states like Egypt and Sudan are enemies of Israel. Throughout the 1960s and until 1973 when Ethiopia broke diplomatic relations with Israel – because of the October 1973 Arab–Israeli War – Israel's aid for Ethiopia was the greatest foreign enterprise at the time, which was promoting Ethiopia's security, health, agriculture, education, construction and finances. Further, by losing Israel, Ethiopia risked losing the US (Johnson 2003: 36–7).

Soviet support for Ethiopia played a considerable role in Ethiopia's fight against its rebels in the north of the country as well as against Somalia. Arms for Ethiopia from Romania, East Germany, Czechoslovakia, mercenaries from Greece, training and troops from Cuba all played a significant role during the Ogaden war between 27 July 1977 and 14 March 1978 (Carment et al. 2006: 93–5). Cuban troop training and armaments from Russia were crucial to Ethiopia's victory over Somalia in 1978 (Lewis 2002: 233). The United States was supporting Ethiopian Emperor Haile Selassie because the Soviets were in Somalia (Johnson 2003: 57). The Somalis were defeated in the Ogaden by a strategy worked out by Soviet generals (Legum 1979: 634). After Ethiopia successfully defeated the Somali irredentists, it was further supported by the Soviets, Cubans and East Germans to suppress the Eritrean guerrilla movements by supplying arms (Winrow 1990: 132, 145; Buzan and Waever 2006: 250). With an enormous advantage in sophisticated weapons and manpower, the Ethiopian army had routed the ELF, lifted the siege around the Red Sea port of Massawa and captured Keren/Eritrea; and the EPLF had been driven back to its original mountain strongholds, and victory claimed by government (Legum 1979: 635), whereas Eritrea was supported by the Arabs against Ethiopia.

However, the support, such as from the Soviets for Ethiopia, was not necessarily because of Communist ideology or a desire for the freedom of the oppressed. For example, it supported nasty tyrannies like Idi Amin in Uganda and President Macias of Equatorial Guniea (Legum 1979: 641). Francisco Macias Nguema banned opposition parties and in 1970 appointed himself 'President for Life' – the first of a string of self-decreed titles that included 'Leader of Steel', 'Implacable Apostle of Freedom' and 'The Sole Miracle of Equatorial Guinea'. As many as 50,000 people, roughly 10 per cent of the population, were murdered during the Macias years and 80,000 fled the country. In 1979 he was overthrown and subsequently executed by his nephew Obiang.

Even after the end of the Cold War, Sudan brought various global actors together. The catastrophic humanitarian situation first in South Sudan and then in Darfur, Sudan's oil, and its relationships with global terrorists such as Osama Bin Laden made Sudan an area of interest for China, Russia, the EU, the US, etc. Whereas the US is mainly interested in fighting terror threats coming from

Sudan, Russia and China have been interested in making business with Sudan. The European Union has been engaged in the Sudan since the early 1990s by supporting the IGAD peace initiative for South Sudan, such as financial support through the IGAD Partners Forum (IPF). Since civil war broke out in western Sudan in early 2003, the EU has been supporting the AU's peacekeeping mission in Darfur and has committed about €300 million to the African Mission in Sudan (AMIS). However, for example, the role of France in Chad is structurally affecting Sudan as well.

Increasing terror threats in the Horn of Africa have drawn the attention of the US. Under pressure from the US, the United Nations imposed the first sanctions on Sudan in April 1996. At the same time the US government intensified its relations with Uganda, Ethiopia and Eritrea by providing them with 'non-offensive' military equipment, such as tents, uniforms and shoes worth $20 million to address the Islamist terrorists in Sudan as well as in the Horn region in general (Prunier 1997).

Since October 2002 Djibouti hosts the US military base in the Horn of Africa – with which forces from France, Germany, Spain and Italy coalesce and is known as the Combined Joint Task Force-Horn of Africa (CJTF-HOA). In 2002 the Combined Task-Force Horn of Africa was established and based in Djibouti to fight terrorism in the region with its 1,700 troops. In 2003 it allocated $100 million to an East African CounterTerrorism initiative. Since the rise of the Union of Islamic Courts in 2005, the United States became directly involved in Somalia by creating the Alliance for Restoration of Peace and CounterTerrorism and by collaborating with a group of Somali warlords who aided the US by snatching – allegedly to fight Islamic 'extremists' – al-Qaeda operatives taking refuge in Somalia. The US paid the warlords around $150,000 a month to support these operations. The US intelligence support, military targeting and logistical support for Ethiopia during the invasion was certainly invaluable. Moreover, the US military attempted to attack Al-Shabab using aircraft, warships and special forces on at least five occasions (BBC 2009). Now the Ethiopians have left; because the US has contributed its enemies hate it more than before; and the Horn of Africa is more instable than before.

Some 1,800 American military and civilian personnel occupy a former French Foreign Legion facility at Camp Lemonier outside the capital city of Djibouti. The force is responsible for fighting terrorism in Djibouti, Ethiopia, Eritrea, Sudan, Kenya, Somalia and Yemen, and in the coastal waters of the Red Sea, the Gulf of Aden and the Indian Ocean to fight terrorism and to enhance long-term stability in the region. CJTF-HOA has been training Ethiopian anti-terrorism battalions in Hurso, in the south-eastern part of the country. In 2002, the US gave Ethiopia $2.25 million in foreign military financing (FMF) and $445,000 in international military education and training (IMET). For 2003, Ethiopia was given $4 million in FMF, $210,000 in IMET, and $1 million in economic support fund (ESF) assistance. For 2004, Ethiopia received an estimated $2.4 million in FMF, $570,000 in IMET and nearly $5 million in ESF. For 2005, Ethiopia was promised $2 million in FMF, $600,000 in IMET and $5 million in ESF.

Probably, the carrot and stick policy of the US Sudan Peace Act of October 2002 had more influence than the IGAD or the EU. The US earmarked $300 million for South Sudan, and it threatened the Khartoum government with sanctions. Until the US intervened in the Somali conflict in the early 1990s, Sudan considered the north–south conflict as an internal matter and did not show any serious interest in negotiating. In order to deflect any mighty foreign intervention like that of Somalia, it accepted negotiations orchestrated by a weak organization, IGAD, under the banner of 'African solutions to African problems', without substantial change. During the IGAD summit on 11 January 2002 – against the backdrop of the US allegations against Sudan that it supports terrorist organizations and fear of severe military consequences from the US after the 11 September 2001 attacks – Al Bashir confessed his willingness to implement the Declaration of Principles of 1994 and declared his willingness to negotiate with the SPLA/M, as foreseen in the IGAD peace process.

Washington was active in Sudan in different parts in different times. Washington's support for South Sudan did not deter the violent policies of Khartoum against the South (Deng and Morrison 2001: 6). For example, when Sadiq al-Mahdi as Prime Minister (1986–89) moved the country closer to Libya and Iran, the US military aid to Sudan was suspended (Johnson 2003: 60), and US (and Israeli) relations with the SPLA was re-strengthened. However, when Sudan supported the Camp David Accord, this move resulted in the break-off of Israel support for the SPLM. Mainly since the early 1990s, Sudan again got the special attention of the US.

Between 1998 and 2002 Sudan was the sixty-sixth-largest recipient of major conventional weapons and was the forty-fourth-largest recipient between 2003 and 2007, whereas armed non-state actors in Darfur were provided with arms and military equipment among others by Eritrea, Chad and Libya (Holtom et al. 2008: 314). A ship loaded in Tunisia with some 680 tons of explosives and 8,000 detonators headed for Sudan in 2003; both can have civilian uses and artillery shell manufacture. In 2000 Bulgaria and Sudan had signed a secret arms deal worth hundreds of millions of dollars. Sudan has been enabled by Chinese licence and support to manufacture small arms; it had been cherishing secret contacts with Iraq and Iran and with military experts from Kazakhstan. In the first half of 2001 Sudan concluded a deal with Russia, estimated to be worth more than $600 million, to manufacture battle tanks (ARB 2001, 38(3): 14340). Oil revenues have also enabled Sudan to acquire surface-to-surface missiles in its war against southern rebels. Oil and conflicts in the region have interlinked the regional and global actors.

French troops in Chad under a 1986 agreement were ready to support Chad when its territorial integrity was under threat. France's support to Chad has had an impact not only on the conflict system within Chad but also on the region as a whole. When the Chad rebels in April 2006 advanced to the capital to overthrow

President Déby and his government, France flew in an extra 150 troops from Gabon on 12 April 2006 to supplement the 1,200 troops already in the country; and the French Defence Ministry acknowledged that one of its mirage jets in Chad fired warning shots towards rebels that had been supported by Sudan. Early in 2008, while the European Union was preparing the deployment of its EUFOR Chad/CAR troops, French President Sarkozy expressed France's willingness and readiness to fly some more French troops (adding the 1,400 already stationed there) to support the Chad government, which undermined the credibility of neutrality of the EUFOR force. Because of such support the Chadian rebels, such as the UFDD, threatened to attack the French forces in particular and EUFOR Chad/CAR in general. Since Chad has been exporting oil (since 2003) significantly, the control of an estimated 1 billion barrels' oil reserve is worth fighting for. The oil has been financing the war instead of financing poverty alleviation, as agreed with the World Bank when it lent money for the oil exploration project.

Because of this oil factor, French interest in Chad is not limited to the historical (colonial) relations between France and Chad, but oil also plays a key role because various global players are involved in Chad. When French oil giant Elf walked out of the oil project in 1999 because the relations between Paris and N'Djamena were poor (though Déby assumed power through the help of France), Exxon arrived as lead player to replace Elf. This was felt in Paris as an embarrassing defeat in the geopolitical and economic competition with the United States. Exxon got 40 per cent of the consortium shares, Chevron 25 per cent and Petronas 35 per cent (Marchal 2007a: 190). Besides the United States, China as Chad's supporter of oil exploration and export has been contributing to reduce Déby's need to depend on the World Bank's dictation.

Global actors, mainly the Permanent UN Security Council Members, have been playing major roles by transferring arms into the Horn of Africa. Between 2003 and 2007, Russia accounted for about 87 per cent of arms transfers to Sudan, including 20 combat helicopters and 12 MiG-29S combat aircraft, whereas China delivered 8 per cent, consisting of, among others six K-8 and three A-5 light combat aircraft. In breach of the arms embargo based on UN Resolutions 1556 (2004) and 1591 (2005), containers were offloaded by Sudanese army soldiers from an Antonov aircraft onto military trucks at the military apron of El Geneina airport in July 2007; and a Russian-supplied Mi-17 military helicopter and Mi-24 attack helicopter were delivered to the Sudanese Air Force at El Geneina. It has to be noted that Russia signed a deal to supply at least 15 such helicopters for delivery in 2005 and 2006. Russia supplied 12 Mi-24 attack helicopters to Sudan in 2005 (Amnesty International 2007). Weapons flowing from different parts of the world, such as Mi-24 helicopters piloted by Ukrainian mercenaries or Russian T-55 tanks, were valuable for the Army of Chad in early 2008 to ward off the rebels' advance from the east of the country (Prunier 2008). Moreover, Chinese, Iranian and Russian arms companies have helped Sudan to assemble and produce small arms, artillery and armoured vehicles (Holtom et al. 2008: 315).

China abstained from the vote on Resolutions 1556 and 1591, and Russia abstained from 1591, both resolutions about arms embargo. But on the other hand, the UN in general and the UN peacekeeping force in particular is seen as the only power to make sustainable peace in Sudan's Darfur region. When UNAMID was created in 2007 the UN mission was hoped to be bigger, better equipped and more capable of aggressively responding to Darfur's unknown numbers of armed groups than the ill-equipped and weak AMIS.

In order to complete the regional and global conflict cycle, the US conflict with the Sudan benefited the Ethiopian government. The strategy of the US during the 1990s was to weaken and contain Sudan and Sudanese-sponsored insurgencies of Ethiopia and Eritrea. In late 1995, President Clinton authorized the transfer of $15 million in non-lethal defensive military assistance to Eritrea and Ethiopia for 1996. A transfer of $4.75 million of such equipment took place in 1997. The aid consisted of boots, backpacks, field radios and tents. The first shipments of this assistance for Eritrea and Ethiopia arrived in February 1997. However, the war that erupted between Ethiopia and Eritrea in May 1998 complicated the strategy. US cooperation with Eritrea and Ethiopia was seen as a key strategy in the Horn of Africa to carefully observe the behaviour and intentions of Sudan and the terrorist developments between the Arab world and Somalia.

Even where the global actors have the power and the will to do the 'right thing', the success of an intervention depends not only on the will and power of the intervening global powers, but also on the readiness of the conflict parties, on the regional actors, and on the global powers that are not intervening in the conflict maybe because of the rivalry and conflict between the intervening and non-intervening global powers. For example, the south of Somalia, where the Somali Patriotic Movement was acting, was heavily affected by the war between this faction and Aideed's fighters around the middle of 1992. Mutual devastation continued until the US 'Operation Restore Hope' arrived in December 1992. In addition to the humanitarian aspect, the US wanted to secure the international waters in the region of oil and trade route (Talentino 2005: 112). Nevertheless most of the population welcomed (at least at the beginning) the US initiative, the United Nations International Task Force (UNITAF) and the UN-sponsored peace conference in March 1993 in Addis Ababa, and Aideed of USC, the SNA (Somali National Alliance), Abdullahi Yusuf of the SSDF (Puntland) and the voted-out president of the SNM (Somaliland) all attended the conference, which ultimately failed. But Aideed objected to the intervention of the UNOSOM, which was trying to rebuild an administration, and claimed he only had the right to nominate administration in the areas under his control. His stubborn rejection and dissemination of accusations of UNOSOM's imperialistic intentions resulted in a UN raid of his radio stations. Based on UN Resolution 814 of 26 March 1993, a multinational force of 20,000 peacekeeper troops, 8,000 logistical staff and 3,000 civilian personnel was envisaged. But because of the continuing US tactical support, the provision of US logistical support and the availability of US special forces, UNOSOM II, which formally took over from UNITAF/UNOSOM

I on 1 May 1993, was marred by a power struggle between Washington (US) and New York (UN); and the Italians had their own agendas (Lewis 2002: 271, 278). Bill Clinton announced that, until its departure by 31 March 1994, the US force would remain under the command of the US, not of the UN. In the face of this global power struggle, escalating violence and perplexity of the situation, the UN was paying $150,000 a day for the hotel bills of 60 Somali participants of a peace conference. At the same time the warlords had enough access to arms in spite of the arms embargo of 1992 (based on UN Resolution 733 of 23 January 1992) and the disarmament programme was without any progress. Aideed's militiamen retaliated by killing 24 Pakistani UN soldiers on 5 June 1993 and 18 US peacekeepers on 3 October of the same year. After these losses the US peacekeepers changed their stance and began to support Aideed's demands as leader of a new government, especially in 1994 (Ahmed 2001). Resolution 953 of 31 October 1994 set the deadline to withdraw UNOSOM II by 31 March 1995, facilitated by the locally recruited Somali staff, costing the UN $1 million per day, which could not save the UN peacekeepers' facilities from massive looting right after the UN left (Lewis 2002: 275). One important aspect about how UNOSOM structurally contributed to the conflict was that out of the $1.6 billion allocated for UNOSOM's military operation up to the end of 1993, probably only 4 per cent entered the Somali economy, mostly to the benefit of warlords and armed entrepreneurs (Lewis 2002: 280).

The more aid came in, the more it fed the competition for resources, and Mogadishu as distribution point attracted many displaced; as a consequence it became an intensified conflict zone (Talentino 2005: 106). One important weakness was that there was no conflict resolution agenda; instead there was only an humanitarian aid agenda and most of the outside actors focused on the drought as a cause of famine and malnutrition (Talentino 2005: 109). Aideed, who occupied Mogadishu (though partly), got considerable international legitimacy. Moreover, he got support from Sudan, Libya and various international Islamic organizations (Ahmed 2001). Until they left Somalia in March 1995, the UN peacekeepers limited their task to humanitarian aid rather than to achieving political stability. Continuous fighting in Mogadishu further destabilized after the UN left (Brons 2001: 219).

Currently, possible global terror threats from Somalia attract the attentions of the global actors like the US to be directly or indirectly militarily involved in Somalia. Global actors like the US argue they know of links to Al-Qaeda in and through Somalia. This leads to actions against it and to conduct dialogue with the frontline nations such as Eritrea, Kenya, Djibouti and Ethiopia (Mentan 2004: 262). That is, the 'war on terrorism' engulfed the 'failed' state of Somalia in the confrontation between the US and its allies on the one hand, and Islamists and its allies in Somalia and worldwide. On 15 November 2001, Susan Rice, the Clinton administration's Africa policymaker, in her speech to the House of Representatives, called Somalia a 'terrorist haven', the 'continent's proverbial black hole, an ungoverned lawless, radicalized, heavily armed country with one of the longest undefended coastlines in the region'. Increasing Wahhabist influence, possible Al Itihad Al Islamiya's (AIAI) activities and Somalia's extreme poverty are feared to make Somalia an ideal site for recruitment into terrorist cells for organizations such as Al-Qaeda (Menkhaus 2005: 23). A violent deadly conflict emerged in May 2006 in Mogadishu between various warlords and Islamist organizations that are believed to have links to Al-Qaeda. Various warlords calculated that the West would come to their aid and help them control Mogadishu against the Al-Qaeda-infiltrated Islamist groups.

Islamist terrorism plays the central role for the presence of regional and global actors. Senior AIAI leader Aweys was accused of involvement in bombings in Ethiopia in 1995–96. Moreover, he is linked to the bombings of the US embassies in Nairobi and Dar-es-Salaam in 1998 and suggested a relationship with bin Laden himself as well as Al-Qaeda and terrorism (ICG 2006e: 10). In order to take advantage of this international interest in fighting such personalities and terrorist organizations, the ARPCT (Alliance for the Restoration of Peace and CounterTerrorism) militia leaders sought to portray all Islamists as terrorists and use American counterterrorism for parochial aims (ICG 2006: 16). In 1997 and 1998, Ethiopia had already entered at various times into the western Somalia, mainly in the Gedo region, in support of the Somali National Front, which was fighting the Al-Ittihad Al Islamia. A key component of the global conflict system in this region consists of Ethiopia's anxiety to protect itself from attacks by Islamic terrorists based in Somalia; the support Ethiopia gets from the US in particular and the West in general in this anti-terrorist project; the lack of effective government in Somalia, a potential breeding ground for fundamentalism; the attempts of these regional and global actors to discourage the emergence of Islamic organizations; the increasing migration flow from Somalia to the Western world, which is being exploited by the right-wing parties in the West to gain votes (Lewis 2002: 309–10); and the security-related threats from Somalia, such as in the EU, in the form of terrorism involving immigrated Somalis like in the London terrorist attacks of July 2005; and piracy.

The TFG leadership's dependence on the historical arch-enemy Ethiopia delegitimizes the credibility of the TFG. As the conflict system in Somalia demonstrates, foreign interveners have been escalating tensions and political competition between the Islamic Courts and clans (ICG 2006e: 1). Because of the growing Islamic terrorism, global and regional actors are increasingly involved in the region; and this increasing presence of the global and regional actors in Somalia is further supporting the rise of Islamic fundamentalism, with destabilizing effects on Somalia and on the region as a whole. This means that the US insistence on counterterrorism policies is paradoxically going to increase the presence and influence of radical Islamic organizations and their destabilizing activities in Somalia and to fuel anti-Western, Islamic sentiments and create a self-fulfilling prophecy (Menkhaus 2005: 23–4). The takeover of Mogadishu by the ICU in June 2006 is partially a reaction to the US support for warlords that have devastated the country since 1991.

The Islamic Courts' success since June 2006 alarmed the TFG, the key regional actor Ethiopia, the US, various states of the EU, the EU itself and the UN. Ethiopia is worried because of the possible links between the ICU and al-Itihaad al-Islaami (AIAI). Kenya is alarmed by the growing importance of Islamism, which could hit its own territory where ethnic Somalis live. The US is still looking for those Islamists responsible for bombing two of its embassies in 1998. The bombing of the American embassies in Kenya and Tanzania by Al-Qaeda in 1998 – attacks that left hundreds dead and thousands wounded – initiated the re-engagement of the US after it left Somalia in 1993. All are attempts to prevent Somalia becoming an African version of Taliban-ruled Afghanistan. Through Eritrean assistance to the ICU, a demonstration of power between Ethiopia and Eritrea has been taking place in Somalia (ICG 2006e: i).

It is important to emphasize that the crisis is and will remain a direct product of ill-conceived foreign interventions. Ethiopia's attempts to supplant the earlier Transitional National Government (2000–03) with a new one dominated by its allies alienated large sections of the Hawiye clan, leaving the TFG with a support base too narrow to operate in and near Mogadishu. The calls of the African Union and the Intergovernmental Authority on Development (IGAD) for foreign peacekeepers, intended to bolster the TFG, have instead cast it as ineffectual and dependent on foreign support, and provided a rallying cry for diverse opposition groups. US counterterrorism efforts meant to contain foreign al-Qaeda operatives have accelerated the expansion of jihad Islamist forces and produced the largest potential safe haven for al-Qaeda in Africa (ICG 2006e: i).

The external involvement would also exacerbate the internal divisions and rivalries, increasing the likelihood of additional schisms, realignments and instability caused by competing interests are complex, reflecting clan cleavages, external alliances, ideology, business competition and personal rivalries. The military involvement also increased the systematic violations of the embargo by a wide range of political factions, merchants, Islamists and a number of countries, such as Ethiopia, Eritrea, Italy and Yemen, etc., as well as the TFG itself (ICG 2006e: 22).

The CIA and US support for various militia groups led to widespread acceptance of the ICU (Dealey 2006: 38). Syria is reported to have equipped and trained the ICU military. For example, on 27 July 2006 fighters from ICU were transported by aircraft to Syria to undergo military training, and a large quantity of arms including surface-to-air-missiles were delivered to the ICU in early September. In the same way Iran supplied the ICU with arms, including a shipment on 25 July and mid-August of 1,000 machine guns and grenade launchers, an unknown quantity of mines and ammunitions, 45 shoulder-fired surface-to-air missiles and medical supplies (Dealey 2006: 38). Somalia is feared to be an ideal conduit through which al-Qaeda could advance into the region from the Arabian peninsula. The ICU have made clear that they would consider the deployment of the African peacekeepers as an act of war and invasion, and the Somalis are not ready to be colonized again by any sort of troops in the world (Dealey 2006: 39).

Sudan initially wanted to accept only African peacekeepers in Darfur instead of the suggested UN peacekeepers. On the domestic front, Khartoum's newspapers led a government-backed anti-Western campaign. President Omar al-Bashir promised to make Sudan a graveyard for foreign intervention and government-backed militia say they are preparing for a holy war. According to Jamal Ibrahim, the foreign ministry spokesman, intervention in Darfur would mean intervention in Sudan's internal affairs. Similarly, Somalia considers regional states like Ethiopia and Kenya as invaders; whereas Sudan considers the non-African peacekeepers as invaders.

Global conflict system consists of economic, strategic and security interests of the intervening powers. Even through their peace initiatives the global powers can produce short-term successes but with long-term structural dangers. It has been argued by various politicians as well as academics that soft and hard power of the foreign actors is indispensable for a successful peace process. For example, in January 2006 the UN Development Programme promised to give each of 275 parliamentarians a daily 'sitting allowance' and per diem payment of $60 for 120 days ($7,200). Similarly, deployment of representatives to key posts abroad was heavily supported by a UNDP project called 'global engagement' fund of $500,000 for diplomatic travel and presence (ICG 2006e: 8). But such a strategy is questionable because those parliamentarians were interested in earning so much money in an economically devastated Somalia, where hundreds of thousands are suffering from hunger and malnutrition. When the money supply dried up, the parliamentarians were not interested in the peace and political process.

International law constitutes an integral part of the global conflict system. For example, international law contributes to the emergence and maintenance of the global conflict system in Somalia. Though it is the only functioning part of Somalia., the global actors are not ready to recognize Somaliland as a state, though a scrupulously conducted referendum, through 97 per cent of the votes in favour, validated the new constitution and Somaliland's independence. The unwillingness of the UN and its international law to recognize this basic democracy suggested that this international organ is not necessarily interested in democracy (Lewis 2002: 303). Certainly, this stance of the global actors will contribute to the future conflict between Somaliland and Somalia, when the latter solves its internal problems and begins to function as a state. So far Somalilanders have been setting an example for the rest of Somalia: traditional grassroots negotiation involving clan elders are working; the state enjoys stability and commitment to free market economies; import–export businesses compete, market streets are stocked with imported goods. At least partially, the success of Somaliland can be traced back to its low-cost, local clan-based peace initiatives against high-profile, internationally sponsored and highly unsuccessful conferences, and absence of power-hungry military leaders (Lewis 2002: 266). International recognition does not always solve problems, but its non-recognition is not going to prevent future conflicts either.

Besides state collapse and terrorist threats, Somalia has got other problem areas that have attracted international actors. Piracy rose in 2008 by 200 per cent from 2007. In 2007, there were 31 actual or attempted attacks off the Somali coast (Westcott 2008). The pirates attacked more than 90 vessels between January and November 2008. Piracy is regarded as a joint threat. UN Resolution 1816 of June 2008 gave authorization to states co-operating with Somalia's transitional government for a period of six months to enter the territorial waters of Somalia. Besides hijacking vessels heading west from Asia across the Gulf of Aden to reach Europe, Somali pirates took, in September 2008, 30 hostages on a French luxury yacht who were released for a ransom of about $2 million (Westcott 2008). For example, China sent two destroyers and one supply ship in late December 2008 to the Indian Ocean, through which four or five Chinese ships pass every day; despite this, seven Chinese ships or crews were attacked in 2008. UN Resolution 1838 was passed in October 2008 to use 'all necessary means' to stop acts of piracy and armed robbery at sea; and the mandate was extended in early December 2008 for another year.

EU NAVCO, established by the Council of the European Union in September 2008, has been transferred to EU NAVFOR and reached its initial operational capability on 13 December 2008. As the Somali pirates are poised for continued attacks on shipping off the coast of Somalia, the EU's anti-piracy taskforce, Operation Atalanta (EU NAVFOR Somalia), with naval officers based at Northwood near London, was deployed in December 2008. The operation is conducted in support of UN Security Council Resolutions 1814 (2008), 1816 (2008), 1838 (2008), 1846 (2008) and 1851 (2008) to protect vessels of the WFP (World Food Programme) delivering food aid to displaced persons in Somalia; to protect vessels cruising off the Somali coast; and to deter, prevent and repress piracy and armed robbery off the Somali coast (Consilium 2009). The EU taskforce warships and maritime patrol aircraft from Britain, France, Germany and Greece are pouring into the Gulf of Aden and Indian Ocean. Due to the deterrent effect on the pirates, the patrolling by coalition warships has resulted in a decrease in hijackings. As well as the EU's, there are warships joining from Russia, China, India and the US anti-piracy taskforce CTF151 (based in Bahrain) to patrol the western Indian Ocean (Gardner 2009). On 23 September 2008, an Indian warship sank a suspected pirate in the Gulf of Aden. One of the hijackings that attracted world attention was the hijack of Saudi Arabian super tanker *Sirius Star*, on 28 September 2008, along with 25 crew and loaded with oil worth $100 million. Towards the end of January 2009, Japanese Defence Minister Yasukazu Hamada ordered the dispatch of ships to join the multinational force. War ships from South Korea and Iran are joining the mission. France, India, South Korea, Russia, Spain, Turkey, Malaysia, the US and Nato, among others, are already present in the region.

Somalis are increasingly complaining that the world only cares about pirates and forgets the causes of the rising piracy. First, according to ex-Somali Army Colonel Mohamed Nureh Abdulle, Somalis living in the coastal areas are facing two problems: dumping of toxic waste causing mysterious illnesses, and overfishing by international ships. The animals in the area are getting sick from drinking the water and the people are suffering from illnesses after having washed themselves in the water. Secondly, overfishing has heavily affected the local communities used to relying on fishing. Former fishermen are turning into pirates, arguing that international ships are illegally threatening or destroying their business. When hijacking became a lucrative business, businessmen and former fighters for the Somali warlords moved in (Westcott 2008). International ships with big nets wipe out everything, pushing the people in the areas to farming and keeping greater numbers of livestock. Somalis are already affected by three successive years of drought, besides chronic insecurity, which has decimated their animals, heavily exposing the population to malnutrition. Soaring world food prices have exacerbated it, leading to a catastrophic combination of circumstances. As a result, the humanitarian crisis had affected 3.5 million people by the end of 2008.

Sudanese and Somali conflicts are the predominant conflicts in the Horn of Africa. But they have transcended the mere intrastate and interstate levels and have directly or indirectly encompassed all states in the region. Various rebel groups from the neighbouring states have been operating from the Sudanese territory or the Sudanese rebels operate from the neighbouring states. The Ugandan rebels, such as the LRA or the ADF, have been operating from southern Sudanese territories; the SPLM/A was operating from the northern Ugandan territories; Darfur rebels are operating from Chadian territory, and Chadian rebels are operating from the Sudanese territory; former Ethiopian rebels were operating from the Sudanese territory, and the SPLM/A are operating from the Ethiopian territory.

The collapsed state of Somalia has become the scene for proxy war and competition mainly between Ethiopia and Eritrea, Ethiopia and Egypt, and even between Ethiopia and Sudan. Whereas in the Sudanese case respective rebels cross borders, in the Somali case states cross borders officially (Ethiopia and Eritrea). Ethiopia, as a regional military and diplomatic power as well as because of various rebellions in the Oromo areas in the south and Ogaden and Afar areas in the east, is a key actor in the regional conflict system.

The interaction of the three conflict system levels in the Horn of Africa will have a negative impact on the chance for peace in the region. This means that the Horn of Africa will increasingly be the scene of the global conflict system. Ethiopia will be increasingly targeted as a proxy of the United States; Eritrea is left isolated from the international community and declared as a state supporting terrorists; and piracy in and around Somali waters have become a rising global concern. In Sudan peace will not depend on how well armed the UNAMID forces are, but instead on how the regional and global actors pursue a genuine and coordinated policy.

Conclusion: Ending Conflict and Violence as System Transformation

Various peace researchers such as Lederach (2003), Galtung (2004), Dietrich (2008) and Kriesberg (1998) have suggested conflict transformation rather than conflict resolution. Moreover, a positive dimension of conflict is also being underlined. Accordingly, conflict can be seen as something positive, normal in human relationships or a motor of change, an opportunity, a gift (Lederach 2003: 4, 14, 18).

Indeed, depending on the context, non-violent interpersonal, intergroup or interstate conflict can be an opportunity out of which something positive can emerge. But the conflicts in DR Congo, Somalia and Sudan have crossed the red line of positive aspects of conflict. These conflicts ended in brutal violence leading to something more destructive than constructive. It is no longer a conflict which can be an opportunity; instead it is the rape, killing, maiming and displacement of millions of people.

Policy makers as well as academics have proffered different solutions for ending Africa's conflicts. In the face of the desperate situation in conflict-ridden states a call for multifaceted intervention in the form of "just intervention" is being suggested. Such intervention implies "justice beyond borders" (Caney 2005). Caney pleads for moral universalism because moral principles should apply to all, if all persons are similar in morally relevant ways and persons throughout the world share common morally relevant properties, and there is no fundamental morally significant difference between the domestic and global realm (Caney 2005: 57). Therefore, the principles to be applied in the global realm should be continuous with those we think appropriate in the domestic (Caney 2005: 265; Beitz 1999: 288). Cosmopolitanism rejects nationalist conclusions that obligations cannot extend further than the group of people with whom one identifies; that one cannot have global obligations; that morality is relevant only on the national level; that political institutions cannot function without their members being bound together by a national identity (Caney 2005: 278). Hence, humanitarian intervention is an intervention to protect the welfare of the members of another political regime and to secure civil, political and economic (distributive) justice based on the concept of universal principles; and it should be based on five principles: just cause (such as violation of human rights); proportionate response to internal wrongs; consideration of less awful measures; reasonable chance of meeting objectives; and authorization by a legitimate body (Caney 2005: 248–50). Similarly, the document "Responsibility to Protect" came in 2001 to circumvent the arguments by crisis states (based on Chapter 2(4) of the UN Charter) that "no international actor is entitled to violate the territorial integrity of sovereign states" if a state fails to fulfil its responsibility.

The failure of intervention in Somalia in the early 1990s showed that intervention alone is insufficient. Many conflicts are too complex to be addressed by mere military intervention. Indeed inward-looking solutions have got a great deal of attention, but that is not enough either. Richard Jackson suggests that power brokers and power holders characterized by diffuse actors have to be identified if a conflict is to end. There are highly complex conflict networks based on political loyalty, ethnic or religious identity, war economies, scarcity, militarized population, ethnic cleansing, etc. Therefore, Jackson suggests that conflict-ridden African states have to be transformed through addressing the malformation and underdevelopment of the African states (Jackson 2006: 25–6). For Bruce Baker, the fundamental issue is good governance, fairness, participation, transparency, effectiveness, openness, responsiveness, accountability, legitimacy, rule of law, etc. Baker considers these as a social capital without which Africa's conflicts cannot be ended (Baker 2006: 42). Rigby underscores that civil societies are essential in the deepening of post-settlement peace processes; and they play a vital role as advocates, catalysts, partners and monitors of nation-building in post-settlement states (Rigby 2006: 58–9). Peter Woodward points out that the IGAD peace process for South Sudan without dealing with Darfur and Sudan's deep regional problems across the country and not just the north–south relationship was a mistake and disappointment (Woodward 2006: 177–8). Indeed, the points raised above by various authors are important components of conflicts but they constitute only one aspect: the intrastate aspect.

In this study I have tried to underline that too little attention has been given to the systemic problems of conflicts and violence in Africa. Conflicts and violence in DR Congo, Somalia and Sudan cannot be separated from regional and global contexts. Understanding the interaction of root causes, trigger causes and tertiary causes is decisive for understanding the conflict and violence system; however, this does not downplay the fact that their root causes are internal. I would like to end my analysis by underling three key points of conflict system.

First, root causes are often artificially constructed, instrumentalized, used and abused to exacerbate conflicts. In DR Congo, owing to at least 200 different ethnic groups, politics and violence were ethnicized right after independence. The colonial power was not interested in creating competent Congolese political elites on the eve of independence, even pursuing divisive politics which resulted in the independence of Katanga in July 1960. The post-independence era was marred by social fragmentation. Further, the contradicting interests and ideologies of political elites such as Tshombe, Kasavubu, Lumumba and Mobutu showed there was no concerted plan for Congo. In the eastern part of Congo a distinction emerged between "indigenous" and "non-indigenous" groups in the 1970s. The Mobutu regime was not able to address interethnic tensions in the area. The new citizenship law of 1981, which superseded that of 1972, ignited conflict between 'immigrant' Tutsis and "indigenous" Congolese. Mobutu constructed an artificial ideology (authenticity) to consolidate the state. However, corruption, patrimonialism, clientelism, kleptocracy and brutal war followed instead (Tetzlaff and Jakobeit 2005: 138–40). Gradually, because of this economic mismanagement and state weakness, the police and army started to harass and extort from the local population. As animosities and violence increased between Congolese Tutsis and "indigenous" Zairians; the state became unable to provide security and address rebel threats from Angola in the 1970s and from the east in the 1990s. Armed rebellion started mainly in the eastern part of the country in the mid-1990s. Zaire was not able or willing to address Rwandan, Ugandan and Burundian rebels operating from its territory in the east. These root problems together with the spill-over effect of the Rwandan genocide triggered the conflict that led to the collapse of DR Congo in 1997.

For sustainable peace, besides extending the negotiations to non-armed actors, it is essential to include actors such as Rwanda and Uganda. External actors (as facilitators or mediators) can support the non-armed groups in playing a constructive role; they can compel the belligerent parties to commit to the negotiation process, and prevent neighbouring states from jeopardizing the peace process with "carrots and sticks". Until the fall of Mobutu, Angola pursued its armed rebels in Zaire with his support. Moreover, these neighbouring states, rebels and global players exploited natural resources. During the Cold War as well as in the post-Cold War era, economic and strategic interests were pursued at the cost of development and democracy. The UN-peacekeeping operation between 1960 and 1964 was undermined by the insufficient mandate and contradicting interests of the global players. Even the success of the current MONUC (since 1999) has been modest, and its reputation is overshadowed by UN-peacekeeping troops' involvement in sexual abuse and gold smuggling. This is one of the most important tertiary causes of conflict and violence in DR Congo. Other global players like the US, France, and China have been part of the problem rather than the solution.

Though homogenous from a religious–cultural point of view (99.8 per cent Muslim), Somalia collapsed in 1991 and descended into violence. Various root causes have contributed to this. In pre-colonial Somalia the north was dominated by pastoral-nomadic groups, whereas sedentary groups lived in the south. Somalia was a politically akephalous (without a central political head) and segmentary society with socio-political organization along clan descendence characterized by fusion (interclan cooperation) and fissure (interclan dissent) depending on imminent threats or interests. Clan and sub-clan family ties determined the colonial and post-colonial party politics as an upper limit of corporate political action. The socio-political groups were stabilized by kinship structures, not by a centralized power. During the preparation for and after independence, political elites attempted to centralize power, to reconcile the "old" kinship system with the "new" state system. But since this was not easy to implement they tried to suppress the "particularistic" clan interests in favour of the "holistic" state interests. Moreover, they had attempted to reunite the Somalis dispersed in Djibouti, Ethiopia and Kenya with those colonized by Italy and Britain. This Somali irredentism (pan-

Somalism) resulted in war against Kenya between 1963 and 1968 and against Ethiopia between 1963 and 1966.

When Barre took power through a military coup in October 1969, he attempted to de-tribalize Somalia, fight patrimonialism, reintroduce democracy, enable universal education, etc. On the one hand, he pursued "scientific socialism" in order to "correct" the old sociopolitical system based on clan structures, but on the other hand, he himself cemented his power on those structures. Barre failed. Nevertheless, the clan structures have continued to play a role in the peace processes and transitional institutions since the state collapse in 1991. Since the mid-1990s, Somali Islamic radicals have intensified their activities in Somalia, attempting to overcome clan-based divisions through overarching Islamism and rebuild Somalia through radical Islamization.

The decisive beginning of the end of Somalia as a functioning state was the lost irredentist war and pan-Somalism. Supported by Somalia in 1977, the Western Somali Liberation Front (WSLF) officially started war against Ethiopia. When in August 1977 the USSR decided to support the Ethiopian side, Somalia lost her most important external ally and supporter. When pan-Somalism failed, national unity started to fail too. The refugees who fled Ethiopia during the Ogaden war, the economic crisis partly triggered by the USSR and the West withdrawing support, internal political crises, etc., led to the emergence and intensification of armed opposition in Somalia.

Sudan has been devastated by wars and rebellions in different parts of the country with various root causes. With 570 distinct peoples, an unprepared pre-independence period and failure to address the political, social and economic interests of differing ethnic groups, Sudan fulfilled the best preconditions for post-independence conflicts and violence. There was no pre-independence gradual integration of the South which was not an integral part of the central state; there was no gradual integration of Darfur (which was a political entity of its own until 1916) into the central state. Concentration of natural resources such as water, oil and grazing land in the culturally, political and economically marginalized western and southern parts of Sudan is an important aspect of the root problem. These have sparked conflicts in different parts of Sudan.

Trigger causes include Islamization, abrogation of the Addis Ababa Agreement (AAA) of 1972, Islamic dominance since 1983, abolition of the autonomy of the south in 1983, and structural economic marginalization of the south and west. In order to address the state crisis, there has been more military coups than democratic change. Khartoum exacerbated conflicts and violence by igniting intra-southern and intra-Darfurian splits and violence, ideological contradictions and hampering coordination of the interests of the rebels. The Government started to arm Arab militias such as the Baggara, Janjawiid, etc. in the border areas between north and south as well as in the Darfur region. In general, Sudan failed in its democratization process. Now it seems that it is going to fail in implementing the Comprehensive Peace Agreement (CPA) of 2005.

However, the failure of Sudan and Somalia cannot be separated from the regional context, and the global dimension, too, has to be taken into account. IGAD as a regional institution is incapable of solving the problem. The success of the AMIS in Darfur was modest. Uganda and Ugandan rebels played their part in the continuation of conflict and violence in southern Sudan. In the 1970s and 1980s Ethiopia also contributed to the conflict dynamic. Between the mid-1990s and mid-2006, Eritrea did the same. Confrontation between Eritrea and Ethiopia has benefited different groups in Sudan. Now the mutual destabilization between Chad and Sudan is deep-rooted. Competition between Saudi Arabia and Libya has extended the scope of the regional system.

In the Somali case, Ethiopia and Somalia had already been at war in the 1960s and 1970s. Kenya and Somalia fought in the 1960s. Ethiopia and Eritrea are conducting their proxy wars in Somalia; Ethiopia has undermined the Transitional National Government, whereas Eritrea has been undermining the Transitional Federal Government. The position of Djibouti has been mostly unclear. Since the collapse of Somalia, the radical Islamic threat has made Ethiopia nervous. Arab-Islamic identity has caused many Arab states to side with radical Islamists at least partly through opposition to Ethiopia.

Global players have been active in the region since the colonial period and the Cold War. The French role in Chad has certainly had negative impacts on Sudan. China is an ally of both Chad and Sudan, and seems more interested in exploiting natural resources and exporting weapons rather than supporting an end to the conflicts. Russia is becoming a heavy-weight in Sudan. In spite of its support for AMIS, etc., the EU is undermined by the contradictory role of member states such as France. The US was active during the Cold War and early 1990s in Somalia, but, although its interest in fighting Islamic terrorism is not negligible, its role is not that substantial. The UN is undermined by the contradicting interests of the global heavy-weights. Therefore, the role of the global players as tertiary causes of conflicts cannot be considered any less important than regional concerns.

The case of Somalia is a good example. The UN-peacekeeping operation in Somalia in the early 1990s was one of the most criticized in UN history. The task of UNOSOM I (UN-Resolution 751 of 24 April 1992) was a massive infusion of food aid, expansion of supplementary feeding, provision of basic health services and mass measles immunization, urgent provision of clean water, sanitation and hygiene, provision of shelter materials, blankets and clothes, simultaneous delivery of seeds, tools and animal vaccines with food rations, prevention of further refugee outflows and the promotion of returnee programmes, institution-building and the rehabilitation of civil society. But because of its weak mandate, the UN facilities, aid deliveries, and UNOSOM convoys were continuously attacked by warlords and armed groups. This forced UNOSOM to withdraw from Mogadishu and the Special Representative of the UN Secretary General to resign.

Based on UN-Resolution 794 of 3 December 1992, the Unified Task Force (UNITAF) was created in order to correct the weaknesses of UNOSOM I. Its task was to create a secure framework for the delivery of aid and to facilitate and extend the work of UNOSOM with robust peacekeeping and humanitarian intervention, a mixture of diplomacy and use of force in case of a threat to peace, to rebuild national political order, feed the starving, protect the defenceless and prepare the way for political economic and social reconstruction. However, one of the weak sides of the UNITAF and the decisive reason for its failure was that the disarmament campaign was not strict enough and did not focus on all armed groups.

UNITAF was replaced by UNOSOM II with the task of observing the ceasefire agreement between Somali factions, political consolidation and reconstruction, peace enforcement, disarmament, securing humanitarian aid delivery, protecting UN personnel and humanitarian organizations, mine clearance and the return of refugees. The confrontation between UNOSOM II and the Somali National Alliance led to the killing of 25 Pakistani soldiers in June 1993 and 18 US soldiers in October of the same year. Ultimately, the UNOSOM and UNITAF operations were successful only in the humanitarian sector. Moreover, the UN legitimized some warlords as negotiating partners in a society where there is no political culture of centralized representation, meaning the marginalization of the clan system. The Transitional National Council was created, and this led to rivalry between the warlords for power. The Somaliland issue was not addressed. Mogadishu became the focal point of the UN. This again resulted in intensification of the war for Mogadishu. Moreover, the UN disarmament attempt in a non-existent state monopoly could not work. The UN weakness was exacerbated by regional (interstate) actors and factors. This regional dimension is true for Sudan and DR Congo as well as Somalia.

Conflicts within and between states have substantial regional implications. OLF, ONLF rebellion in Ethiopia, conflicts between Ethiopia and Eritrea, tensions between Somalis in Kenya and the Kenyan state, conflicts within Chad, between Chad and Sudan, and within Sudan, cannot be reduced to intrastate or interstate conflicts. Successful peaceful transition in Burundi and the disappearance of UNITA contributed to gradual peaceful developments in DR Congo. Cooperation between Rwanda and DR Congo resulted in the capture of Nkunda in early 2009. Similarly, the interests of ethnic groups are interconnected. Peace in Congo will depend on transforming the interethnic isolationism and hatred into mutual interest. Power sharing, wealth sharing and security sector reform (SSR) are essential. It is also important to acknowledge that hatred and marginalization of some ethnic groups was the result of colonial powers and weak political elites. This recognition is essential for system transformation. What does this mean?

States systems can be *differentiated*, but cannot be *separated*. Conflicts are systems in state systems. The identities and interests of Somalia, Eritrea, Ethiopia, Sudan, Chad, DR Congo, Uganda, Rwanda, etc., can of course be differentiated but not separated. Their security and interests are interdependent. Further, their long-term security and interests have substantial impacts on the long-term interests of the global players directly or indirectly acting in these states. This is about transforming the system of isolationism into the system of togetherness.

Only through this transformation (power sharing, wealth sharing and political and economic cooperation) can we address the destructiveness of conflicts and violence.

System transformation is the process of overcoming the contradiction between the realities of the social constructs on the one hand and the subjective perception of these realities on the other. This means, to acknowledge this reality (though constructed) without forgetting that this reality depends on historically constructed and subjective perceptions.

Second, it is important to deal with the fact that root causes change in the course of system transformation. It is about deconstructing the constructed ethnic and national identities and interests discussed in the chapters above. A key component of system transformation is transforming the system of exclusive violence against other groups into a system of mutual interests and togetherness. Power sharing, wealth sharing and security sector reform are cornerstones of system-transforming peace processes in DR Congo, Somalia and Sudan. Power sharing was the cornerstone of the Inter-Congolese Dialogue of December 2002 as a result of which armed groups and civil society got the posts of four vice presidents. In Sudan, the Comprehensive Peace Agreement of 2005 has been attempting to transform the system of northern dominance and southern marginalization into a system in which the south is given a post of vice president, among other posts like foreign minister, etc. Though so far unsuccessful, the Somali peace processes have been attempting to transform the system of inter-clan antagonism into inter-clan power sharing. Similarly, especially in the case of Sudan and DR Congo through the system of wealth sharing (the product of CPA of 2005 and Inter-Congolese Dialogue of 2002) the system of economic marginalization and exploitation is being pursued though its implementation is modest. Reforming the security sector is a key for ending the system of violence. Integrating the army of north and south Sudan is an essential component of the CPA of 2005. Similarly, without security sector reform, conflict and violence systems in DR Congo cannot be ended. A specialized unit of the DRC National Police (PNC) was created in 2002 under Congolese command. The 'Memorandum on Security and the Army' of 29 June 2003 provided for the establishment of an Integrated Police Unit (IPU). The support of the UN and EU has been crucial for this. On 12 April 2005 EUPOL Kinshasa was established by the EU to monitor, mentor and advise the IPU. In June 2005 the EU established EUSEC RD Congo to provide practical support for the integration of the Congolese Army and good governance in the field of security. The UN SSR focused on demobilization, disarmament and reintegration of the Congolese ex-combatants, training of the Congolese armed forces, and reforming the police, judicial and prison systems.

Third, it is important to underline that some factors and actors of the three system levels are easier to change and transform than others. It is easier to change the intrastate actors and factors than regional and global actors and factors. Further, it is easier to change regional actors and factors than the global ones. The global actors and factors are conceived to be too distant from the conflict states; and they are too mighty to be coerced. In the worst case it could be theoretically possible to intervene militarily in Rwanda or Uganda within the context of the Congolese conflict system, but it is categorically impossible to threaten economically or militarily those global players in DR Congo. There is no authority to do this. Similarly, it is possible to threaten Chad militarily in the context of the Sudanese conflict system or threaten Sudan militarily in the context of the Chad conflict system, but it is certainly impossible to threaten global players like China or France in the region. But their role is substantial in the conflict dynamic. This reality makes solving conflicts difficult. Some actors and factors can be addressed, whereas others not. However, to address all remains essential for sustainable peace. Conflicts persist because of this asymmetry. A peace process which focuses only on democratization, state-building, security sector reform, rebuilding financial institutions, humanitarian military intervention, arms embargoes, etc., cannot achieve sustainable peace if regional and global powers pursue contradicting interests. It is in the interests of the global players not to see themselves as closed systems (accumulating power and resources at the cost of their rivals or at the cost of the weak and collapsed African states).

At any rate, peace processes should not limit their focus to these material aspects of conflict causes (economic interests, natural resources, etc.) but instead include also the psychological, relational and intercultural aspects. System transformation is about a relationship where the conflicting parties, although acknowledging that there is no rational solution to their conflict, nevertheless recognize the legitimacy of their opponents (Mouffe 2005). Wolfgang Dietrich suggests "energetic peaces" in which the human existence is embedded in a relationship of all in all. He pleads for peaces (plural) than peace (singular) since there are different concepts and forms of peace in different cultures. According to Dietrich, these peaces are relational and dynamic (Dietrich 2008: 408). Such a relational and dynamic concept of peace is an integral part of the systemic character of the world in a globalized world society (Dietrich 2008: 285), whereas for positivist-realist conflict managers the actors and the interests that initiate the conflict are "given", "real" and "objective". Their systems are antagonistic rather than relational. But we can deal with conflicts sustainably only when we move away from a systemically metaphysical and isolationist position to a systemically relational and dynamic position wherein identity and interests are relationally differentiated, not categorically separated.

In his extensive research Lederach has suggested that instead of conflict resolution, conflict transformation creates a dynamic process that involves key people who together focus on the realities of the conflict in their context (Lederach 2006: 109). Further he proposes that a transformative approach is focusing on strategic linkage, i.e. about how the activity impacts the broader setting and building relationship (Lederach 2006: 109–10). However, this transformative action presupposes the identification of all those decisive actors of conflict dynamic. This is usually the main challenge.

As Lederach argues, conflict transformation must focus on relationship building. Relationship building is system transformation. Usually states see themselves as isolated systems. On the grounds of national sovereignty and territorial integrity they perceive themselves as separated from other state systems. States are sovereign only *in relation* to other states. Their interests, identity and integrity cannot be separated from those of other states. System transformation is about transforming the system of isolationism to a system of relationship. As much as states are interrelated systems regionally as well as globally, conflicts in state systems are interrelated regionally as well as globally. Even if many African states are weak states, they are part of the global system.

The principle of transformation is inclusivity: the "both/and", the "inside and outside", deconstructing the metaphysical categorization of good and evil. This is often the problem with interstate, interethnic, intercultural or inter-religious conflicts. System transformation means seeing your own perceptions and convictions as incomplete insights, looking at them from above, detaching the Self from perceptions. As Galtung says, the willingness to exert just half of the strength the warring parties harness to wage the war would surely be enough to find a common peaceful solution.

This study has attempted to show that the social (ethnic, religious composition, etc.), economic (natural resources, economic marginalization, etc.) and political (democratization, security, etc.) actors and factors of the internal conflict system condition each other. Indeed these actors and factors in their interaction in a failed or collapsed state constitute conflict and violence systems, but as the DR Congo, Somali and Sudanese cases show, they cannot be separated from the regional and global systems of interests and strategies. Therefore, the issue at stake is not only to prevent, manage, resolve or even transform a conflict in an intra-state system, instead to transform the system and the relationships, interests and the very constitution of society (internally, regionally and globally) that supports the continuation of violent conflicts. As Lederach says, transformation must actively envision, include, respect, and promote the human and cultural resources from within a given setting. This involves a new set of lenses through which we do not primarily 'see' the setting and the people in it as the 'problem' and the outsider as the 'answer'. Rather, we understand the long-term goal of transformation as validating and building on people and resources within the setting.

For conflict system transformation first the actors and factors of a conflict system have to be identified. On which actors and factors do the interveners, mediators or facilitators focus? Can they identify that there are actors that essentially constitute the conflict system but they are not immediately conflict parties? Do those who mediate or intervene have the *power*, the *will* and the *intention* to address all actors and factors that constitute a conflict system? In DR Congo, Somalia and Sudan the conflicts became protracted because the main focus was on the intrastate level. In some cases, the regional dimension has been addressed to some extent as in Congo. But in Sudan and Somalia this has not been the case. The intrastate approach is only one part of the problem. Conflicts in and between Ethiopia, Eritrea, Somalia, Djibouti, Kenya, Sudan, Chad, Central African Republic, Uganda, Rwanda, DR Congo, Angola, Burundi, etc., are systemically interrelated as the case studies have shown. Similarly, competition and conflicts between the US, China, Russia, the UK, France, the EU, India, Iran, Syria, Israel, etc., have essential global systemic impacts on DR Congo, Somalia and Sudan. Therefore, it is an illusion to expect a positive outcome from military interventions in DR Congo, Somalia or Sudan, and at the same time give less attention to the regional and global dimensions.

The successes and failures of peacekeeping are contingent on the domestic, regional, and global dynamics of conflict systems (Adebayo and Landsberg 2006: 162). As long as the current global system persists, global actors in the long run multiply their own enemies and the enemies of their allies in this region. It is in the interests of all levels to acknowledge the regional and global systemic impacts. Conflicts in Africa must be seen as long-term destabilizing factors even to those who believe they are geographically far from them.

References

Abay-Asmerom, G. (1999), Super Race Mentality? Who is guilty of it? In http://www.denden.com/Conflict/newscom/superrace.html, accessed 26 July 2007.

Abegunrin, O. (2007), The Soviet Involvement in Southern African Liberation Struggles, 1960–1990. In Matusevich, M. (Ed.), *Africa in Russia, Russia in Africa: Three centuries of encounters* (Trenton, NJ: Africa World Press), pp. 315–30.

Abgaje, A. (2004), Nigeria: Prospects for the Fourth Republic. In Gyimah-Boadi, E. (Ed.), *Democratic Reform in Africa: The quality of progress* (Boulder, CO: Lynne Rienner), pp. 201–33.

Abrahamsen, R. and Williams, P. (2001), Ethics and Foreign Policy: The antinomies of New Labour's 'Third Way' in Sub-Saharan Africa. *Political Studies*, 49, 249–64.

Acharya, A. (2005), Beyond Anarchy: Third World instability and international order after the Cold War. In Neuman, S. (Ed.), *International Relations Theory and the Third World* (New York: St. Martin's Press), pp. 159–211.

Achieng, J. (1998), Horn of Africa: Flow of weapons fuels conflict. http://www.oneworld.org/ips2/sept98/12_52_040.html.

Adebajo, A. (2004), From Congo to Congo: United Nations Peacekeeping in Africa after the Cold War. In Taylor, I. and Williams, P. (Eds), *Africa in International Politics: External involvement in the Continent* (London: Routledge), pp. 195–212.

Adebajo, A. and Landsberg, C. (2006), Back to the Future: UN peacekeeping in Africa. In Adebajo, A. and Sriram, C. L. (Eds), *Managing Armed Conflicts in the 21st Century* (London: Cass), pp. 161–88.

Adelman, K. L. (1975), The Recourse to Authenticity and Négritude in Zaire. *Journal of Modern African Studies*, 13:1, 134–9.

Aho, J. A. (1981), *Religious Mythology and the Art of War: Comparative religious symbolisms of military violence* (Westport: Greenwood Press).

Alden, C. (2007), *China in Africa* (London: Zed Books).

Alford, R. R. (1994), Paradigms of Relations between State and Society. In Hall, J. A. (Ed.), *The State: Critical concepts* (London: Routledge), pp. 63–76.

Ambafrance. (2008), State Visit to South Africa: Speech by M. Nicolas Sarkozy, President of the Republic, to the Parliament of South Africa. In http://ambafrance-us.org/spip.php?article857, accessed 15 March 2008.

Amnesty International. (2005), Spain: The southern border: The State turns its back on the human rights of refugees and migrants, 41/008/2005. In http://web.amnesty.org/library/pdf/EUR410082005ENGLISH/$File/EUR4100805.pdf, accessed, 29 December 2005).

Amnesty International. (2007) Sudan: New photographs show further breach of UN arms embargo on Darfur. In: <http://www.amnesty.org/en/library/asset/AFR54/045/2007/en/dom-AFR540452007en.html>, accessed 9 January 2009.

Amoo, S. K. (1997), *The Challenge of Ethnicity and Conflicts in Africa: The necessity for a new model* (New York: PNUD).

Anderson, B. R. (2000), *Imagined Communities: Reflections on the origin and spread of nationalism* (London: Verso).

ARB. (1998a), *Africa Research Bulletin*, 35:4, 13077–8.

ARB. (1998b), *Africa Research Bulletin*, 35:1, 12979.

ARB. (1998c), *Africa Research Bulletin*, 35:2, 12996.

ARB. (2002), Somalia: RRA factions fight. *Africa Research Bulletin*, 39:7, 14943.

ARB. (2006), Somalia: SICC declares war on Ethiopia. *Africa Research Bulletin*, 43:10, 16834–6.

ARB. (2009), Democratic Republic of Congo: Peace Talks Flounder. *Africa Research Bulletin*, 45:12, 17789B-17791A.

Arop Madut Arop. (2008), The Founding of the Sudan People's Revolution: SPLM/SPLA (1983–2005). <http://www.cmi.no/sudan/doc/?id=992>, accessed 11 December 2008.

Asylumlaw.org. (2000), Report on Minority Groups in Somalia. <http://www.asylumlaw.org/docs/somalia/dis00_somalia_minoritygroups.pdf>, accessed 12 September 2006.

Ayittey, G. B. N. (1999), *Africa in Chaos* (New York: St. Martin's Griffin).

Ayittey, G. B. N. (2005), *Africa Unchained: The blueprint for Africa's future* (New York: Palgrave Macmillan).

Ayoade, J. A. A. (1986), The African Search for Democracy: Hopes and reality. In Ronen, D. (Ed.), *Democracy and Pluralism in Africa* (Boulder, CO: Lynne Rienner), pp. 19–33.

Ayoob, M. (2005), Subaltern Realism: International relations theory meets the Third World. In Neuman, S. (Ed.), *International Relations Theory and the Third World* (New York: St. Martin's Press), pp. 31–54.

Baker, B. (2006), Post-settlement Governance Programmes: What is being built in Africa? In Furley, O. and May, R. (Eds), *Ending Africa's wars* (Aldershot: Ashgate), pp. 31–45.

Barash, D. P. and Webel, C. P. (2002), *Peace and Conflict Studies* (Thousand Oaks, CA: Sage).

Barber, B. (1996), Foundationalism and Democracy. In Benhabib, S. (Ed.), *Democracy and Difference: Contesting the boundaries of the political* (Princeton: Princeton University Press), pp. 348–60.

Barnett, M. (2001), Authority, Intervention, and the Outer Limits of International Relations Theory. In Callaghy, T. M., et al. (Eds), *Intervention and Transnationalism in Africa: Global–local networks of power* (Cambridge: Cambridge University Press), pp. 47–65.

Barnett, M. (2003), Bureaucratizing the Duty to Aid: The United Nations and Rwandan genocide. In Lang, A. (Ed.), *Just Intervention* (Washington, D.C.: Georgetown University Press), pp. 174–91.

Bartos, O. J. and Wehr, P. (2002), *Using Conflict Theory* (Cambridge: Cambridge University Press).

Baudrillard, J. (2003), *The Spirit of Terrorism: And requiem for the Twin Towers* (London: Verso).

Bayart, J.-F. et al. (2001), *The Criminalization of the State in Africa* (Oxford: The International African Institute).

BBC. (2009), US Fails to Break Somali Islamists, http://news.bbc.co.uk/1/hi/world/africa/7807291.stm, accessed 17 February 2009.

Beehner, L. (2005), Violence in the Horn of Africa: Ethiopia vs. Eritrea, http://www.cfr.org/publication/9214/, accessed 21 February 2007.

Beinart, P. (2007), America's Foreign Policy Trap, *TIME*, 169:26, June 25–July 2, 2007, p. 29.

Beitz, C. (1999), International Liberalism and Distributive Justice: A survey of recent thought. *World Politics*, 51:2, 269–96.

Bellinger, C. K. (2001), *The Genealogy of Violence: Reflections on creation, freedom, and evil* (Oxford: Oxford University Press).

Benhabib, S. (1996), Toward a Deliberative Model of Democratic Legitimacy. In Benhabib, S. (Ed.), *Democracy and Difference: Contesting the boundaries of the political* (Princeton: Princeton University Press), pp. 67–94.

Berman, B., et al. (2004), Ethnicity and Democracy in Historical and Comparative Perspective. In Berman, B., et al. (Eds), *Ethnicity and Democracy in Africa* (Oxford: Currey), pp. 1–21.

Bernard, J. (1977), The Sociological Study of Conflict. In Hoffmann, S. (Ed.), *Contemporary Theory in International Relations* (Westport: Greenwood Press), pp. 124–36.

Bertalanffy, L. von (1968), *General System Theory: Foundations, development, applications* (New York: George Braziller).

Billig, M. (1995), *Banal Nationalism* (London: Sage Publications).

Birch, A. H. (1993), *The Concepts and Theories of Modern Democracy* (London: Routledge).

Biscop, S. (2005), *The European Security Strategy* (Aldershot: Ashgate).

Bjorgo, T. (2002), Gewalt gegen ethnische und religiöse Minderheiten. In Heitmeyer, W. and Hagan, J. (Eds), *Internationales Handbuch der Gewaltforschung* (Wiesbaden: Westdeutscher Verlag), pp. 981–99.

Blair, D. (2005), Spectre of War Haunts Horn of Africa. <http://www.telegraph.co.uk/news/main.jhtml?xml=/news/2005/12/19/wafr19.xml&sSheet=/news/2005/12/19/ixworld.html>, accessed 19 December 2005.

Blair, T. (1999), Doctrine of the International Community, http://www.pm.gov.uk/output/Page1297.asp, accessed 25 November 2000.

Blair, T. (2007), A Battle for Global Values. *Foreign Affairs*, 86, 1, 79–90.

Boas, M. and Dunn, K. C. (2007), African Guerilla Politics: Raging against the machine? In Boas, M. and Dunn, K. C. (Eds), *African Guerrillas: Raging against the machine* (Boulder, CO: Lynne Rienner), pp. 9–37.

Bratton, M. and Masunungure, E. (2007), Popular Reactions to State Repression: Operation Murambatsvina in Zimbabwe. *African Affairs*, 106, 422, 21–45.

Broadman, H. G. (2007), *Africa's Silk Road: China and India's new economic frontier* (Washington, D.C.: World Bank).

Brons, M. H. (2001), *Society, Security, Sovereignty and the State in Somalia: From statelessness to statelessness?* (Utrecht: International Books).

Brzezinski, Z. (2007), How to Avoid a New Cold War, *TIME*, 169:25, June 18, 2007, pp. 28–9.

Bull, H. (1995), *The Anarchical Society: A study of order in world politics* (New York: Columbia University Press).

Buzan, B. (2005), System versus Units in Theorizing about the Third World. In Neuman, S. (Ed.). *International Relations Theory and the Third World* (New York: St. Martin's Press), pp. 213–34.

Buzan, B., et al. (1998), *Security: A new framework for analysis* (Boulder, CO: Lynne Rienner).

Buzan, B. and Gonzalez-Pelaez, A. (2005), International Community after Iraq. *International Affairs*, 81:1, 31–52.

Buzan, B. and Little, R. (2000), *International Systems in World History: Remaking the study of international relations* (Oxford: Oxford University Press).

Buzan, B. and Waever, O. (2006), *Regions and Powers: The structure of international security* (Cambridge: Cambridge University Press).

Bwenge, A. M. (2005), Researching Ethno-political Conflicts and Violence in the Democratic Republic of Congo. In Porter, E. (Ed.), *Researching Conflict in Africa: Insights and experiences* (Tokyo: United Nations University Press), pp. 90–109.

Caldwell, L. T. (2007), Russian Concepts of National Security. In Legvold, R. (Ed.), *Russian Foreign Policy in the Twenty-first Century and the Shadow of the Past* (New York: Columbia University Press), pp. 242–79.

Campbell, D. (1998), *Writing Security: United States foreign policy and the politics of identity* (Minneapolis: University of Minnesota Press).

Caney, S. (2005), *Justice Beyond Borders: A global political theory* (Oxford: Oxford University Press).

Caplan, R. (2003), Humanitarian Intervention: Which way forward? In Lang, A. (Ed.), *Just Intervention* (Washington, D.C.: Georgetown University Press), pp. 131–44.

Carlson, A. (2006), More Than Just Saying No: China's evolving approach to sovereignty and intervention since Tiananmen. In Johnston, A. I. (Ed.), *New Directions in the Study of China's Foreign Policy* (Stanford: Stanford University Press), pp. 217–41.

Carment, D., et al. (2006) *Who Intervenes? Ethnic conflict and interstate crisis* (Columbus: Ohio State University Press).

Carneiro, R. L. (1970), A Theory of the Origin of the State: Traditional theories of state origins are considered and rejected in favor of a new ecological hypothesis. *Science*, 169, 733–8.

Carothers, T. (2003) Promoting Democracy and Fighting Terror. *Foreign Affairs*, 82: 1, 84–98.

CFR. (2002), Somalia's Clan Families. <http://www.cfr.org/content/publications/attachments/cia_somalia.pdf>, accessed 25 October 2008.

Chabal, P. (1998), A Few Considerations on Democracy in Africa. *International Affairs*, 74:2, 289–303.

Cheadle, D. and Prendergast, J. (2005), The Darfur Genocide. *The Wall Street Journal*, 24 March.

Chimanikire, D. P. (2003), Foreign and Security Policy of Zimbabwe: From independence to the DRC. In Darnolf, S. and Laakso, L. (Eds), *Twenty Years of Independence in Zimbabwe: From liberation to authoritarianism* (London: Palgrave Macmillan), pp. 179–95.

Citizens for Global Solutions. (2005), The United Nations Responds to Sexual Abuse by Peacekeepers in the Democratic Republic of Congo. <http://oldsite.globalsolutions.org/programs/peace_security/peace_ops/conflicts/DRC_Articles/abuse_in_congo.pdf>, accessed 23 February 2009.

Clapham, C. (1996), *Africa and the International System: The politics of state survival* (Cambridge: Cambridge University Press).

Clapham, C. (2001), Rethinking African States. *African Security Review*, 10:3, 7–16.

Clapham, C. (2003), Notes on the Ethio–Eritrean Boundary Demarcation. <http://www.aigaforum.com/clapham0311.htm>, accessed October 2003.

Clapham, C. (2005), Introduction. *International Affairs*, 81:2, 275–9.

Clark, J. F. (Ed.). (2002), *The African Stakes of the Congo War* (New York: Palgrave).

Clark, J. and Mungbalemwe, K. (2002), The Economic Impact of the Congo War. In Clark, J. F., et al. (Eds), *The African Stakes of the Congo War* (New York: Palgrave Macmillan), pp. 201–23.

Cohen, M. J. (2007), Food Security: Vulnerability Despite Abundance. <http://www.ipacademy.org/asset/file/192/mgs-fsec.pdf>, accessed 18 March 2008.

Collier, P. (2007) *The Bottom Billion: Why the poorest countries are failing and what can be done about it* (Oxford: Oxford University Press).

Collier, P., et al. (2005) The Collier–Hoeffler Model of Civil War: Onset and the case study project research design. In Collier, P. and Sambanis, N. (Eds), *Understanding Civil War: Evidence and analysis* (Washington, D.C.: World Bank), pp. 1–33.

Connell, D. (2005), Eritrea: On a slow fuse. In Rotberg, R. I. (Ed.), *Battling Terrorism in the Horn of Africa* (Cambridge: World Peace Foundation), pp. 64–92.

Consilium. (2005a), The EU And Africa: Towards a strategic partnership, 15961/05 (Presse 367), Brussels, 19 December 2005.

Consilium. (2005b), EU Support for Peace and Security in Africa, February 2005.

Consilium. (2005c), EU Council Secretariat Darfur – Consolidated EU package in support of AMIS II, October 2005.

Consilium. (2006), Council Conclusions on Strengthening African Capabilities for the Prevention, Management and Resolution of Conflicts 2760th General Affairs, Council meeting Brussels, 13 November 2006 <http://www.consilium.europa.eu/uedocs/cmsUpload/Capabilities_Africa_20.11.pdf>, accessed 21 August 2007.

Consilium. (2009), EU NAVFOR Somalia. <http://consilium.europa.eu/cms3_fo/showPage.asp?id=1518&lang=en>, accessed 11 February 2009.

Control Arms Campaign. (2006), The Call for Tough Arms Controls: Voices from the Democratic Republic of the Congo. <http://www.oxfam.de/download/DRC_Bericht.pdf>, accessed 12 January 2007.

Cook, M. L. (2003), 'Immaculate War': Constraints on humanitarian intervention. In Lang, A. (Ed.), *Just Intervention* (Washington, D.C.: Georgetown University Press), pp. 145–54.

Cook, T. I. and Moos, M. (1954), *Power through Purpose: The realism of idealism as a basis for foreign policy* (Baltimore: The John Hopkins Press).

Cornwell, R. (2004), Somalia: Fourteenth time lucky? http://www.iss.co.za/pubs/papers/87/Paper87.htm, accessed 12 March 2006.

Cottam, M. L., et al. (2004), *Introduction to Political Psychology* (Mahwah, NJ: Erlbaum).

Cousens, E. M. (2004), Conflict Prevention. In Malone, D. M. (Ed.), *The UN Security Council.* (Boulder, CO: Lynne Rienner), pp. 101–15.

Cramer, C. (2005), Angola and the Theory of War. In Chabal, P., et al. (Eds), *Is Violence Inevitable in Africa? Theories of conflict and approaches to conflict prevention* (Leiden: Brill), pp. 17–34.

Crawford, J. (2006), *The Creation of States in International Law* (Oxford: Oxford University Press).

Crenshaw, M. (2004), The Pschology of Political Terrorism. In Jost, J. T. and Sidanius, J. (Eds), *Political Psychology* (New York: Psychology Press), pp. 411–31.

Creveld, M. van (1999), *The Rise and Decline of the State* (Cambridge: Cambridge University Press).

Crocker, C. (1992), *High Noon in Southern Africa: Making peace in a rough neighborhood* (New York: Norton).

Crocker, C. A. (2003), Engaging Failing States. *Foreign Affairs*, 82:5, 32–45.

Dahl, R. A. (1989), *Democracy and Its Critics* (New Haven, CT: Yale University Press).

Daly, M. W. (2007), *Darfur's Sorrow: A history of destruction and genocide* (Cambridge: Cambridge University Press).

Darfur Consortium. (2008), Putting People First: The Protection Challenge Facing UNAMID in Darfur. <http://www.darfurconsortium.org/darfur_consortium_

actions/reports/2008/Putting%20People%20First%20UNAMID%20report.pdf>, accessed 25 January 2009.

Darley, J. M. (2004), Social Organisation for the Production of Evil. In Jost, J. T. and Sidanius, J. (Eds), *Political Psychology* (New York: Psychology Press), pp. 383–410.

David, S. R. (2005), The Primacy of Internal War. In Neuman, S. (Ed.), *International Relations Theory and the Third World* (New York: St. Martin's Press), pp. 77–101.

De Waal, A. (2004), Introduction. In De Waal, A. (Ed.), *Islamism and Its Enemies in the Horn of Africa* (London: Hurst and Company), pp. 1–20.

De Waal, A. (2005), Who are the Darfurians? Arab and African identities, violence and external engagement. *Affrican Affairs*, 104:415, 181–205.

De Waal, A. (2007), Sudan: The turbulent state. In De Waal, A. (Ed.), *War in Darfur: And the search for peace* (Harvard: Harvard University Press), pp. 1–38.

De Waal, A. and Salam, A. (2004), Islamism, State Power and Jihad in Sudan. In De Waal, A. (Ed.), *Islamism and Its Enemies in the Horn of Africa* (London: Hurst and Company), pp. 71–113.

Dealey, S. (2006), Terror's Playground. *TIME*, 168:23, November 27, 2006, pp. 36–9.

Deng, F. M. and Morrison, J. S. (2001), U.S. Policy to End Sudan's War: Report of the CSIS Task Force on U.S.–Sudan Policy. February.

Deutsch, K. W. (1974), *Politics and Government: How people decide their fate* (Boston, Mass: Houghton Mifflin).

Deutsch, K. W. (1979), *Tides Among Nations* (New York: The Free Press).

Deutsch, M. (1986), The Malignant (Spiral) Process of Hostile Interaction. In White, R. K. (Ed.), *Psychology and the Prevention of Nuclear War* (New York: New York University Press), pp. 131–54.

Deutsch, M. (2006) *Handbook of Conflict Resolution* (San Francisco: Jossey-Bass).

Dews, P. (2003), Postmodernism: Pathologies of modernity from Nietzsche to the post-structuralists. In Ball, T. and Bellamy, R. (Eds), *The Cambridge History of Twentieth-century Political Thought* (Cambridge: Cambridge University Press), pp. 343–67.

DFID. (2006), Reducing Conflict in Africa: progress and challenges. Africa Conflict Prevention Pool Performance Report, 2001–05. In http://www.dfid.gov.uk/pubs/files/ACPP_performance_report.pdf.

DFID. (2007), Africa Conflict Prevention Pool Annual Report 2006/07. In http://www.dfid.gov.uk/Pubs/files/ACPP-annual-report-0607.pdf, accessed 19 September 2008.

DFID. (2008), DFID Humanitarian Allocations in Africa 2007/8. <http://www.dfid.gov.uk/pubs/files/2007-8-Africa-humanitarian-spend-analysis.pdf>, accessed 23 October 2008.

Diamond, L. (2004), Promoting Real Reform in Africa. In Gyimah-Boadi, E. (Ed.), *Democratic Reform in Africa: The quality of progress* (Boulder, CO: Lynne Rienner), pp. 263–92.

Dietrich, W. (2008), *Variationen über die vielen Frieden*, Band 1, Deutungen, (Wiesbaden: VS).

Dixon, M. (2006), *Textbook on International Law* (Oxford: Oxford University Press).

Donaldson, R. H. and Nogee, J. L. (2005), *The Foreign Policy of Russia: Changing systems, enduring interests* (Armonk, NY: Sharpe).

Dörner, D. (1989), *Die Logik des Mißlingens* (Reinbek: Rowohlt).

Dowden, R. (2007), Endgame in Zimbabwe. *TIME*, 169:14, 15.

Dunleavy, P. (2000), The State. In Goodin, R. E. and Pettit, P. (Eds), *A Companion to Contemporary Political Philosophy* (Cambridge: Blackwell Publishers), pp. 611–21.

Dunn, F. S. (1977), The Scope of International Relations. In Hoffmann, S. (Ed.), *Contemporary Theory in International Relations* (Westport: Greenwood Press), pp. 13–16.

Easterly, W. (2006), *The White Man's Burden: Why the West's efforts to aid the rest have done so much ill and so little good* (New York: The Penguin Press).

Eckstein, H. (2004), Theoretical Approaches to Explaining Collective Political Violence. In Jost, J. T. and Sidanius, J. (Eds), *Political Psychology* (New York: Psychology Press), pp. 332–48.

Edgerton, R. B. (2002), *The Troubled Heart of Africa: A history of the Congo* (New York: St. Martin's Press).

Edkins, J. (2002), After the Subject of International Security. In Finlayson, A. and Valentine, J. (Eds), *Politics and Post-structuralism* (Edinburgh: Edinburgh University Press), pp. 66–80.

EISA (Electoral Institute of Southern Africa). (2008), Angola: Towards peace and democracy (2002–2007). <http://www.eisa.org.za/WEP/angoverview10.htm>, accessed 4 March 2009.

Eizenstat, S., et al. (2005), Rebuilding Weak States. *Foreign Affairs*, 84:1, 134–46.

El-Din, A. K. (2007), Islam and Islamism in Darfur. In De Waal, A. (Ed.), *War in Darfur: And the search for peace* (Harvard: Harvard University Press), pp. 92–112.

Elwert, G. (2002), Sozialanthropologisch erklärte Gewalt. In Heitmeyer, W. and Hagan, J. (Eds), *Internationales Handbuch der Gewaltforschung* (Wiesbaden: Westdeutscher Verlag), pp. 330–67.

Emmott, B., et al. (1997), *Managing the International System over the Next Ten Years: Three essays*. Trilateral Commission.

Ethiopian News Agency. (2008), Ethiopia Welcomes Russian Oil, Gas Companies' Plan to Develop Its Gas Deposits. <http://www.ena.gov.et/EnglishNews/2008/Feb/14Feb08/50302.htm>, accessed 20 February 2008.

Etzioni, A. (2004), A Self-restrained Approach to Nation-building by Foreign Powers. *International Affairs*, 80:1, 1–17.

EUR LEX (2005), Council Joint Action 2005/557/CFSP of 18 July 2005 on the European Union civilian-military supporting action to the African Union mission in the Darfur region of Sudan.

Evans, P. (1994), The State as Problem and Solution: Predation, embedded autonomy, and structural change. In Hall, J. A. (Ed.), *The State: Critical concepts. Vol. 1* (London: Routledge), pp. 386–423.

Ezekwesili, O. K. (2007), Sudan: Public Expenditure Review, 41840-SD. <http://siteresources.worldbank.org/INTSUDAN/Resources/SD_PER_synthesis_report.pdf>, accessed, 19 February 2009.

Fan, S., Johnson, M., Saurkar, A. and Makombe, T. (2008), Investing in African Agriculture to Halve Poverty by 2015. In http://www.ifpri.org/pubs/dp/IFPRIDP00751.pdf, accessed, 17 March 2008.

Fanon, F. (2005), *The Wretched of the Earth* (New York: Grove Press).

Federal Research Division. (1993), *Somalia: A country study* (Washington, D.C.: Federal Research Division).

Fink, C.F. (1968), Some Conceptual Difficulties in the Theory of Social Conflict. *Journal of Conflict Resolution*, 12:4, 412–60.

Flint, J. and Waal, A. de (2006) *Darfur: A short history of a long war* (Cape Town: Zed Books).

Fox, G. H. (2004), Democratization. In Malone, D. M. (Ed.), *The U.N. Security Council: From the Cold War to the 21st century* (Boulder, CO: Lynne Rienner), pp. 69–84.

Freedom House. (2007), Congo, Democratic Republic of (Kinshasa) (2007), http://www.freedomhouse.org/template.cfm?page=363&year=2007&country=7308, accessed 11 January 2008.

Freedom House. (2008), Sudan 2008. <http://www.freedomhouse.org/template.cfm?page=251&year=2008>, accessed 12 December 2008.

Fruchart, D., et al. (2007), United Nations Arms Embargoes: Their impact on arms flows and target behaviour. *Stockholm International Peace Research Institute and Department of Peace and Conflict Research Uppsala University*. <http://books.sipri.org/files/misc/UNAE/SIPRI07UNAE.pdf>, accessed 18 March 2008.

Fukuyama, F. (2004), *State-building: Governance and world order in the 21st century* (London: Cornell University Press).

Galtung, J. (1975), *Peace, Research, Education, Action* (Copenhagen: Ejlers).

Galtung, J. (2004), *Transcend and Transform: An introduction to conflict work* (London: Pluto Press).

Gardner, F. (2009), Taking on Somalia's Pirates. <http://news.bbc.co.uk/go/pr/fr/-/1/hi/world/africa/7882618.stm>, accessed 10 February 2009.

Gebrewold, B. (2005), Civil Militias and Militarization of Society in the Horn of Africa. In Francis, D. (Ed.), *Civil Militia: Africa's intractable security menace* (Aldershot: Ashgate), pp. 187–212.

Genocide Intervention Network. (2009), Democratic Republic of Congo. <http://www.genocideintervention.net/educate/crisis/democratic_republic_of_congo>, accessed 9 March 2009.

Girard, R. (1986a), *Violence and the Sacred* (Baltimore: Johns Hopkins University Press).

Girard, R. (1986b), *The Scapegoat* (London: Athlone Press).

Girard, R. (1987), *Things Hidden Since the Foundation of the World* (London: Athlone Press).

Glasersfeld, E. von (2003), Konstruktion der Wirklichkeit und des Begriffs der Objektivität. In Foerster, H. V., et al. (Eds), *Einführung in den Konstruktivismus* (Muenchen: Piper), pp. 9–39.

Gleijeses, P. (2003), *Conflicting Missions: Havana, Washington, and Africa, 1959–1976* (Chapel Hill, NC: University of North Carolina Press).

Global Security. (2008), Sudan – First Civil War http://www.globalsecurity.org/military/world/war/sudan-civil-war1.htm, accessed January 2008.

Goemans, H. E. (2006), Bounded Communities: Territoriality, territorial attachment, and conflict. In Kahler, M. and Walter, B. F. (Eds), *Territoriality and Conflict in an Era of Globalization* (Cambridge: Cambridge University Press), pp. 25–61.

Goldsmith, J. L. and Posner, E. A. (2005), *The Limits of International Law* (Oxford: Oxford University Press).

Gourevitch, P. (1994), The Second Image Reversed: The international sources of domestic politics. In Hall, J. A. (Ed.), *The State: Critical concepts. Vol. 1* (London: Routledge), pp. 203–34.

Green, W (2007), Where Investors Fear to Tread. *TIME*, 170:1, 9 July 2007, p. 48.

Griffiths-Fulton, L. (2002), Small Arms and Light Weapons in the Horn of Africa. *Ploughshares Monitor*, http://www.ploughshares.ca/CONTENT/MONITOR/monj02a.html, accessed 15 May 2009.

Grosse-Kettler, S. (2004), *External Actors in Stateless Somalia: A war economy and its promoters*. Bonn International Center for Conversion, Paper 39. <http://www.bicc.de/publications/papers/paper39/paper39.pdf>, accessed 11 November 2005.

Guggenberger, W. (1998), *Niklas Luhmanns Systemtheorie: eine Herausforderung der christlichen Gesellschaftslehre* (Innsbruck: Tyrolia-Verlag).

Guo, S. and Hua, S. (2007), New Dimensions of Chinese Foreign Policy. In Guo, S. and Hua, S. (Eds), *New Dimensions of Chinese Foreign Policy* (Lanham: Lexington Books), pp. 1–6.

Gurr, T. R. (1970), *Why Men Rebel* (Princeton: Princeton University Press).

Guyomarch, A., et al. (2002), *France in the European Union* (Basingstoke: Palgrave MacMillan).

Gyimah-Boadi, E. (2004), Africa: The quality of political reform. In Gyimah-Boadi, E. (Ed.), *Democratic Reform in Africa: The quality of progress* (Boulder, CO: Lynne Rienner), pp. 5–27.

Haas, E. B. (1977), The Challenge of Regionalism. In Hoffmann, S. (Ed.), *Contemporary Theory in International Relations* (Westport: Greenwood Press), pp. 223–40.

Haggar, A. (2007), The Origins and Organization of the Janjawiid in Darfur. In De Waal, A. (Ed.), *War in Darfur: And the search for peace* (Harvard: Harvard University Press), pp. 113–39.

Hanagan, M. (2002), Gewalt und die Entstehung von Staaten. In Heitmeyer, W. and Hagan, J. (Eds), *Internationales Handbuch der Gewaltforschung* (Wiesbaden: Westdeutscher Verlag), pp. 153–76.

Hansen, A. (2007), The French Military in Africa, Council on Foreign relations. <http://www.cfr.org/publication/12578/#8>, March 22, 2007.

Hanson, E. O. (2006), *Religion and Politics in the International System Today* (Cambridge: Cambridge University Press).

Harbeson, J. W. and Rothchild, D. (Eds). (2000), *Africa in World Politics: The African state system in flux* (Boulder, CO: Westview Press).

Hardin, R. (1995), *One for All: The logic of group conflict* (Princeton, NJ: Princeton University Press).

Hardt, M. and Negri, A. (2001), *Empire* (Cambridge: Harvard University Press).

Hayes, C. J. H. (1966), *Essays on Nationalism* (New York: Russel and Russel).

Heelas, P. (1982), Anthropolgy, Violence and Catharsis. In Marsh, P. (Ed.), *Aggression and Violence* (Oxford: Blackwell), pp. 47–61.

Heitmeyer, W. (2004), Kontrolverluste: zur Zukunft der Gewalt. In Heitmeyer, W. and Soeffner, H. G. (Eds), *Gewalt: Entwicklungen, Strukturen, Analysprobleme* (Frankfurt: Suhrkamp), pp. 86–106.

Heitmeyer, W. and Hagan, J. (2002), Gewalt: zu den Schwierigkeiten einer systematischen internationalen Bestandsaufnahme. In Heitmeyer, W. and Hagan, J. (Eds), *Internationales Handbuch der Gewaltforschung* (Wiesbaden: Westdeutscher Verlag), pp. 15–25.

Heitmeyer, W. and Soeffner, H. G. (2004), Gewalt: Entwicklungen, Strukturen, Analyseprobleme. In Heitmeyer, W. and Soeffner, H. G. (Eds), *Gewalt: Entwicklungen, Strukturen, Analyseprobleme* (Frankfurt: Suhrkamp), pp. 11–17.

Held, D. (1989), *Political Theory and the Modern State* (Stanford: Stanford University Press).

Helly, D. (2006), Developing an EU Strategy for Security Sector Reform. *European Security Review*, 28, February 2006, International Security Information Service.

Hendricks, C. (2004), The Burdens of the Past and Challenges of the Present: Coloured IDENTITY and the 'Rainbow Nation'. In Berman, B., et al. (Eds), *Ethnicity and Democracy in Africa* (Oxford: Currey), pp. 113–28.

Hentz, J. J. (2004), The Contending Currents in United States Involvement in Sub-Saharan Africa. In Taylor, I. and Williams, P. (Eds), *Africa in International Politics: External involvement on the continent* (London: Routledge), pp. 23–40.

Herbst, J., et al. (2004), *Africa's Big States: Toward a new realism.* Carnegie Endowment for International Peace.

Hinde, R. (1990), Human Aggression: Biological propensities and social forces. In Smoker, P., et al. (Eds), *Reader in Peace Studies* (Oxford: Pergamon Press), pp. 172–81.

Hintze, O. (1994), Military Organization and the Organization of the State. In Hall, J. A. (Ed.), *The State: Critical concepts. Vol. 1* (London: Routledge), pp. 181–202.

Hippler, J. (Ed.). (2004), *Nation-Building: ein Schlüsselkonzept für friedliche Konfliktbearbeitung?* (Bonn: Dietz).

Hobson, J. M. (2000), *The State and International Relations* (Cambridge: Cambridge University Press).

Holsti, K. J. (2005), International Relations Theory and Domestic War in the Third World: The limits of relevance. In Neuman, S. (Ed.), *International Relations Theory and the Third World* (New York: St. Martin's Press), pp. 103–32.

Holtom, P. et al. (2008), International Arms Transfers. In *SIPRI Yearbook: Armaments, disarmament and international security* (Oxford: Oxford University Press), pp. 293–317.

HRW. (1993), Somalia. *Human Rights Watch*, 5:2, March 7.

HRW. (2003), Democratic Republic of Congo. *Human Rights Watch*, 15:11(a).

HRW. (2007), Shell-Shocked: Civilians under siege in Mogadishu. *Human Rights Watch*, 19:12(a).

Huband, M. (2001), *The Skull beneath the Skin: Africa after the Cold War* (Boulder, CO: Westview Press).

Huth, P. K. and Allee, T. L. (2002), *The Democratic Peace and Territorial Conflict in the Twentieth Century* (Cambridge: Cambridge University Press).

IANSA. (2008), Somalia: Continuing inflow of arms worsens a human rights catastrophe. <http://www.iansa.org/un/documents/CaseStudy-Somalia.doc>, accessed 22 January 2009.

ICG. (2000), Scramble for the Congo: Anatomy of an ugly war. *International Crisis Group*, 26, 20 December 2000.

ICG. (2003), Ethiopia and Eritrea: War or peace? *International Crisis Group*, 68, 24 September 2003.

ICG. (2004), Sudan's Dual Crisis: Refocusing on IGAD. *International Crisis Group*, 76, 5 October 2004.

ICG. (2005a), Darfur: The failure to protect, 89. <http://www.crisisgroup.org/library/documents/africa/horn_of_africa/089_darfur_the_failure_to_protect.pdf>, accessed 11 September 2008.

ICG. (2005b), Ethiopia and Eritrea: Preventing war. *International Crisis Group*, 101, 22 December 2005.

ICG. (2006a), Escaping the Conflict Trap: Promoting good governance in the Congo. *International Crisis Group*, 114, 20 July 2006.

ICG. (2006b), Security Sector Reform in the Congo. *International Crisis Group*, 104, 13 February 2006.

ICG. (2006c) Congo's Elections: Making or breaking the peace. <http://www.crisisgroup.org/library/documents/africa/central_africa/108_congo_elections_making_or_breaking_the_peace.pdf>, accessed 3 March 2009.

ICG. (2006d), Can the Somali Crisis be Contained? <http://www.crisisgroup.org/library/documents/africa/horn_of_africa/116_can_the_somali_crisis_be_contained.pdf>, accessed 12 March 2007.

ICG. (2006e), Can the Somali Crisis be Contained? <http://www.crisisgroup.org/library/documents/africa/horn_of_africa/116_can_the_somali_crisis_be_contained.pdf>, accessed 12 December 2008.

ICG. (2007a), Congo: Consolidating the peace. <http://www.crisisgroup.org/library/documents/africa/central_africa/128_congo_consolidating_the_peace_web.pdf>, accessed 21 January 2008.

ICG. (2007b), Darfur's New Security Reality, http://www.crisisgroup.org/library/documents/africa/horn_of_africa/134_darfur_s_new_security_reality.pdf, accessed 11 October 2008.

ICG. (2007c), Central African Republic: Anatomy of a phantom state, http://www.crisisgroup.org/library/documents/africa/central_africa/136_central_african_republic___anatomy_of_a_phantom_state.pdf>, accessed 22 April 2008.

ICG. (2008a), Sudan's Southern Kordofan Problem: The next Darfur? http://www.crisisgroup.org/library/documents/africa/horn_of_africa/145_sudans_southern_kordofan_problem___the_next_darfur.pdf, accessed 21 March 2009.

ICG. (2008b), Northern Uganda: The road to peace, with or without Kony. <http://www.crisisgroup.org/library/documents/africa/central_africa/146_northern_uganda___the_road_to_peace__with_or_without_kony.pdf>, accessed 23 January 2009.

ICG. (2008c), Chad: A new conflict resolution framework, Africa Report No.144, 24 September 2008. <http://www.crisisgroup.org/library/documents/africa/horn_of_africa/144_chad_a_new_conflict_resolution_framework.pdf>, accessed 21 November 2008.

Idris, A.H. (2005), *Conflict and Politics of Identity in Sudan* (New York: Palgrave Macmillan).

IGAD. (2002), Machakos Protocol, 20 July 2002.

Ikambana, P. (2007), *Mobutu's Totalitarian Political System: An Afrocentric analysis* (New York: Routledge).

ILO. (2004) *Global Employment Trends 2004* (Geneva: International Labour Office).

ILO. (2005), *A Global Alliance against Forced Labour* (Geneva: International Labour Office).

Imbusch, P. (2002), Der Gewaltbegriff. In Heitmeyer, W. and Hagan, J. (Eds), *Internationales Handbuch der Gewaltforschung* (Wiesbaden: Westdeutscher Verlag), pp. 26–57.

IMF. (2009), Impact of the Global Financial Crisis on Sub-Saharan Africa. <http://www.imf.org/external/pubs/ft/books/2009/afrglobfin/ssaglobalfin.pdf>, accessed 11 March 2009.

IRIN. (2003), Ethiiopia: Interview with Prime Minister Meles Zenawi. <http://www.irinnews.org/report.aspx?reportid=46954>, accessed 23 July 2007.

IRIN. (2004), Somalia: Fighting in Bulo Hawa Displaces Hundreds of Families. http://www.irinnews.org/Report.aspx?ReportId=49879#somfig, accessed 22 August 2005.

ISS. (2005), Somalia: Security information. <http://www.iss.co.za/AF/profiles/Somalia/SecInfo.html#top>, 22 February 2008.

Iyob, R. (1995), *The Eritrean Struggle for Independence: Domination, resistance, nationalism 1941–1993* (Cambridge: Cambridge University Press).

Jack, A. El (2002), *Gender Perspectives on the Management of Small Arms and Light Weapons in the Sudan* (Bonn: Bonn International Centre for Conversion, Small arms in the Horn of Africa).

Jackson, R. (2006), Africa's Wars: Overview, causes and the challenges of conflict transformation. In Furley, O. and May, R. (Eds), *Ending Africa's Wars* (Aldershot: Ashgate), pp. 15–29.

Jackson, R. and Sorensen, G. (2003) *Introduction to International Relations: Theories and approaches* (Oxford: Oxford University Press).

Jackson, R. H. and Roseberg, C. G. (1982) Why Africa's Weak States Persist: The empirical and the juridical in statehood. *World Politics*, 35:1, 1–24.

Janzekovic, J. (2006), *The Use of Force in Humanitarian Intervention: Morality and practicalities* (Aldershot: Ashgate).

Jeong, H.-W. (2005), *Peace and Conflict Studies: An introduction* (Aldershot: Ashgate).

Johnson, C. (2004), *The Sorrows of Empire: Militarism, secrecy, and the end of the republic* (London: Verso).

Johnson, D. H. (2003), *The Root Causes of Sudan's Civil Wars* (Oxford: James Currey).

Jost, J. T. and Sidanius, J. (2004), Political Psychology: An INTRODUCTION. In Jost, J. T. and Sidanius, J. (Eds), *Political Psychology* (New York: Psychology Press), pp. 1–21.

Kabongo, I. (1986), Democracy in Africa: Hopes and prospects. In Ronen, D. (Eds), *Democracy and Pluralism in Africa* (Boulder, CO: Lynne Rienner), pp. 35–39.

Kagan, R. (2003), *Macht und Ohnmacht. Amerika und Europa in der neuen Weltordnung* (Berlin: Siedler).

Kahler, M. (2006), Territoriality and Conflict in an Era of Globalization. In Kahler, M. and Walter, B. F. (Eds), *Territoriality and Conflict in an Era of Globalization* (Cambridge: Cambridge University Press), pp. 1–21.

Kant, I. (1900), *Beantwortung der Frage: Was ist Aufklärung?* (Frankfurt am Main: Neuer Frankfurter Verlag).

Kaplan, M. (1968), The Systems Approach to International Politics. In Kaplan, M. (Ed.), *New Approaches to International Relations* (New York: St. Martin's Press), pp. 381–404.

Kaplan, M. (1977), System and Process in International Relations. In Hoffmann, S. (Ed.), *Contemporary Theory in International Relations* (Westport: Greenwood Press), pp. 104–23.

Karstedt, S. (2004), Typen der Sozialintegration und Gewalt: Kollektivismus, Individualismus und Sozialkapital. In Heitmeyer, W. and Soeffner, H. G.

(Eds), *Gewalt: Entwicklungen, Strukturen und Analyseprobleme* (Frankfurt: Suhrkamp), pp. 269–92.

Kasfir, N. M. (2005), Sudan's Darfur: Is it genocide? *Current History*, 104:682, 195–202.

Kaufmann, D., et al. (2007), *Governance Matters VI: Aggregate and Individual Governance Indicators 1996–2006*. The World Bank, July 2007.

Keller, E. J. (2006), Africa and the United States: Meeting Challenges of Globalisation. In Rothchild, D. and Keller, E. J. (Eds), *Africa–US Relations: Strategic encounters* (Boulder, CO: Lynne Rienner), pp. 1–20.

Kelman, H. C. and Fisher, R. J. (2003), Conflict Analysis and Resolution. In Sears, D. O., et al. (Eds), *Oxford Handbook of Political Psychology* (Oxford: Oxford University Press), pp. 315–53.

Kern, K. (2007), The Human Cost of Cheap Cell Phones. In Hiatt, S. (Ed.), *A Game as Old as Empire: The secret world of economic hit men and the web of global corruption* (San Francisco: Berrett-Koehler), pp. 93–112.

Kirby, W. C. (2006), Traditions of Centrality, Authority, and Management in Modern China's Foreign Relations. In Robinson, T. W. and Shambaugh, D. (Eds), *Chinese Foreign Policy: Theory and practice* (Oxford: Clarendon Press), pp. 13–29.

Klomegah, K. K. (2006), Russia–Africa: Putin promises development, but promotes business. <http://en.civilg8.ru/allnews/2334.php>, accessed 22 April 2008.

Koepf, T. (2005), Die französische Sicherheits- und Militärpolitik in Subsahara-Afrika nach dem Ende des Kalten Krieges Das Ende einer Ära oder Business as Usual? Nr. 43/2005, Forschungsstelle Dritte Welt, am Geschwister-Scholl-Institut für Politische Wissenschaft der Ludwig-Maximilians-Universität München.

Koslowski, R. (2006), The Mobility Money Can Buy: Human smuggling and border control in the European Union. In Messina, A. and Lahav, G. (Eds), *The Migration Reader: Politics and policies* (Boulder, CO: Lynne Rienner), pp. 571–87.

Kpundeh, S. J. (2004) Corruption and Corruption Control. In Gyimah-Boadi, E. (Ed.), *Democratic Reform in Africa: The quality of progress* (Boulder, CO: Lynne Rienner), pp. 121–39.

Krause, K. (2007), Small Arms and Light Weapons: Towards global public policy. In http://www.ipacademy.org/asset/file/154/CWC_Working_Paper_SMALL_ARMS_KK.pdf, accessed 18 March 2008).

Krech, V. (2002), Opfer und Heiliger Krieg: Gewalt aus religionswissenschaftlicher Sicht. In Heitmeyer, W. and Hagan, J. (Eds), *Internationales Handbuch der Gewaltforschung* (Wiesbaden: Westdeutscher Verlag), pp. 1255–75.

Kriesberg, L. (1998), *Constructive Conflicts: From escalation to resolution* (Lanham: Rowman and Littlefield).

Kroslak, D. (2004), France's Policy towards Africa: Continuity or change? In Taylor, I. and Williams, P. (Eds), *Africa in International Politics: External involvement on the continent* (London: Routledge), pp. 61–82.

Laitinen, I. (2007), Frontex and African Illegal Migration to Europe. In Gebrewold, B. (Ed.), *Africa and Fortress Europe: Threats and opportunities* (Aldershot: Ashgate), pp. 127–38.

Lake, A., et al. (2006) *More than Humanitarianism: A strategic U.S. approach toward Africa* (New York: Council on Foreign Relations).

Lal, R. (2006), *Understanding China and India: Security implications for the United States and the world* (Westport: Praeger Security International).

Land Mine Monitor. (1999), Somalia. <http://www.icbl.org/lm/1999/somalia>, accessed 11 January 2001.

Lang, A. F. (2003), Humanitarian Intervention: Definitions and debates. In Lang, A. F. (Ed.), *Just Intervention* (Washington, D.C.: Georgetown University Press), pp. 1–7.

Latham, R. et al. (2001), Transboundary Formations, Intervention, Order, and Authority. In Callaghy, T. M., et al. (Eds), *Intervention and Transnationalism in Africa: Global–local networks of power* (Cambridge: Cambridge University Press), pp. 1–22.

Lederach, J. P. (2003), *The Little Book of Conflict Transformation* (Pennsylvania: Good Books).

Lederach, J. P. (2006), *Building Peace: Sustainable reconciliation in divided societies* (Washington, D.C.: United States Institute of Peace Press).

Legum, C. (1979), The African Crisis. *Foreign Affairs*, 632–51.

Lemarchand, R. (2000), The Crisis in the Great Lakes. In Harbeson, J. W. and Rothchild, D. (Eds), *Africa in World Politics: The African state system in flux* (Boulder, CO: Westview Press), pp. 324–52.

Lemarchand, R. (2003), The Democratic Republic of Congo: From failure to potential reconstruction. In Rotberg, R. I. (Ed.), *State Failure and State Weakness in a Time of Terror* (Washington, D.C.: Brookings Institution Press), pp. 29–70.

Lemarchand, R. (2007), Consociationalism and Power Sharing in Africa: Rwanda, Burundi, and the Democratic Republic of the Congo. *African Affairs*, pp. 1–21.

Lesch, A. M. (1998), *The Sudan: Contested national identities* (Oxford: James Curry).

Levi, M. (1994), The Predatory Theory of Rule. In Hall, J. A. (Ed.), *The State: Critical concepts. Vol. 1* (London: Routledge), pp. 146–75.

Lewis, I. M. (2003), *Modern History of the Somali* (Ohio: Ohio University Press).

Lewis, I. M. (2002), *A Modern History of the Somali: Nation and state in the Horn of Africa* (Oxford: James Currey).

Lockwood, M. (2006), *The State They Are In: An agenda for international action on poverty in Africa* (Bourton on Dunsmore: Practical Action Publishing).

Loeser, B. (2007), The EU's Peace and Security Strategy in the DRC. In Gebrewold, B. (Ed.), *Africa and Fortress Europe: Threats and opportunities* (Aldershot: Ashgate), pp. 159–70.

Luhmann, N. (1972), *Rechtssoziologie* (Reinbeck bei Hamburg: Rowolt).

Luhmann, N. (1974), *Soziologische Aufklärung. I* (Wiesbaden: VS Verlag für Sozialwissenschaften).

Luhmann, N. (1984), *Soziale Systeme: Grundriß einer allgemeinen Theorie* (Frankfurt am Main: Suhrkamp).

Luhmann, N. (1986), *Ökologische Kommunikation: Kann die moderne Gesellschaft sich auf ökologische Gefährdung einstellen?* (Opladen: Westdteutscher Verlag).

Luhmann, N. (1990a), *Essays on Self-reference* (New York: Columbia University Press).

Luhmann, N. (1990b), *Soziologische Aufklärung. V* (Wiesbaden: VS Verlag für Sozialwissenschaften).

Luhmann, N. (1993), Zeichen als Form. In Baecker, D. (Ed.), *Probleme der Form* (Frankfurt am Main: Suhrkamp), pp. 49–65.

Luhmann, N. (1999), *Social Systems* (Stanford: Stanford University Press).

Luhmann, N. (2004), *Einführung in die Systemtheorie* (Heidelberg: Carl-Auer).

Lutundula Report. (2005), Rights and Accountability in Development. Unofficial Translation of Lutundula Report, http://www.raid-uk.org/docs/Lutundula/Unofficial_Translation.pdf, accessed 10 April 2007.

Lyman, P. N. and Morrison, J. S. (2004), The Terrorist Threat in Africa. *Foreign Affairs*, 8:1, 75–86.

Lyons, T. (2006), Diasporas and Homeland Conflict. In Kahler, M. and Walter, B. F. (Eds), *Territoriality and Conflict in an Era of Globalization* (Cambridge: Cambridge University Press), pp. 111–29.

Mackie, G. (2003), *Democracy Defended* (Cambridge: Cambridge University Press).

Magang, D. N. (1986), Democracy in African tradition: The case of Botswana. In Ronen, D. (Ed.), *Democarcy and pluralism in Africa* (Boulder, CO: Lynne Rienner), pp. 103-108

Malone, D. M. (2004), Introduction. In Malone, D. M. (Ed.), *The U.N. Security Council* (Boulder, CO: Lynne Rienner), pp. 1–15.

Mamdani, M. (2002), *When Victims Become Killers: Colonialism, nativism, and the genocide in Rwanda* (Princeton: Princeton University Press).

Mann, M. (1977), States, Ancient and Modern. *Archives Européennes de Sociologie*, 18, 262–98.

Mann, M. (1985), The Autonomous Power of the State: Its origins, mechanisms and results. *European Journal of Sociology*, 25, 185–213.

Mann, M. (2005), *The Dark Side of Democracy: Explaining ethnic cleansing* (Cambridge: Cambridge University Press).

Mansfield, E. D. and Snyder, J. (1995), Democratization and the Danger of War. *International Security*, 20, 1, 5–38.

Marchal, R. (2004), Islamic Political Dynamics in the Somali Civil War. In De Waal, A. (Ed.), *Islamism and Its Enemies in the Horn of Africa* (London: Hurst and Company), pp. 114–45.

Marchal, R. (2007a), The Unseen Regional Implications of the crisis in Darfur. In De Waal, A. (Ed.), *War in Darfur: And the search for peace* (Harvard: Harvard University Press), pp. 173–98.

Marchal, R. (2007b), Warlordism and Terrorism: How to obscure an already confusing crisis? The case of Somalia. *International Affairs*, 83:6, 1091–106.

Marks, S. (2004), The Dog that Did Not Bark, or Why Natal Did Not Take Off: Ethnicity and democracy in South Africa – KwaZulu-Natal. In Berman, B., et al. (Ed.), *Ethnicity & Democracy in Africa* (Oxford: Currey), pp. 183–99.

Marks, S. (2007), Introduction. In Manji, F. (Ed.), *African Perspectives on China in Africa* (Oxford: Fahamu), pp. 1–14.

Martin, G. (1995), Continuity and Change in Franco–African Relations. *Journal of Modern African Studies*, 33:1, 1–20.

Matthies, V. (2005), *Kriege am Horn von Afrika* (Berlin: Köster).

Maturana, H. J. (1987), Biologie der Sozialität. In Schmidt, S. J. (Ed.), *Der Diskurs des radikalen Konstruktivismus* (Frankfurt am Main: Suhrkamp), pp. 287–302.

Mazov, S. (2007), Soviet Policy in West Africa: An episode of the Cold War, 1956–1964. In Matusevich, M. (Ed.), *Africa in Russia, Russia in Africa: Three centuries of encounters* (Trenton: Africa World Press), pp. 293–314.

Mazrui, A. A. (1967), *On the Concept of 'We are All Africans', Towards a Pax-Africana: A study of ideology and ambition* (London: Weidenfeld and Nicolson).

Mazula, B. (2004), Mozambique: The challenge of democratisation. In Gyimah-Boadi, E. (Ed.), *Democratic Reform in Africa: The quality of progress* (Boulder, CO: Lynne Rienner), pp. 183–200.

McDermott, R. (2006), *Political Psychology in International Relations* (Ann Arbor, MI: University of Michigan Press).

Mearsheimer, J. J. (2006), Structural Realism. In Dunne, T., et al. (Ed.), *International Relations Theories: Discipline and diversity* (Oxford: Oxford University Press), pp. 71–88.

Medhanie, S. (2007), *Is EPRDF Veridical on Ethiopian Nationalism Issues?* <http://www.ethiopiafirst.com/news2007/May/EPRDF_Veridical.html>, accessed 03 August 2007.

Men, J. (2007), Changing Ideology in China and its Impact on Chinese Foreign Policy. In Guo, S. H. and Hua, S. (Eds), *New Dimensions of Chinese Foreign Policy* (Lanham: Lexington Books), pp. 7–39.

Mengisteab, K. and Yohannes, O. (2005), *Anatomy of an African tragedy: Political, economic, and foreign policy crisis in post-independence Eritrea* (Trenton, NJ: Red Sea Press).

Menkhaus, K. (2005), Somalia and Somaliland: Terrorism, political Islam, and state collapse. In Rotberg, R. I. (Ed.), *Battling Terrorism in the Horn of Africa* (Cambridge: World Peace Foundation), pp. 23–47.

Menkhaus, K. (2006/7), Governance without Government in Somalia: Spoilers, state building, and the politics of coping. *International Security*, 31:3, 74–106.

Mentan, T. (2004), *Dilemmas of Weak States: Africa and transnational terrorism in the twenty-first century* (Aldershot: Ashgate).

Meredith, M. (2003), *Our Votes, Our Guns: Robert Mugabe and the tragedy of Zimbabwe* (New York: PublicAffairs).

Meredith, M. (2005), *The Fate of Africa: A history of fifty years of independence* (New York: PublicAffairs).

Milkias, P. (1999), Ethiopia and Eritrea at War: Saga of triumph and tragedy at the dawn of the millennium. *Horn of Africa Bulletin*, 17:1–4, 33–71.

Mohamed, A. (2001), *The Peace Process in Somalia.* <www.civiwebs.com/ cwvlib/>, accessed 15 September 2004.

Molutsi, P. (2004), Botswana: The path to democracy and development. In Gyimah-Boadi, E. (Ed.), *Democratic Reform in Africa: The quality of progress* (Boulder, CO: Lynne Rienner), pp. 159–81.

Montague, D. (2002), Stolen Goods: Coltan and conflict in the Democratic Republic of Congo. *SAIS Review*, 22:1, 103–18.

Mouffe, C. (2005), *On the Political* (London: Routledge).

Münkler, H. and Llanque, M. (2002), Die Rolle der Eliten bei der Legitimation von Gewalt. In Heitmeyer, W. and Hagan, J. (Eds), *Internationales Handbuch der Gewaltforschung* (Wiesbaden: Westdeutscher Verlag), pp. 1215–32.

Murphy, S. D. (2006), *Principles of International Law* (St. Paul: Thompson-West).

Mustapha, A. R. (2004), Ethnicity and the Politics of Democratisation in Nigeria. In Berman, B., et al. (Eds), *Ethnicity & Democracy in Africa* (Oxford: Currey), pp. 257–75.

Nardin, T. (2003), The Moral Basis for Humanitarian Intervention. In Lang, A. (Ed.), *Just Intervention* (Washington, D.C.: Georgetown University Press), pp. 11–27.

Negash, T. and Tronvoll, K. (2000), *Brothers at War: Making sense of the Eritrean–Ethiopian war* (Oxford: James Currey).

Nest, M., F. Grignon, et al. (2006), *The Democratic Republic of Congo: Economic dimensions of war and peace* (Boulder, CO: Lynne Rienner).

Nettl, J. P. (1994), The State as a Conceptual Variable. In Hall, J. A. (Ed.), *The State: Critical concepts* (London: Routledge), pp. 9–36.

Neuman, S. (2005), International Relations Theory and the Third World: An oxymoron? In Neuman, S. (Ed.), *International Relations Theory and the Third World* (New York: St. Martins's Press), pp. 1–29.

New York Times. (1992), Chaotic Somalia Starves as Strongmen Battle. <http://query.nytimes.com/gst/fullpage.html?res=9E0CEEDC1538F937A35753C1A964958260&sec=&spon=&pagewanted=2>, accessed 12 September 2007.

Newman, D. (2006), The Resilience of Territorial Conflict in an Era of Globalization. In Kahler, M. and Walter, B. F. (Eds), *Territoriality and Conflict in an Era of Globalization* (Cambridge: Cambridge University Press), pp. 85–110.

Nicholson, M. (1995), *Rationality and the Analysis of International Conflict* (Cambridge: Cambridge University Press).

Nunner-Winkler, G. (2004) Überlegungen zum Gewaltbegriff. In Heitmeyer, W. and Soeffner, H.-G. (Eds), *Gewalt: Entwicklungen, Strukturen* (Frankfurt: Suhrkamp), pp. 21–61.

Nzongola-Ntalaja, G. (2004a), *From Zaire to the Democratic Republic of the Congo* (Uppsala: Nordiska Afrikainstitutet).

Nzongola-Ntalaja, G. (2004b), *The Politics of Citizenship in the DRC: States, borders and nations: negotiating citizenship in Africa.* Annual International Conference (University of Edinburgh: Centre of African Studies).

O'Ballance, E. (2000), *The Congo–Zaire Experience: 1960–1998* (Basingstoke: MacMillan).

Obiorah, N. (2007), Who is Afraid of China in Africa? Towards an African civil society perspective on China–Africa relations. In Manji, F. and Marks, S. (Eds), *African Perspectives on China in Africa* (Oxford: Fahamu), pp. 35–55.

Offe, C. (1994), Structural Problems of the State: Class rule and the political system. On the selectiveness of the political institutions. In Hall, J. H. (Ed.), *The State: Critical concepts. Vol. 1* (London: Routledge), pp. 104–29.

Oliver, R. and Fage, J. D. (1966), *A Short History of Africa* (Harmondsworth: Penguin Books).

Omar, M. J. (2002), Management and Control of Small Arms: The Somaliland experience. In *Small Arms in the Horn of Africa* (Bonn: Bonn International Centre for Conversion), pp. 18–24.

Onuf, N. (2003), Normative Frameworks for Humanitarian Intervention. In Lang, A. (Ed.), *Just Intervention* (Washington, D.C.: Georgetown University Press), pp. 28–45.

Ottaway, M. et al. (2004), *Africa's Big States: Toward a New Realism.* <http://www.carnegieendowment.org/files/PolicyOutlookOttaway.pdf>, accessed 26 January 2008.

Oxfam. (2007), *Africa's Missing Billions: International arms flows and the cost of conflict.* <http://www.oxfam.org.uk/resources/policy/conflict_disasters/downloads/bp107_africa_missing_billions.pdf>, accessed 11 January 2008.

Pankhurst, R. (2007), The Russians in Ethiopia: Aspirations of progress. In Matusevich, M. (Ed.), *Africa in Russia, Russia in Africa: Three centuries of encounters* (Trenton: Africa World Press), pp. 219–38.

Parry, G. and Moran, M. (1994), *Democracy and Democratization* (London: Routledge).

Parsons, T. (1964), *The Social System* (London: Routledge and Kegan Paul).

Parsons, T. (1968), System Analysis: Social systems. *International Encyclopaedia of Social Sciences*, 15, 458–73.

Parsons, T. (1969), *Politics and Social Structure* (New York: Free Press).

Peninou, J. L. (1998), Why the Eritrean–Ethiopian Conflict? *Review of African Political Economy*, 25:77, 508–26.

Peres, Y. and Hopp, M. (1990), Loyalty and Aggression in Human Groups. In Denner, J. V. D. and Falger, V. (Eds), *Sociobiology and Conflict: Evolutionary perspectives on competition, cooperation, violence, and warfare* (London: Chapman and Hall), pp. 123–30.

Perlez, J. (1988), *Sudan's Premier Is Visiting Libya after Proposal for Unity Is Signed.* http://query.nytimes.com/gst/fullpage.html?res=940DE4DF1531F937A25752C1A96E948260, accessed 24 June 2008.

Perry, A. (2007), Africa's Oil Dreams. *TIME*, 169:24, 11 June, 22–9.

Perry, A. (2008), *Falling off the Edge: Globalization, world peace and other lies* (London: Macmillan).

Ploughshares. (1999) *Military Profiles 1998/99 (Horn of Africa Countries).* <www.ploughshares.ca/content/BUILD%20PEACE/MilitaryProfile.htm>, accessed 12 March 2004.

Porteous, T. (2005), British Government Policy in Sub-Saharan Africa under New Labour. *International Affairs*, 81:2, 281–97.

Prahalad, C. K. (2007), Poverty: The world for sale. *Foreign Policy*, May/June, 50.

Prendergast, J. (1999), *Building for Peace in the Horn of Africa: Diplomacy and beyond.* http://www.usip.org/pubs/specialreports/sr990628.htm, accessed 15 May 2009.

Prunier, G. (1995), *Somalia: Civil war, intervention and withdrawal 1990–1995.* <http://www.asylumlaw.org/docs/somalia/country_conditions/Prunier.pdf>, accessed 22 April 2008.

Prunier, G. (1997), *Sudan's Regional War.* http://mondediplo.com/1997/02/02sudan#nb5#nb5, accessed 9 April 2008.

Prunier, G. (2002), *Oil, Islam, Chinese Weapons and American Money. Sudan: Irreconcilable differences.* http://mondediplo.com/2002/12/06sudan, accessed 8 April 2008.

Prunier, G. (2007a), *Africa, Continent of Organised Pillage. Sudan: genocide in Darfur.* <http://mondediplo.com/2007/03/08darfur>, accessed 8 March 2008.

Prunier, G. (2007b), *Darfur: The Ambiguous Genocide* (Ithaca, NY: Cornell University Press).

Prunier, G. (2008), *Chad: Caught in the Darfur crossfire.* <http://mondediplo.com/2008/03/05chad>, accessed 10 June 2008.

Rae, H. (2002), *State Identities and the Homogenisation of Peoples* (Cambridge: Cambridge University Press).

Ramsbotham, A., et al. (2005), Enhancing African Peace and Security Capacity: A useful role for the UK and the G8? *International Affairs*, 81:2, 325–39.

Reliefweb. (2009), Somalia Food Security Outlook January–June 2009. http://www.reliefweb.int/rw/RWFiles2009.nsf/FilesByRWDocUnidFilename/CJAL-7PMT69-full_report.pdf/$File/full_report.pdf, accessed 30 March 2009.

Renan, E. (1995), *Was ist eine Nation? und andere politische Schriften* (Wien: Folio-Verlag).

Reno, W. (1998), *Warlord Politics and African States* (Boulder, CO: Lynne Rienner).

Report of the Secretary-General (2006), Report of the UN Secretary-General on the situation in Somalia, S/2006/122, 2006.

Reporters Without Borders. (2008), 2008 Annual Report, http://www.rsf.org/IMG/pdf/rapport_en-2.pdf, accessed 19 March 2009.

Reuters. (2007), China to Lend Congo $5 billion in Latest Africa Foray. <http://www.reuters.com/article/worldNews/idUSL181569372007 0918>, accessed 22 March 2008.

Rigby, A. (2006), Civil Society, Reconciliation and Conflict Transformation in Post-war Africa. In Furley, O. and May, R. (Eds), *Ending Africa's Wars* (Aldershot: Ashgate), pp. 47–62.

Roberts, B. (1992), Democracy and World Order. In Roberts, B. (Ed.), *US Foreign Policy after the Cold War.* 2nd ed. (Cambridge: MIT Press), pp. 293–307.

Robinson, S. (2006), Time Running Out. *TIME*, 168:13, 18, 2006, 15–16.

Rocha, J. (2007), A New Frontier in the Exploitation of Africa's Natural Resources: The emergence of China. In Manji, F. and Marks, S. (Eds), *African Perspectives on China in Africa* (Oxford: Fahamu), pp. 15–34.

Rogier, E. (2006), Democratic Republic of Congo: Problems of the peacekeeping process. In Furley, O. and May, R. (Eds), *Ending Africa's Wars* (Aldershot: Ashgate), pp. 99–113.

Ronning, H. (2003), The Media in Zimbabwe: The struggle between state and civil society. In Darnolf, S. and Laakso, L. (Eds), *Twenty Years of Independence in Zimbabwe: From liberation to authoritarianism* (Basingstoke: Palgrave Macmillan), pp. 196–221.

Rotberg, R. I. (Ed.). (2003), *State Failure and State Weakness in a Time of Terror* (Cambridge: World Peace Foundation).

Rotberg, R. I. (Ed.). (2004), *When States Fail: Causes and consequences* (Princeton: Princeton University Press).

Rotberg, R. I. (2005), The Horn of Africa and Yemen: Diminishing the threat of terrorism. In Rotberg, R. I. (Ed.), *Battling Terrorism in the Horn of Africa* (Cambridge: World Peace Foundation), pp. 1–22.

Rotberg, R. I., et al. (2000), *Peacekeeping and Peace Enforcement in Africa: Methods of conflict prevention* (Cambridge: The World Peace Foundation).

Rothchild, D. and Harbeson, J. W. (2000), The African State and State System in Flux. In Harbeson, J. W. and Rothchild, D. (Eds), *Africa in World Politics: The African state system in flux* (Boulder, CO: Westview Press), pp. 3–20.

Russel, R. (1999), Eritrean–Ethiopian Conflict. *Horn of Africa Bulletin*, 11:1, 8–12.

RWB. (2008), 2008 Annual Report Africa.

Sachs, J. (2005) *The End of Poverty: How we can make it happen in our lifetime* (London: Penguin).

Saltanov, A. (2008), *Russian Relations with the Sub-Saharan African Countries in 2008.* <http://www.ln.mid.ru/brp_4.nsf/itogi/600123F8FC5F74B6C325752E00373694>, accessed 10 March 2009.

Sa'oudi, M. A.-G. (2001), An Overview of the Egyptian–Sudanese Jonglei Canal Project. *International Politics Journal (Al-Siyassa Al-Dawliya)*, 143, January 2001.

Sbacchi, A. (1985), *Ethiopia under Mussolini: Fascism and the colonial experience* (Totowa: Zed Books)

Sbacchi, A. (1996), *The Concept of the Political* (Chicago: University of Chicago Press).

Sbacchi, A. (2004), *Der Wert des Staates und die Bedeutung des Einzelnen* (Berlin: Duncker & Humblot).

Schmitt, C. (1996), *The Concept of the Political* (Chicago: University of Chicago Press).

Schmitt, C. (2005), *Political Theology: Four chapters on the concept of sovereignty* (Chicago: University of Chicago Press).

Schraeder, P. J. (1994), *United States Foreign Policy toward Africa: Incrementalism, crisis and change* (Cambridge: Cambridge University Press).

Sen, A. (2008), Violence, Identity and Poverty. *Journal of Peace Research*, 45:5, pp. 5–15.

Shambaugh, D. (1994), Patterns of Interaction in Sino–American Relations. In Robinson, T. W. and Shambaugh, D. (Eds), *Chinese Foreign Policy: Theory and practice* (Oxford: Clarendon Press), pp. 197–223.

Shaxson, N. (2005), New Approaches to Volatility: Dealing with the 'resource' in sub-Saharan Africa. *International Affairs*, 81: 2, 311–24.

Shinn, D. (2003), The Horn of Africa: How does Somaliland fit? *The Somalilaand Times*.

Shinn, D. H. (2005), Ethiopia: Governance and terrorism. In Rotberg, R. I. (Ed.), *Battling Terrorism in the Horn of Africa* (Cambridge: World Peace Foundation), pp. 93–118.

Shinn, D. H. (2006), *Somalia: Regional involvement and implications for US policy*. <http://www.gwu.edu/~elliott/news/transcripts/shinn11.html>, accessed 6 December 2006.

Shubin, V. (2004), Russia and Africa: Moving in the right direction? In Taylor, I. and Williams, P. (Eds), *Africa in International Politics: External involvement in the continent* (London: Routledge), pp. 102–5.

Shubin, V. (2007), Beyond the Fair Tales: The reality of Soviet involvement in the liberation of Southern Africa. In Matusevich, M. (Ed.), *Africa in Russia, Russia in Africa: Three centuries of encounters* (Trenton, NJ: Africa World Press), pp. 331–52.

Simmel, G. (1964), *Conflict and the Web of Group Affiliations* (New York: Free Press).

Simon, B. and Klandermans, B. (2004), Politicised Collective Identity. In Jost, J. T. and Sidanius, J. (Eds), *Political Psychology* (New York: Psychology Press), pp. 449–65.

Singer, J. D. (1960), International Conflict: Three levels of analysis. *World Politics*, 12:3, 453–61.

Singer, J. D. (2006), *A Culture of Corruption: Everyday deception and popular discontent in Nigeria* (Princeton, NJ: Princeton University Press).

Smith, A. D. (1994), State-making and Nation-building. In Hall, J. A. (Ed.), *The State: Critical concepts. Vol. 2* (London: Routledge), pp. 59–89.

Smith, A. D. (2000), *Nationalism and Modernism: A critical survey of recent theories of nations and nationalism* (London: Routledge).

Smith, A. D. (2001), *Nationalism: Theory, ideology, history* (Oxford: Polity).

Smith, D. J. (2006), *A Culture of Corruption: Everyday deception and popular discontent in Nigeria* (Princeton, NJ: Princeton University Press).

Smith, J. (2005), *Dinner with Mobutu: A chronicle of my life and times* (Philadelphia: Xlibris Corporation).

Snow, P. (1994), China and Africa: Consensus and camouflage. In Robinson, T. W. and Shambaugh, D. (Eds), *Chinese Foreign Policy: Theory and practice* (Oxford: Clarendon Press), pp. 283–321.

Soeffner, H.-G. (2004), Gewalt als Faszinosum. In Heitmeyer, W. and Soeffner, H.-G. (Eds), *Gewalt: Entwicklungen, Strukturen, Analyseproblem* (Frankfurt: Suhrkamp), pp. 62–85.

Spear, J. (2006), Disarmament, Demobilization, Reinsertion and Reintegration in Africa. In Furley, O. and May, R. (Eds), *Ending Africa's Wars* (Aldershot: Ashgate), pp. 63–80.

Stark, C. (1994), *Autopoiesis und Integration: eine kritische Einführung in die Luhmannsche Systemtheorie*, (Hamburg: Kovac).

Sutter, R. G. (2005), *China's Rise in Asia: Promises and perils* (Lanham: Rowman & Littlefield).

Tadesse, M. (1999), *The Eritrean–Ethiopian War: Retrospect and prospects. Reflections on the making of conflicts in the Horn of Africa, 1991–1998* (Addis Ababa: Mega Printing Enterprise).

Talentino, A. K. (2005), *Military Intervention after the Cold War: The evolution of theory and practice* (Ohio: Ohio University Press).

Taylor, I. (2004), The 'All-weather Friend'? In Taylor, I. and Williams, P. (Eds), *Africa in International Politics: External involvement in the continent* (London: Routledge), pp. 83–101.

Taylor, I. (2005), Advice is Judged by Results, Not by Intentions: Why Gordon Brown is wrong about Africa. *International Affairs*, 81:2, 299–310.

Taylor, I. (2006), *China and Africa: Engagement and compromise* (London: Routledge).

Teigist Lemma. (2006), United Nations Conference on Trade and Development: Ad Hoc Expert Meeting in preparation for the Mid-term Review of the Programme of Action for the Least Developed Countries for the Decade 2001–2010; Case study on Ethiopia. <http://www.unctad.org/en/docs/ldcmisc20065_en.pdf>, accessed 12 February 2009.

Terlinden, U. (2004), IGAD – Paper Tiger Facing Gigantic Tasks. In http://www.ulf-terlinden.de/docs/Terlinden2004-IGAD-engl.pdf, accessed 10 January 2005.

Tetzlaff, R. (1993), *Staatswerdung im Sudan: ein Bürgerkriegsstaat zwischen Demokratie, ethnischen Konflikten und Islamisierung* (Münster: Lit-Verl).

Tetzlaff, R. and Jakobeit, C. (2005), *Das nachkoloniale Afrika* (Wiesbaden: VS).

Thompson, K. W. (1977), Toward a Theory of International Politics. In Hoffmann, S. (Ed.), *Contemporary Theory in International Relations* (Westport: Greenwood Press), pp. 17–28.

Thomson, A. (2007), *An Introduction to African Politics* (London: Routledge).

Tilly, C. (1985), War Making and State Making as Organized Crime. In Evans, P. B. et al. (Ed.), *Bringing the State Back In* (Cambridge: Cambridge University Press).

Tilly, C. (2001), *Coercion, Capital, and European States, AD 990–1992* (Cambridge: Blackwell).

Tilly, C. (2002), Großdimensionale Gewalt als Konfliktive Politik. In Heitmeyer, W. and Hagan, J. (Eds), *Internationales Handbuch der Gewaltforschung* (Wiesbaden: Westdeutscher Verlag), pp. 547–69.

Tilly, C. (2003), *The Politics of Collective Violence* (Cambridge: Cambridge University Press).

Tocqueville, A. de (1956), *Democracy in America* (New York: New American Library of World Literature).

Todd, B. (2006), Congo, Coltan, Conflict. *The Heinz School Review*, 3:1, 15 March 2006.

Tow, W. T. (1994), China and the International Strategic System. In Robinson, T. W. and Shambaugh, D. (Eds), *Chinese Foreign Policy: Theory and practice* (Oxford: Clarendon Press), pp. 115–57.

Transparency International. (2008) *Annual Report 2007.*

Trivelli, R. M. (1998), Divided Histories, Opportunistic Alliances: Background notes on the Ethiopian Eritrean conflict. *Afrika Spectrum*, 33:3, 257–89.

Tsegaye. (1999), *The Dynamics of the Ethio–Eritrea Conflict*. http://www.geocities.com/~dagmawi/NewsApr99/Apr18_Viewpoint_Tsegaye.html, accessed 24 April 2007.

Tubiana, J. (2007), Darfur: A conflict for land? In De Waal, A. (Ed.), *War in Darfur: And the search for peace* (Harvard: Harvard University Press), pp. 68–91.

Tull, D. (2007), The Democratic Republic of Congo: Militarized politics in a ‘failed state’. In Boas, M. and Dunn, K. C. (Eds), *African Guerrillas: Raging against the machine* (Boulder, CO: Lynne Rienner), pp. 113–30.

Tull, D. (2008), *The Chad Crisis and Operation EUFOR Chad/CAR*. <http://www.swp-berlin.org/en/common/get_document.php?asset_id=4766>, accessed 26 February 2009.

UN. (2001), *Interim Report of the United Nations Expert Panel on the Illegal Exploitation of Natural Resources and Other Forms of Wealth of the Democratic Republic of the Congo*. <http://www.globalsecurity.org/military/library/report/2001/49e.pdf>, accessed 22 December 2006.

UN. (2006), Report of the Monitoring Group on Somalia pursuant to Security Council Resolution 1676 (2006).

UN. (2008a), Report of the Secretary-General on Children and Armed Conflict in the Democratic Republic of the Congo, S/2008/693.

UN. (2008b), Report of the Secretary-General on the Situation in Somalia, S/2008/709.

UN. (2008c), Report of the Secretary-General on Children and Armed Conflict in Somalia, S/2008/352.

UNCTAD. (2004), Trade Performance and Commodity Dependence, Geneva.

UNDP. (2008), *2007/2008 Human Development Report.* <http://hdrstats.undp.org/countries/data_sheets/cty_ds_SOM.html>, accessed 22 February 2009.

UNECA. (2005), Meeting the Challenges of Unemployment and Poverty in Africa. *UNECA.* <http://www.uneca.org/era2005/overview.pdf, accessed>, accessed 20 March 2008.

UNHCR. (2008), *Supplementary Appeal for Darfur 2008.* http://www.reliefweb.int/rw/RWFiles2008.nsf/FilesByRWDocUnidFilename/LSGZ-7DDFVX-full_report.pdf/$File/full_report.pdf, accessed 8 April 2008.

UNHCR. (2009a), *DRC: Growing insecurity causing more displacement in North Kivu.* <http://www.unhcr.org/news/NEWS/49a3c9a82.html>, accessed 3 March 2009.

UNHCR. (2009b), More than 40,000 Somalis Return to Mogadishu Despite Renewed Fighting. <http://www.unhcr.org/news/NEWS/49a8070b2.html>, accessed 2 March 2009.

United Nations. (2005), *Report of the International Commission of Inquiry on Darfur to the United Nations Secretary-General.* <http://www.un.org/news/dh/sudan/com_inq_darfur.pdf>, accessed 22 March 2007.

United Nations. (2006a), Report of the Secretary-General on the Situation in Somalia, 23 October 2006, S/2006/838.

United Nations. (2006b), Report of the Panel of Experts Established Pursuant to Paragraph 3 of Resolution 1591 (2005) Concerning the Sudan, S/2006/65.

United Nations. (2008), Report of the Panel of Experts Established Pursuant to Resolution 1591 (2005) Concerning the Sudan, S/2008/647.

Utley, R. (2002), Not To Do Less but To Do Better. French military policy in Africa. *International Affairs*, 78:1, 129–46.

Van Der Dennen, J. and Falger, V. (Eds), (1990), *Sociobiology and Conflict: Evolutionary perspectives on competition, cooperation, violence and warfare* (London: Chapman and Hall).

Vines, A. (2005), Combating Light Weapons Proliferation in West Africa. *International Affairs*, 81:2, 341–60.

Vines, A. and Middleton, R. (2008), Options for the EU to Support the African Peace and Security Architecture. In http://www.europarl.europa.eu/activities/committees/studies.do?language=EN, accessed 23 August 2008,

Waal, A. De (2005), Defining Genocide. *Index on Censorship*, 34:1, 6–13.

Wagner, J. (2004), Afrika im Fadenkreuz. Vom vergessenen Kontinent zum Objekt der Begierde. *Blätter für deutsche und internationale Politik*, 6, 703–11.

Waldmann, P. (2002), Bürgerkriege. In Heitmeyer, W. and Hagan, J. (Eds), *Internationales Handbuch der Gewaltforschung* (Wiesbaden: Westdeutscher Verlag), pp. 368–89.

Wallander, C. A. (2007), Global Challenges and Russian Foreign Policy. In Legvold, R. (Ed.), *Russian Foreign Policy in the Twenty-first Century and the Shadow of the Past* (New York: Columbia University Press).

Wallensteen, P. and Johansson, P. (2004), Security Council Decisions in Perspective. In Malone, D. M. (Ed.), *The U.N. Security Council: From the Cold War to the 21st century* (Boulder, CO: Lynne Rienner), pp. 17–33.

Wallerstein, I. M. (2006), *World-systems Analysis: An introduction* (Durham, NC: Duke University Press).

Walt, S. M. (2006), *Taming American Power: The global response to U.S. primacy* (New York: Norton).

Walt, V. (2008), A Furious Hunger. *TIME*, vol. 171, no. 11, 2008: 42–3.

Walter, B. F. (2006), Conclusion. In Kahler, M. and Walter, B. F. (Eds), *Territoriality and Conflict in an Era of Globalization* (Cambridge: Cambridge University Press), pp. 288–96.

Waltz, K. N. (1979), *Theory of International Politics* (New York: McGraw-Hill).

Waltz, K. N. (2001), *Man, the State and War: A theoretical analysis* (New York: Columbia University Press).

Walzer, M. (2002), The Argument about Humanitarian Intervention. In Mills, N. and Brunner, K. (Eds), *The New Killing Fields: Massacre and the politics of intervention* (Albany: State University of New York Press).

Walzer, M. (2004), *Arguing about War* (New Haven: Yale University Press).

Warburg, G. (2003), *Islam, Sectarianism and Politics in Sudan since the Mahdiyya* (London: Hurst and Company).

Wassara, S. (2002), *Conflict and State Security in the Horn of Africa: Militarization of civilian groups*. <www.aaps.co.zw/Publications/AIJP/Wassara.html>, accessed 12 March 2005.

Weber, M. (1972), *Wirtschaft und Gesellschaft* (Tübingen: Mohr).

Weiss, T. G. (2003), Principles, Politics and Humanitarian Action. In Lang, A. (Ed.), *Just Intervention* (Washington, D.C.: Georgetown University Press), pp. 84–103.

Weiss, T. G. (2004), The Humanitarian Impulse. In Malone, D. M. (Ed.), *The U.N. Security Council: From the Cold War to the 21st century* (Boulder, CO: Lynne Rienner), pp. 37–54.

Wendt, A. (1992), Anarchy Is What the States Make of It: The social construction of power politics. *International Organization*, 46:2, 391–425.

Weschler, J. (2004), Human Rights. In Malone, D. M. (Ed.), *The U.N. Security Council: From the Cold War to the 21st century* (Boulder, CO: Lynne Rienner), pp. 55–68.

Westcott, K. (2008), *Somalia's Pirates Face Battles at Sea*. <http://news.bbc.co.uk/go/pr/fr/-/1/hi/world/africa/7358764.stm>, accessed 29 October 2008.

WFP. (2007), WFP in Africa: 2006 Facts, figures and partners. *World Food Programme*. http://www.wfp.org/policies/Introduction/other/Documents/pdf/WFP_in_Africa_Briefing_2006.pdf, accessed 18 March 2008.

Wheeler, N. J. (2003), Humanitarian Intervention after September 11, 2001. In Lang, A. (Ed.), *Just Intervention* (Washington, D.C.: Georgetown University Press), pp. 193–216.

WHO. (2006), HIV Prevention in the African Region: A strategy for renewal and acceleration. Report of the Regional Director. *World Health Organization*, AFR/RC56/8, 17 June 2006 <http://www.afro.who.int/aids/publications/afr_rc56_8_hiv_prevention.pdf>, accessed 26 March 2008.

Williams, P. (2004), Britain and Africa after the Cold War: Beyond damage limitation? In Taylor, I. and Williams, P. (Eds), *Africa in International Politics: External involvement on the continent* (London: Routledge), pp. 41–60.

Willke, H. (1993), Kontingenz und Notwendigkeit des Staates. In Baecker, D. (Eds), *Probleme der Form* (Frankfurt am Main: Suhrkamp), pp. 212–29.

Willke, H. (1997), *Supervision des Staates* (Frankfurt am Main: Suhrkamp).

Willke, H. (2005), *Systemtheorie II: Interventionstheorie: Grundzüge einer Theorie der Intervention in komplexe Systeme* (Stuttgart: Lucius und Lucius).

Willke, H. (2006), *Systemtheorie I: Grundlagen* (Stuttgart: UTB).

Wimmer, A. and Schetter, C. (2002), Ethnische Gewalt. In Heitmeyer, W. and Hagan, J. (Eds), *Internationales Handbuch der Gewaltforschung* (Wiesbaden: Westdeutscher Verlag), pp. 313–29.

Winrow, G. M. (1990), *The Foreign Policy of the GDR in Africa* (Cambridge: Cambridge University Press).

Woods, E. (1997), *Somalia.* <http://www.fpif.org/pdf/vol2/19ifsoma.pdf>, accessed 6 March 2009.

Woods, N. (2005), The Shifting Politics of Foreign Aid. *International Affairs*, 81:2, 393–409.

Woodward, P. (2003), *The Horn of Africa: Politics and international relations* (London: Tauris).

Woodward, P. (2006), Peacemaking in Sudan. In Furley, O. and May, R. (Eds), *Ending Africa's Wars* (Aldershot: Ashgate), pp. 167–78.

World Bank. (2008), *The Little Data Book on Africa 2007*. The World Bank.

Wrong, M. (2006), *In the Footsteps of Mr. Kurtz: Living on the brink of disaster in Mobutu's Congo* (New York: Harper Perennial).

Xing, G. (2007) 'Peaceful Rise': China's public diplomacy and international image cultivation. In Guo, S. and Hua, S. (Ed.), *New Dimensions of Chinese Foreign Policy* (Lanham: Lexington Books), pp. 133–47.

Zenawi, M. (2004), Speech to Members of House of Peoples Representatives on the Ethiopia–Eritrea border issue, 25th November 2004.

Zhang, W. (2007), Tapping Soft Power: Managing China's 'peaceful rise' and the implications for the world. In Guo, S. and Hua, S. (Ed.), *New Dimensions of Chinese Foreign Policy* (Lanham: Lexington Books), pp. 109–31.

Zweig, D. and Jianhai, B. (2005) China's Global Hunt for Energy. *Foreign Affairs*, 84:5, 25–38.

Index